Coastal Geomorphology in Australia

Coastal Geomorphology in Australia

Edited by
B. G. Thom
Department of Geography
University of New South Wales
Duntroon, A. C. T.

ACADEMIC PRESS
(Harcourt Brace Jovanovich, Publishers)
Sydney Orlando San Diego Petaluma New York
London Toronto Montreal Tokyo
1984

ACADEMIC PRESS AUSTRALIA
Centrecourt, 25–27 Paul Street North
North Ryde, N.S.W. 2113

United States Edition published by
ACADEMIC PRESS INC.
Orlando, Florida 32887

United Kingdom Edition published by
ACADEMIC PRESS, INC. (LONDON) LTD.
24/28 Oval Road, London NW1 7DX

Printed in Australia

National Library of Australia Cataloguing-in-Publication Data

---

Coastal geomorphology in Australia.

Includes index.
ISBN 0 12 687880 3.

1. Coasts - Australia. 2. Landforms - Australia.
I. Thom, B. G. (Bruce Graham), 1939- .

551.4'5'0994

Library of Congress Catalog Card Number: 84-72293

Academic Press Rapid Manuscript Reproduction

CONTENTS

3 Morphodynamics of High Energy Beaches: An Australian Perspective

A. D. SHORT and L. D. WRIGHT

4 Foredune Formation in Southeast Australia

PATRICK A. HESP

5 New South Wales Estuaries: Their Origin and Evolution

P. S. ROY

6 Facies Variation and Holocene Reef Growth in the Southern Great Barrier Reef

JOHN F. MARSHALL and P. J. DAVIES

7 The Holocene 'High Energy Window' on the Central Great Barrier Reef

DAVID HOPLEY

8 Holocene Sea-Level History of a Tropical Estuary: Missionary Bay, North Queensland

JOHN GRINDROD and E. G. RHODES

9 The Holocene Marine Transgression as a Forcing Function in Episodic Dune Activity on the Eastern Australian Coast

K. PYE and G. M. BOWMAN

10 Chenier Plain Formation in Northern Australia

JOHN CHAPPELL and JOHN GRINDROD

11 Sand Barriers of Eastern Australia: Gippsland — A Case Study

B. G. THOM

12 Subaerial Denudation and Weathering of Vegetated Coastal Dunes in Eastern Australia

C. H. THOMPSON and G. M. BOWMAN

13 Quaternary Stratigraphy and Sea-Level Variation in the Southeast of South Australia

D.A. SCHWEBEL

14 Quaternary Emergent Shorelines of Tasmania

A. R. BOWDEN and E. A. COLHOUN

CONTRIBUTORS

Numbers in parentheses indicate the pages on which the authors' contributions begin.

Bowden, A. R. (313), Australian Groundwater Consultants, St Kilda, Victoria, Australia

Bowman, G. M. (179, 263), Division of Soils, CSIRO, Glen Osmond, South Australia, Australia

Chappell, John (197), Department of Biogeography and Geomorphology, Australian National University, Canberra, Australian Capital Territory, Australia

Colhoun, E. A. (313), Department of Geography, University of Tasmania, Hobart, Tasmania, Australia

Davies, P. J. (123), Bureau of Mineral Resources, Canberra, Australian Capital Territory, Australia

Grindrod, John (151,197), Department of Biogeography and Geomorphology, Australian National University, Canberra, Australian Capital Territory, Australia

Hesp, Patrick A. (69), Division of Resource Management, Department of Agriculture, Jarrah Road, South Perth, Western Australia, Australia

Hopley, David (135), Department of Geography, James Cook University of North Queensland, Townsville, Queensland, Australia

Jenkin, J. J. (23), Soil Conservation Authority, Melbourne, Victoria, Australia

Marshall, John F. (123), Bureau of Mineral Resources, Canberra, Australian Capital Territory, Australia

Pye, K. (179), Department of Earth Sciences, University of Cambridge, Cambridge, United Kingdom

Rhodes, E. G. (151), Gulf Oil Company, Houston, Texas, United States of America

Roy, P. S. (99), Geological Survey of New South Wales, Department of Mineral Resources, Sydney, New South Wales, Australia

Schwebel, D. A. (291), Esso Exploration and Production Australia Ltd, Sydney, New South Wales, Australia

Short, A. D. (43), Coastal Studies Unit, Department of Geography, University of Sydney, Sydney, New South Wales, Australia

Thom, B. G. (1, 233), Department of Geography, University of New South Wales, Duntroon, Australian Capital Territory, Australia

Thompson, C. H. (263), Division of Soils, CSIRO, Cunningham Laboratory, St Lucia, Queensland, Australia

Wright, L. D.[1] (43), Coastal Studies Unit, Department of Geography, University of Sydney, Sydney, New South Wales, Australia

---

Present address:

1. Department of Geological Oceanography, Virginia Institute of Marine Science, Gloucester Point, Virginia, United States of America

FOREWORD

As I write this, it is exactly thirty years since my arrival in Australia — thirty years that have seen the emergence of an indigenous conception of Australian coasts which is now contributing to ideas on a world scale. In the first twenty of these years a small number of people were grappling with questions whose answers have become much clearer in the last ten, when coastal studies have advanced at accelerated pace in universities and government bodies. In 1954 coastal studies was in its infancy and largely confined to looking at rocky shores. There had been notable contributions which remain extremely relevant today: Fairbridge on limestone platforms and coral reefs, and Hills on high shore platforms. Yet the beaches, dunes and tidal flats which make up the greater proportion of the coast had received little attention and ideas about them were either archaic or derived by analogy from inappropriate studies in the northern hemisphere. The intervening years have been increasingly exciting as the nature and history of the long coastline have been discovered progressively. Gradually we are learning what is where, and why and when it got there. Ability to quantify the environmental factors has also steadily grown: Sydney boasts the longest continuous instrumental wave record of anywhere in the world.

As the scientific discovery of the coast took place, it became increasingly apparent that Australia represented an enormous natural laboratory in which the opportunities for formulating and testing generally applicable process models were great indeed. Though the group of Australian workers is small, their work covers an entire continent. Australia has 30 000 or so kilometres of coastline and thus has a range of environments difficult to match elsewhere. The major deficiencies perhaps lie in the absence of storm dominated cold coasts, as were well represented in the early literature from the North Atlantic; and in ice affected coasts, for which it is necessary to travel to Australian bases in the Antarctic. But if one takes a world view, the sandy, swell-dominated, microtidal coast of New South Wales is probably very close to the mean condition and there are very real advantages in working from mean conditions towards less typical or more extreme examples. The chapter by Short and Wright in this book is an admirable demonstration of how to develop a general conceptual model on shores where conditions oscillate about the apparent mean and then to amplify and extend this model by testing it in a number of other situations involving different and less usual combinations of environmental factors.

I see this utilisation of Australia as a big laboratory, within which concepts can be developed and tested, as a major way in which contributions can be made to ideas of coastal development. There is more work to be done on the beaches, but the available variety of dunes, estuaries, cliffs, platforms and other coastal features suggests that many opportunities to develop similar models of general application remain to be exploited. This book succeeds in encapsulating a sample of the acceleration in coastal studies which has taken place over the last decade. It is a timely illustration of things which have been accomplished and contains many signposts for the future.

J. L. DAVIES

ACKNOWLEDGEMENTS

A book of this type involves the cooperation and active support of many individuals and institutions. The basic concept evolved from discussions with many colleagues who considered it was time to produce a collection of Australian papers on coastal geomorphology. John Chappell, Don Wright and Jack Davies gave me considerable encouragement in this endeavour; all the authors, as well as those colleagues who for some reason did not make a contribution in the end, made the book possible by supporting the concept of an edited volume. Many authors have made specific acknowledgements to individuals and institutions who have helped them.

I would like to add my thanks to two members of the staff of the Geography Department at Duntroon. Karl Shaw has assisted many of the authors with his various skills in the field; he has been an invaluable participant in my field programs over the past ten years. This book would not have been possible without the skill and patience of Paul Ballard, the Departmental Cartographer.

Finally, I must acknowledge those institutions which have provided funds, facilities and opportunities for the various authors to undertake such work in this vast continent. Organizations such as Bureau of Mineral Resources, Australian Institute of Marine Science, CSIRO, various universities, Australian Research Grants Scheme and Esso Australia have made much of our work possible.

1

Geomorphic Research on the Coast of Australia: A Preview

B. G. THOM

Department of Geography
University of New South Wales
Duntroon, ACT

Introduction

The last decade has seen major advances in coastal geomorphic research in Australia. In a review in 1976, I considered that the descriptive morphology of wave and wind-built features on temperate coasts, and the reconstruction of relative sea-level movements from shoreline features constituted the two dominant themes in coastal geomorphology in this country (Thom, 1976, 1978a). Neglected areas at that time included systematic studies of processes in various environments and morphostratigraphic investigations of coastal deposits. It is extremely pleasing to see that these are no longer neglected areas, and that in the past few years rigorous attempts have been made to investigate problems of a process and evolutionary nature. This change to a more analytical as distinct from a descriptive approach has been made possible by an influx of scholars trained overseas, by the application of new and often expensive techniques, and by a growing awareness of what Australian environments have to offer in the way of field sites for specific research projects.

This book provides me with an opportunity to select studies which reflect the major trends in coastal geomorphic research of the past few years. It is not designed to provide a comprehensive review of scholarship nor of environments. The selection has been made with the aim of bringing together in one volume the works of many recent contributors to Australian coastal research. It has not been possible to invite all geoscientists active in this field of geomorphology and its

COASTAL GEOMORPHOLOGY
IN AUSTRALIA
ISBN 0 12 687880 3

associated disciplines to participate in the publication. The result is of necessity a varied collection of topics and regions, but the importance of recent work should be quite evident.

The Australian Coastline

Coastal geomorphic studies in Australia are confronted with a lengthy coastline (c. 30,000 km, Galloway pers. comm., 1984), much of which is unpopulated. Although no part of the Australian continent is encompassed by a zone of plate convergence, there is still considerable structural geologic diversity along its coastline embracing a range of lithologic types (Jenkin, this volume).

It extends from latitude 10^0S to 43^0S, and from longitude 114^0E to 154^0E, and experiences several climatic and biotic types. However, this variety is perhaps less clear-cut than any other continent because of the latitudinal position and low topographic relief. Tectonic structure and a generally dry climate combine to produce a coastline which is in general free of rivers continuously supplying large quantities of sediment to the nearshore zone. In fact, oceanographic conditions along western and southern coasts are such as to promote the dominance of continental shelf-derived carbonate sediments over terrigenous materials (Davies, 1977; Bird, 1978). Along the northeast coast, the Great Barrier Reef exemplifies the power of biotic factors in coastal development (Hopley, 1982).

The latitudinal position of the continent places the southern half in sphere of influence of Southern Ocean swell waves, and the north-east corner in the southeasterly trade wind and wave zone (Davies, 1977). Northern coasts experience the impact of tropical cyclones which may dramatically modify coastal terrain. Storm surges of several metres magnitude occur in association with such cyclones (Hopley and Harvey, 1979).

Superimposed on climatic variation is a complex pattern of tidal regimes around Australia. In northwest and northeast Australia there are localities with tidal ranges in excess of 6 m, whilst extensive areas in southern Australia the range is 1-2 m (Easton, 1970; Radok, 1976).

Climatic, topographic, oceanographic and biotic factors combine to produce a variety of ecological systems ranging from salt marshes in the south to mangrove swamps in the north. Landform-plant patterns in dry areas contrast markedly with those of wetter portions (see Clough, 1982).

Coastal Classification and Landscapes

R.J. Russell once commented that too little is known about coasts to line up examples and shove them into pigeonholes: 'much of the literature on sea coasts is concerned with classification of shoreline types, but to me this is both unfortunate and premature' (Russell, 1967, p. 84). As director of the Coastal Studies Institute at Louisiana State University, he instigated reconnaissance and systematic studies of coastal areas in Australia (e.g. Sauer 1965; Russell and McIntire, 1966; Coleman *et al.*, 1966; Wright *et al.*, 1973), and elsewhere. Such studies were considered necessary to build up the 'precise and factual information' from which later classification could be developed.

In Australia, there have been few attempts to 'pigeonhole' the coast. Until recently, this was mainly due to inadequate knowledge coupled with lack of interest by the few coastal specialists. Overseas scientists have felt less concerned about the lack of data, and as part of world-wide studies have attempted to classify Australian coastal types (Valentin, 1952; McGill, 1958; Inman and Nordstrom, 1971). These attempts have incorporated some obvious errors, such as designating the southern part of the east coast as a collision coast 'although it is hard to think of one reason why it should be so categorized' (Davies, 1972, p.12; see also Jenkin, this volume). As part of his global review of geographical variability of coastal features, Davies (1972) has depicted many physical and biologic properties of the Australian coast. His intimate knowledge of this coast has made possible a more detailed categorization of Australian coastal landscapes (Davies, 1977; see also Radok, 1976).

Figure 1 synthesizes several diagrams prepared by Davies (1977, Figs. 7.1, 7.4, 7.5 7.6). Figure 1a shows coastal types occurring in Australia which form a series as the coast becomes progressively more 'low-lying'. Also shown (Fig. 1b) are three other important elements of the coastal landscape (and seascape): distribution of mangroves, dune rocks (and associated 'sandstone reefs') and platform or patch coral reefs. For purposes of description, Davies distinguished four basic coastal segments defined in terms of process environments. These are also shown in Figure 1. They are unequal in length and are unequally known, but this division has the advantage of highlighting major homogenies of process, sediment system and evolution at the present stage of knowledge.

Bird (1978) has subdivided the coast on the basis of the nature and sources of sediments which make up beach materials.

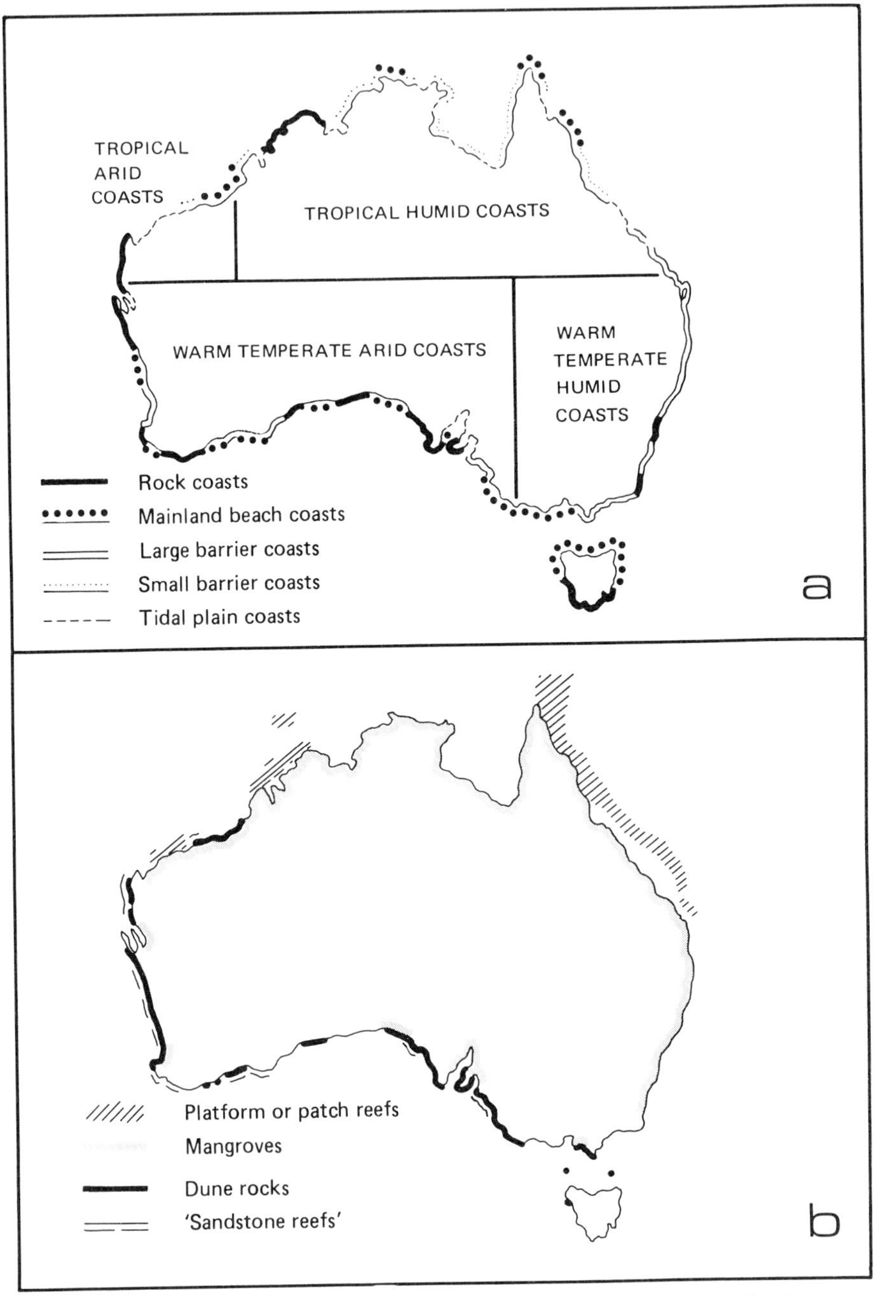

FIGURE 1. Coastal types and selected features of the Australian coast based on Davies (1977).

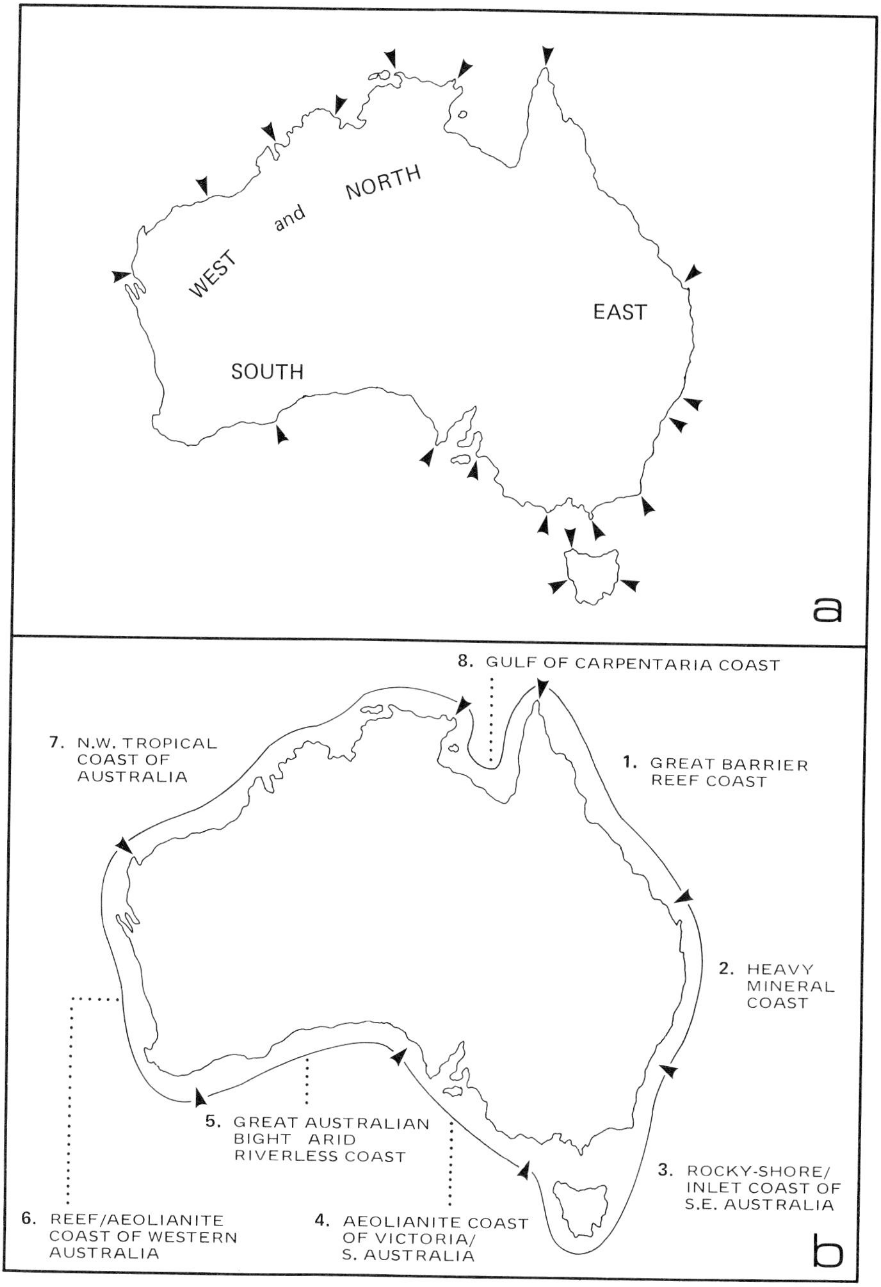

FIGURE 2. (a) Boundaries of coastal sediment systems as delimited by Bird (1978); (b) coastal types of Australia by Gill (1982).

This is a more specialist form of classification. He did not produce a map in his paper, but Figure 2a shows the general boundaries of coastal sediment systems based on descriptions in the text. Bird also stresses the lack of information, but he considers sufficient data exist to highlight major sediment systems. Interested readers should refer to his paper for more information. A more recent attempt to descriptively classify the coast of Australia is by Gill (1982). He subdivided the coast into eight parts on the basis of 'dominant features' (Fig. 2b).

A comparison of these four generalized maps reveals several common points as well as particular discrepancies. However, it is not the purpose of this paper to elaborate on details, but simply to note the existence of certain views as to how this coast can be categorized. Below this broad level of generalization is a more detailed scale of subdivision as seen in the 'provinces' which Wright *et al.*, (1973) delimit between East Kimberley and Darwin. At a still higher level are the attempts to define coastal sediment compartments (Davies, 1974; Chapman *et al.*, 1982).

The search for more information has led to the development of an 'information system' on coastal lands by the Division of Water and Land Resources, CSIRO (Galloway *et al.*, 1980). This system sub-divides coastal land into *sections* 10 km long and 3 km inland from the shoreline (3027 sections for mainland Australia and Tasmania). A grid of sample points (1 per 3 km^2) was marked on each 10 km^2 section on aerial photos; geology, landform, vegetation and land use were determined at each point. Galloway *et al.* (1980) explain the numerous problems and implications of this sample procedure. The result was 41,721 sample sites and providing an extensive inventory on these features. Data from these sample points and 30 km^2 sections can be integrated into smaller scale sections. Examples using two different scales are shown in Figure 3. Figure 3a shows Australia sub-divided into c. 600 km long sections, and Figure 3b depicts part of the east coast of Arnhem Land. The former represents the base for tabulation of data compiled by Dr. Robert Galloway on areas of the coastal zone, as defined above, covered by a particular surface type. Table 1 lists the areas for different surface types. Figure 3b is an example of the most detailed scale used by Galloway plotted initially at 1:1,000,000. From his data I have selected only those 10 km sections which possess mangrove tidal flats covering an area greater than 10 km^2. Such an information base is invaluable in assessing the characteristics of our coastal lands for a multitude of purposes (e.g. Galloway's evaluation of mangrove environments, see Galloway, 1982; and Chappell and

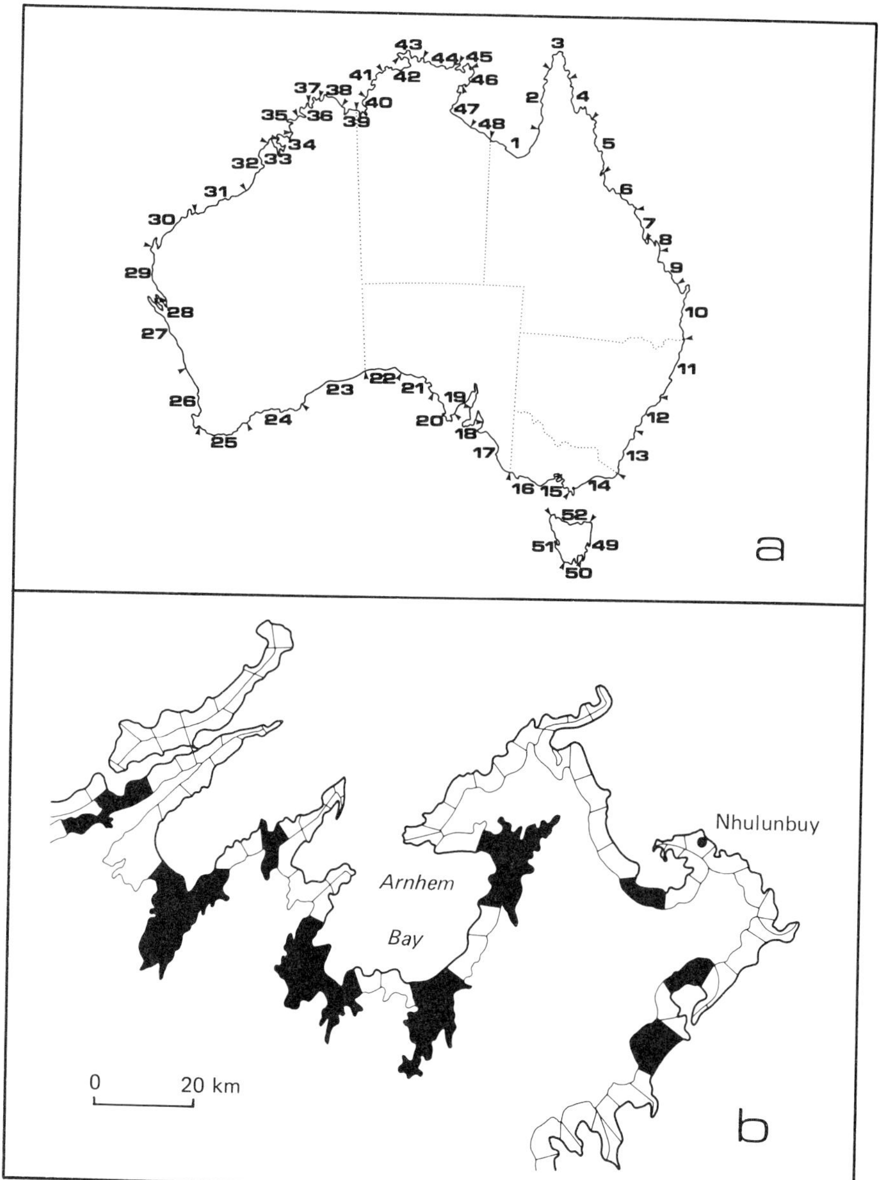

FIGURE 3. (a) Division of Australia into 600 km long sections (Table 1 shows values in km^2 for surface types by numbered sections); (b) an example of the division of part of Australia, northeast Arnhem Land, into 10 km long sections with shaded areas representing sections where mangrove tidal flats cover in excess of 10 km^2 (data from R. Galloway, CSIRO).

TABLE 1. Areas in km^2 *for different coastal surface types based on 600 km long sections as sampled by Galloway* et al. *(1980); Figure 3a shows location of each section.*

Section No	Total area	Bare mud flats	Mangrove mud flats	Holocene sand total area	Holocene sand bare	Holocene parabolic dunes
QLD						
1	6684	3432	795	690	36	0
2	3852	462	411	846	60	0
3	2229	81	855	381	45	129
4	1791	291	249	309	93	69
5	1980	27	234	348	24	72
6	2136	204	477	261	39	9
7	1536	69	249	87	21	3
8	2145	252	498	120	12	51
9	2484	453	393	345	51	84
10 (680km)	2808	81	243	207	21	51
NSW						
11	2772	15	33	267	93	15
12	3027	6	45	357	75	45
13 (520km)	1704	6	9	132	60	15
VIC						
14	3060	159	6	378	93	105
15	1635	18	9	204	42	75
16 (540km)	1515	0	0	204	111	69
SA						
17	3903	150	0	666	225	180
18	1767	81	63	243	45	57
19	2022	108	99	132	21	0
20	1860	63	0	435	168	135
21	1887	84	42	384	183	78
22 (270km)	903	0	0	147	54	33
WA						
23	2055	129	0	387	141	42
24	2280	63	0	864	225	286
25	2490	6	0	828	267	297
26	2841	12	0	1044	144	378
27	1869	51	0	537	90	240
28	1734	72	0	126	39	15
29	4002	2217	15	441	63	147
30 (750 km)	3351	1383	351	252	72	3
31	2214	711	219	207	69	6
32	2301	162	99	267	78	12
33	2682	1059	303	72	30	0
34	3069	153	429	9	6	0
35	1668	9	264	3	3	0
36	1419	60	147	6	6	0
37	1101	30	84	12	6	0
38	1374	18	75	21	12	0
39 (430km)	3160	1215	393	21	3	0
NT						
40	4578	1647	307	48	6	0
41	2115	54	456	105	24	0
42	4395	585	618	78	6	0
43	1290	18	114	75	6	6
44	2757	297	489	114	24	0
45	1845	138	330	69	15	0
46	1452	60	129	270	60	69
47	3288	1098	333	255	27	18
48 (230km)	1743	711	153	138	6	0
TAS						
49	1632	15	0	66	21	3
50	1695	6	0	36	3	0
51	2214	0	0	210	81	51
52 (430km)	1578	24	0	141	48	42

Grindrod's examination of beach and chenier terrain in northern Australia - this volume). In a somewhat similar but more specialized way, Hopley (in prep.) is developing a computerised classification of coral reef types for the Great Barrier Reef Marine Park Authority.

Themes in Australian Coastal Geomorphology

It is possible to identify six themes in research on Australian coastlines over the past 50 years or so. To some extent these themes are not self-contained because individuals often participate in more than one subject area. Specializiation has not been practical until quite recently given the abundance of problems awaiting the attention of the very few field scientists. Nevertheless, there has been a progressive concentration in the two fields of process and morphostratigraphic research which demand more specialized skills and funding to support expensive field programs.

The following is a brief review of the themes. Only a few references have been used to illustrate aspects of the work undertaken in each subject area.

Coral Reefs and Carbonate Sedimentary Environments

The presence of the magnificent Great Barrier Reef provided the basis for the first major thrust in coastal geomorphic research. From last century (Jukes, 1847; Agassiz, 1898) into the early part of this century (Andrews, 1902; Davis, 1917), the reef has fascinated scientists interested in explaining its topographic features (see Hopley, 1982, for an excellent review). The works of Steers (1929) and Spender (1930) as part of the first Great Barrier Reef Expedition of the Royal Society provided more specific details. These were later enhanced by studies based on war-time aerial photography (Fairbridge, 1950). Maxwell (1968) synthesized much data on the Reef and produced a general model of its morphologic evolution. This model was tested and further detailed geomorphic/sedimentologic work undertaken during the second Royal Society Expedition in 1973 (Stoddart *et al.*, 1978). Since then David Hopley, Peter Davies and John Marshall (see papers in this volume), amongst others, have added considerably to our knowledge of the reef region, especially through detailed drilling and dating programs. Morphodynamic work is also being undertaken (e.g. Frith, 1983).

On western and southern coasts, carbonate environments of different types have long attracted attention. Aeolianites have been studied to some degree (Teichert, 1947; Fairbridge and Teichert, 1953; Read, 1974; Warren, 1983 and Schwebel, this volume). Shark Bay in Western Australia (Logan, *et al.*, 1970, 1974) and the Coorong in South Australia (von der Borch, 1977) are areas of carbonate chemical precipitation which have been studied in detail. Research undertaken in these localities has also enhanced our knowledge of coastal depositional sequences and regional sea-levels.

Shore Platforms

A second theme which received early attention was the examination of shore platforms. Before World War 11, the influence of structural and lithological variations in rocks on the coast of New South Wales (N.S.W.) and Victoria received attention (Johnson, 1931; Jutson, 1939; Edwards, 1941). This work led to more comprehensive thinking of processes at work on shore platforms, particularly by Hills (1949) with his analysis of the process of water-layer weathering. Studies of platforms in a variety of rock types (e.g. Bird and Dent, 1966), and an appreciation of the special characteristics of horizontal platforms formed in aeolianites (Gill, 1973), have been more recent contributions to shore platform literature. Gill (1973) has also been able to make estimates on rates of platform evolution. However, it is probably fair to say that interest in platform research, so dominant in the 30's and 40's, has waned over the past two decades: 'rocky coasts have been somewhat eschewed for the more comfortable realm of beach and dune, and the common pre-1951 theme of shore platforms has been neglected with odd exceptions' (Spate and Jennings, 1972, p. 24). Nevertheless, any visit to the striking rock platforms along the N.S.W. or Victorian coast with a group of international coastal scientists is still certain to spark a good argument on platform genesis.

Morphology of Depositional Landforms

The 'explanatory descriptive' approach to understanding the distribution and character of Australia's vast areas of depositional landforms was to a large extent initiated on independent lines by J.N. Jennings, J.L. Davies and E.C.F. Bird in the mid 1950's. A notable earlier exception was the important work of Fairbridge in Western Australia (Fairbridge, 1948). The contribution of the abovementioned trio has been reviewed elsewhere (Thom, 1976). I simply wish to refer to a few of their contributions which have greatly influenced my appreciation of Australia's coast: first, by Jennings there is his examination of coastal dunes of King Island, Tasmania (Jennings, 1957); second, there is the classic paper by Davies (1957) on beach ridges, plus his appreciation of the refraction of swell waves in controlling the orientation of wave-built features (Davies, 1958); and third, the morphological study of barriers and associated lagoons by Bird (1961, 1967) has paved the way for more recent studies.

Others have expanded on the base provided by this trio. In N.S.W., Langford-Smith and his students (e.g. Thom, 1965; Hails, 1968; Langford-Smith and Thom, 1969; Wright, 1970) have described sections of an embayed coastal plain. Twidale (1968) has also encouraged coastal research in South Australia, whilst Hopley and others have conducted 'explanatory descriptive' studies in North Queensland (e.g. Hopley, 1970).

Sea-Level Change

This theme was expounded by Fairbridge (1950) who indicated that the shorelines of Australia exhibited a variety of features which could be interpreted as representing a recent fall in sea level. With the advent of radiocarbon dating, the study of relict shorelines was pursued in earnest, particularly when it was generally considered that Australia provided an apparently stable tectonic platform to which reference could be made. Visitors from the U.S.A., such as R.J. Russell and F. Shepard, came to examine and critically assess the field evidence. Australian work became an important centre for the so-called 'Indo-Pacific school' which asserted a 2-4 m high eustatic sea level(s) in the Holocene. A leader of this school was Edmund Gill who has forcefully tried to document this point over three decades (see Gill and Hopley, 1972). A number of studies supported him, including works by the trio mentioned above (see Jennings, 1959), and the early work of Hopley (1971).

Sceptics arose from two sources, one primarily geological which felt that land movements were equally possible in areas of undisputed 'raised' deposits, and thus the eustatic sea-level hypothesis, although not discounted, should not be asserted dogmatically (Bowler, 1969; Logan *et al.*, 1970). The second source of scepticism came from those who felt that much of the so-called 'raised' phenomena could be explained by present-day processes acting at the contemporary sea level (Russell, 1963; Hails, 1965; Langford-Smith and Thom, 1969; Thom *et al.*, 1969, 1972; Wright *et al.*, 1972).

Over the last decade there have been numerous Australia-wide studies of relative sea-level change. Deposits of Last Interglacial age (and older) have been identified and confirmed by radiometric dating (e.g. Marshall and Thom, 1976; Schwebel, this volume). Holocene features in a few areas have been closely examined and relative sea-level 'envelopes' produced (Thom and Chappell, 1975). A recent review volume edited by Hopley (1983) synthesizes much of this information. Papers in this volume point to local and regional variability in sea-level histories. It is clear that some of our best documented studies are emerging from dated reef and chenier deposits in northern Australia (Chappell *et al.*, 1983; Grindrod and Rhodes, this volume). However, there is still a need for more detailed regional studies of relative sea-level change supported by paleoecologic and radiometric studies in many areas.

Process Studies

In recent years, two major thrusts can be recognized in the advance of the knowledge of coastal geomorphology in Australia. The first concerns process studies of beach and inshore areas. These so-called morphodynamic systems exhibit a wide range of variability around the margins of this continent. Largely through the efforts of L.D. Wright and his colleagues at the University of Sydney, field observations have been made of changing morphologic features, water circulation patterns, wave and current spectra and sediment transport (Wright *et al.*, 1979; Short and Wright, this volume). Interactions of process and form in estuaries (Wright, 1976), high energy, wave-dominated beaches of southeast Australia (see Wright *et al.*, 1982a), across reefs of the Great Barrier Reef (Frith, 1983), and in high tidal areas of the north west (see Wright *et al.*, 1982b) are now better understood as a result of these studies. Undoubtedly, such work shows the benefit of combined physical and mental fortitude as well as modern technology.

During the last decade there have been contributions made by government organizations in coastal process studies. For instance, the Beach Protection Authority in Queensland, and the Department of Public Works, Coastal Engineering Branch, in N.S.W., have been very active in monitoring waves, currents and sediment movement along parts of the eastern coast (e.g. Gordon and Lord, 1980). These studies are quite basic to the long-term understanding of the process regimes and the development of management schemes. The Division of Oceanography, CSIRO, has been involved in estuarine dynamics research.

Detailed studies of beach morphologic change and sediment movement have also been undertaken during this period. This has been approached in several quite different ways. Bryant (1981) has been more traditionally sedimentologic in the examination of beach sand 'fabric' and its variability. Processes of beach liquefaction in response to rising water table have been described and tested experimentally by Chappell *et al.* (1979). At a somewhat expanded time scale, there have been detailed studies of beach changes both on-offshore and alongshore. In N.S.W., examples of such studies are those by Short (1979) at Narrabeen, Eliot and Clarke (1982) at Warilla, and Thom and Bowman (1980) at Moruya. These studies are continuing to provide valuable insights into temporal variability of beaches in relation to wave and atmospheric climate (see Thom, 1978b; Bryant, 1983a and b).

Morphostratigraphic Research

Another recent thrust has increased our knowledge of depositional histories and morphostratigraphies of many coastal areas. To some extent this work has developed from strong interests in carbonate depositional systems on either side of the continent: for example, Shark Bay, Western Australia (see Logan *et al.*, 1970, 1974), and the Great Barrier Reef (Maxwell, 1968; Hopley, 1982). Extensive application of radiometric dating and the use of powered drilling rigs and vibro corers are two developments which have been used on a large scale in the investigation of depositional sequences in Australia. These sequences are linked to an improved understanding of late Quaternary environmental changes, for instance relative sea level for different parts of the continent and paleo-climates.

It is not possible in this brief review to highlight all contributors to the morphostratigraphic theme. Several reef and carbonate studies have already been mentioned. The important work of Peter Cook and colleagues at Broad Sound, Queensland (e.g. Cook and Mayo, 1977), and in southeast South Australia (Cook *et al.*, 1977) must be mentioned. The value of magneto-stratigraphy in defining the Brunhes - Matuyama boundary in coastal sequences is highlighted by the latter study. Also in South Australia, there are the recent works of Hails, Gostin and Belperio on the one hand (e.g. Hails *et al.*, 1983), and Burne (1982) on the other. Several chapters in this book summarize the research of other contributors to the morphostratigraphic theme indicating the value and problems of using radiocarbon dates (see also Thom *et al.*, 1981; Nielsen and Roy, 1982) and the richness of the stratigraphic record below present sea level.

Plan of the Book

The basic approach taken in the presentation of chapters in this book is to provide readers with examples of studies which represent the two modern themes. These themes have dominated much of our coastal research in the last decade as indicated in the previous section. However, the first chapter by Jenkin provides a broader framework in which the present coastal landscape of this continent is seen from an evolutionary perspective.

The next two chapters reflect different aspects of the morphodynamic process approach to coastal depositional systems. Short and Wright synthesize much of their recent work on intermediate energy beaches. The chapter by Hesp represents the monitoring work on the coast, involving in this case detailed measurements of form and vegetation change in association with an assessment of interacting processes responsible for the development of foredunes.

Peter Roy's chapter on estuaries provides some sort of link between the geologic approach to coastal change and changes to estuarine biota. His study involves an understanding of variation in presentday estuarine morphology, sedimentology and dynamics as well as changes through time.

The Great Barrier Reef is featured in two papers. Recent summaries of Reef research appear elsewhere (see Hopley, 1982). However, the papers in this volume by Marshall and Davies on the one hand, and by Hopley on the other, provide some insight into the increased scope of Reef research.

Sea-level studies have become more focused in the last 5 years. Grindrod and Rhodes provide local documentation of Holocene sea-level change, especially the transgression phase. In contrast, Pye and Bowman use the concept of the marine transgression in their explanation of widespread coastal dunes which formed in the early Holocene in eastern Australia.

A set of three papers examine the morphology, stratigraphy, age structure and soils of coastal deposits in Queensland, N.S.W. and Victoria. Two studies are regional in character and examine modes of deposition along prograding coasts of two distinct types: a chenier plain and a complex sand barrier. Thompson and Bowman utilize a knowledge of age of coastal sand deposits to examine various soil features. Soil-geomorphic relations has long been an important topic in geoscience research in Australia.

A broader historical perspective is developed in the final two papers. The well-studied aeolianite belt of southeast South Australia is examined from the point of view of facies analysis. Schwebel utilizes both magneto and radiometric dating techniques to provide some order to this sequence. Northeast Tasmania offers information on possibly quite significant earth movements of the late Quaternary. The meaning of the disposition of marine deposits in this area remains poorly understood. Nevertheless, the paper by Bowden and Colhoun provides some suggestions based on Bowden's detailed field work (Bowden, 1981).

Acknowledgements

The author would like to thank Dr. Robert Galloway for his contribution to this paper.

References

Agassiz, A. (1898). A visit to the Great Barrier Reef of Australia in the steamer Croydon during April and May, 1896. *Bull. Mus. Comp. Zool. Harvard Coll.* 28, 95-148.

Andrews, E.C. (1902). Preliminary note on the geology of the Queensland coast. *Proc. Linn. Soc. N.S.W.* 27, 146-185.

Bird, E.C.F. (1961). The coastal barriers of east Gippsland, Australia. *Geographical Journal* 127, 460-468.

Bird, E.C.F. and Dent, O.F. (1966). Shore platforms on the south coast of New South Wales. *Australian Geographer* 10, 71-80.

Bird, E.C.F. (1967). Coastal lagoons of southeastern Australia. *In* Jennings, J.N. and Mabbutt, J.A. (eds.), *Landform Studies from Australia and New Zealand*, A.N.U. Press, Canberra, 366-385.

Bird, E.C.F. (1978). The nature and source of beach materials on the Australian coast. *In* Davies, J.L. and Williams, M.A.J. (eds.) *Landform Evolution in Australasia*, A.N.U. Press, Canberra, 144-157.

Bowden, A.R. (1981). Coastal Sands of Northeastern Tasmania: Geomorphology and Groundwater Hydrology. Ph.D. thesis, Department of Geogrpahy, University of Tasmania.

Bowler, J.M. (1969). Port Phillip survey, 1957-1963, geology and geomorphology. *National Museum Memoir* 27, Melbourne, 19-67.

Bryant, E. (1982). Behaviour of grain size characteristics on reflective and dissipative foreshores, Broken Bay, Australia. *Journal of Sedimentary Petrology* 52, 431-450.

Bryant, E. (1983a). Coastal erosion and accretion, Stanwell Park beach, N.S.W., 1890-1980. *Australian Geographer* 15, 382-390.

Bryant, E. (1983b). Regional sea level, Southern Oscillation and beach change, New South Wales, Australia. *Nature* 305, 213-216.

Chappell, J., Eliot, I.G., Bradshaw, M.P. and Lonsdale, E. (1979). Experimental control of beach face dynamics by water-table pumping. *Engineering Geology* 14, 29-41.

Chappell, J., Chivas, A., Wallensky, E., Polach, H. and Aharon, P. (1983). Holocene palaeo-environmental changes, central to north Great Barrier Reef inner zone. B.M.R. *Journal of Australian Geology and Geophysics 8*, 223-236.

Clough, B.F. (ed.) (1982). *Mangrove Ecosystems in Australia.* A.N.U. Press, Canberra, 302 pp.

Coleman, J.M., Gagliano, S.M. and Smith, W.G. (1966). Chemical and physical weathering on saline high tidal flats, northern Queensland, Australia. *Geological Society of America, Bulletin* 77, 205-206.

Cook, P.J. and Mayo, W. (1977). Sedimentology and Holocene history of a tropical estuary (Broad Sound, Queensland). *Bureau of Mineral Resources, Geology and Geophysics, Bulletin* 170, 206 pp.

Cook, P.J., Colwell, J.B., Firman, J.B., Lindsay, J.M., Schwebel, D.A. and von der Borch, C.C. (1977). Late Cainozoic sequence of southeast of South Australia and Pleistocene sea level changes. *B.M.R. Journal of Geology and Geophysics* 2, 81-88.

Davies, J.L. (1957). The importance of cut and fill in the development of sand beach ridges. *Australian Journal of Science* 20, 105-111.

Davies, J.L. (1958). Wave refraction and the evolution of shoreline curves. *Geographical Studies* 5, 1-14.

Davies, J.L. (1972). *Geographical Variation in Coastal Development.* Oliver and Boyd, Edinburgh, 204 pp.

Davies, J.L. (1974). The coastal sediment compartment. *Australian Geographical Studies* 12, 139-151.

Davies, J.L. (1977). The coast. *In* Jeans, D.M. (ed.), *Australia, a geography,* University of Sydney Press, Sydney, 134-151.

Davis, W.M. (1917). The Great Barrier Reef of Australia. *American Journal of Science Ser.* 4, 44, 339-350.

Easton, A.K. (1970). The tides of the continent of Australia. *Horace Lamb Centre Research Paper* 37, Flinders University, 326 pp.

Edwards, A.B. (1941). Storm-wave platforms. *Journal of Geomorphology* 4, 223-236.
Eliot, I.G. and Clarke, D.J. (1982). Seasonal and biennial fluctuation in subaerial beach sediment volume on Warilla beach, New South Wales. *Marine Geology* 48, 89-103.
Fairbridge, R.W. (1948). The geology and geomorphology of Point Peron, Western Australia. *Journal of Royal Society of Western Australia* 34, 35-72.
Fairbridge, R.W. (1950). Recent and Pleistocene coral reefs of Australia. *Journal of Geology* 58, 330-401.
Fairbridge, R.W. and Teichert, C. (1953). Soil horizons and marine bands in the coastal limestones of Western Australia. *Proceedings Royal Society of N.S.W.* 86,68-87.
Frith, C.A. (1983). Aspects of lagoon sedimentation and circulation at One Tree Reef, southern Great Barrier Reef. *B.M.R. Journal of Australian Geology and Geophysics* 8, 211-222.
Galloway, R.W. (1982). Distribution and physiographic patterns of Australian mangroves. *In* Clough, B. (ed.), *Mangrove Ecosystems in Australia,* A.N.U. Press, Canberra, 31-53.
Galloway, R.W. and Bahr, M.E. (1979). What is the length of the Australian coast? *Australian Geographer* 14, 244-247.
Galloway, R.W., Story, R., Cooper, R. and Yapp, G.A. (1980). Coastal lands of Australia. *Division of Land Use Research, CSIRO, Technical Memorandum* 80/24, 61 pp.
Gill, E.D. (1973). Rate and mode of retrogradation on rocky coasts in Victoria, Australia, and their relationship to sea level changes. *Boreas* 2, 143-147.
Gill, E.D. (1982). Eight coasts of Australia. *Division of Applied Geomechanics, Technical Report* No. 119, CSIRO, 66 pp.
Gill, E.D. and Hopley, D. (1972). Holocene sea levels in eastern Australia - a discussion. *Marine Geology* 12, 223-233.
Gordon, A.D. and Lord, D.B. (1980). An approach to understanding coastal processes. *Proceedings 17th International Coastal Engineering Conference,* Sydney, March, 1980, 1235-1254.
Hails, J.R. (1965). A critical review of sea-level changes in eastern Australia since the last glacial. *Australian Geographical Studies* 3, 63-78.
Hails, J.R. (1968). The late Quaternary history of part of the mid-north coast, New South Wales, Australia. *Transactions Institute British Geographers* 44, 133-149.
Hails, J.R., Belperio, A.P. and Gostin, V.A. (1983). Holocene sea levels of Upper Spencer Gulf, South Australia. *In* Hopley, D. (ed.) *Australian Sea Levels in the Last 15,000 Years: A Review,* Department of Geography, James Cook University, Monograph Series, Occasional Paper 3, 48-53.

Hills, E.S. (1949). Shore platforms. *Geological Magazine* 86, 137-152.

Hopley, D. (1970). *Geomorphology of the Burdekin Delta, North Queensland*. Department of Geography, James Cook University, Monograph Series, 1, 66 pp.

Hopley, D. (1971). The origin and significance of North Queensland island spits. *Zeitzchr. Geomorph.* 15, 371-389.

Hopley, D. (1982). *The Geomorphology of the Great Barrier Reef: Quaternary Evolution of Coral Reefs*. John Wiley-Interscience, New York, 453 pp.

Hopley, D. (ed.) (1983). Australian sea levels in the last 15,000 years: a review. *Occasional Paper No. 3, Monograph Series, Department of Geography, James Cook University*, 104

Hopley, D. and Harvey, N. (1979). Regional variations in storm surge characteristics around the Australian coast: a preliminary investigation. *In* Heathcote, R.L. and Thom, B.G. (eds.), *Natural Hazards in Australia*, Australian Academy of Science, Canberra, 164-185.

Inman, D.L. and Nordstrom, C.E. (1971). On the tectonic and morphologic classification of coasts. *Journal of Geology* 79, 1-21.

Jennings, J.N. (1957). On the orientation of parabolic or U-dunes. *Geographical Journal* 123, 474-480.

Jennings, J.N. (1959). The coastal geomorphology of King Island, Bass Strait, in relation to changes in the relative level of land and sea. *Rec. Queen Victoria Mus.* (N.S.), 11, 1-39.

Johnson, D.W. (1931). Supposed two metre eustatic bench of the Pacific shores. *Comptes Rendus, International Geographical Congress* (Paris), 2, 158-163.

Jukes, J.B. (1847). *Narrative of the Surveying Voyage of H.M.S. Fly*, T. and W. Boone, London, 2 vols.

Jutson, J.T. (1939). Shore platforms near Sydney, N.S.W. *Journal of Geomorphology* 2, 237-250.

Langford-Smith, T. and Thom, B.G. (1969). New South Wales coastal morphology. *Journal of the Geological Society of Australia* 16, 572-580.

Logan, B.W., Davies, G.R., Read, J.F. and Cebulski, D.E. (1970). Carbonate sedimentation and environments, Shark Bay, Western Australia. *American Association of Petroleum Geologists Memoir* 13, 223 pp.

Logan, B.W., Read, J.F., Hagan, G.M., Hoffman, P., Brown, R.G., Woods, P.J. and Gebelein, C.D. (1975). Evolution and diagenesis of Quaternary carbonate sequences, Shark Bay, Western Australia. *American Association of Petroleum Geologists Memoir* 22, 358 pp.

McGill, J.T. (1958). Map of coastal landforms of the world. *Geographical Review* 48, 402-405.

Marshall, J.F. and Thom, B.G. (1976). The sea level in the last interglacial. *Nature* 263, 120-121.

Maxwell, W.G.H. (1968). *Atlas of the Great Barrier Reef.* Elsevier, Amsterdam, 258 pp.

Nielsen, A.F. and Roy, P.S. (1982). Age contamination of radiocarbon dates on shell hash from coastal sand deposits, southeast Australian examples. *Proceedings 5th Australian Conference on Coastal and Ocean Engineering,* Perth, November, 1981, 177-182.

Radok, R. (1976). *Australia's Coast.* Rigby, Adelaide, 100 pp.

Read, J.F. (1974). Calcrete deposits and Quaternary sediments, Edel Province, Shark Bay, Western Australia. *American Association Petroleum Geologists Memoir* 22, 250-282.

Russell, R.J. (1963). Recent recession of tropical cliffy coasts. *Science* 38, 9-15.

Russell, R.J. (1967). *River Plains and Sea Coasts.* University of California Press, Berkeley, 173 pp.

Russell, R.J. and McIntire, W.G. (1966). Australian tidal flats. *Coastal Studies Series* No. 12, L.S.U. Press, Baton Rouge, 48 pp.

Sauer, J. (1965). Geographic reconnaissance of Western Australian nearshore vegetation. *Australian Journal of Botany* 13, 39-69.

Short, A.D. (1979). Three dimensional beach-stage model. *Journal of Geology* 87, 553-571.

Spate, O.H.K. and Jennings, J.N. (1972). Australian geography, 1951-1971. *Australian Geographical Studies* 9, 197-224.

Spender, M.A. (1930). Island reefs of the Queensland coast. *Geographical Journal* 76, 194-214, 273-297.

Steers, J.A. (1929). The Queensland coast and the Great Barrier Reef. *Geographical Journal* 74, 232-257, 341-370.

Stoddart, D.R., McLean, R.F., Scoffin, T.P., Thom, B.G. and Hopley, D. (1978). Evolution of reefs and islands, northern Great Barrier Reef: synthesis and interpretation. *Phil. Trans. Roy. Soc. London* B, 284, 149-159.

Teichert, C. (1947). Contributions to the geology of Houtman's Abrolhos, Western Australia. *Proc. Linn. Soc. N.S.W.* 71, 145-196.

Thom, B.G. (1965). Late Quaternary coastal morphology of the Port Stephens-Myall Lakes area. *Journal and Proceedings of the Royal Society of New South Wales* 98, 23-36.

Thom, B.G. (1976). An assessment of coastal research in Australia with particular reference to geomorphology. *Geoscience and Man* (Louisiana State University Press) 14, 127-134.

Thom, B.G. (1978a). Future trends in coastal geomorphology. *Australian Geographer* 14, 64-65.

Thom, B.G. (1978b). Coastal sand deposition in southeast Australia during the Holocene. *In* Davies, J.L. and Williams, M.A.J. (eds.), *Landform Evolution in Australasia,* A.N.U. Press, Canberra, 197-214.

Thom, B.G., Hails, J.R. and Martin, A.R.H. (1969). Radiocarbon evidence against higher postglacial sea levels in eastern Australia. *Marine Geology* 7, 161-168.

Thom, B.G., Hails, J.R., Martin, A.R.H. and Phipps, C.V.Q. (1972). Postglacial sea levels in eastern Australia - a reply. *Marine Geology* 12, 233-242.

Thom, B.G. and Chappell, J. (1975). Holocene sea levels relative to Australia. *Search* 6, 90-93.

Thom, B.G. and Bowman, G.M. (1980). Beach erosion and accretion at two time scales on a high energy coast, N.S.W. Australia. *Proceedings 17th International Coastal Engineering Conference,* Sydney, March, 1980, 934-945.

Thom, B.G., Bowman, G.M., Gillespie, R., Temple, R. and Barbetti, M. (1981). Radiocarbon dating of Holocene beach-ridge sequences in southeast Australia. *Monograph No. 11, Geography Department, University of N.S.W. Duntroon,* 36 pp.

Twidale, C.R. (1968). *Geomorphology, with special reference to Australia.* Ch. 8 Coastal Process and Forms, Nelson, Melbourne.

Valentin, H. (1952). *Die Kusten der Erde.* Supplement to *Petermanns Geogr. Milt.,* No. 246.

von der Borch, C.C. (1977). Stratigraphy and formation of Holocene dolomite carbonate sediments of the Coorong area, South Australia. *Sedimentology* 46, 952-966.

Warren, J.K. (1983). Pedogenic calcrete as it occurs in Quaternary calcareous dunes in coastal South Australia. *Journal of Sedimentary Petrology* 53, 787-796.

Wright, L.D. (1970). The influence of sediment availability on patterns of beach ridge development in the vicinity of the Shoalhaven River delta, N.S.W. *Australian Geographer* 11, 336-348.

Wright, L.D. (1976). Nearshore wave-power dissipation and the coastal energy regime of the Sydney-Jervis Bay region, New South Wales: a comparison. *Australian Journal of Marine and Freshwater Research* 27, 633-640.

Wright, L.D., Coleman, J.M. and Thom, B.G. (1972). Emerged tidal flats in the Ord River estuary, Western Australia. *Search* 3, 339-341.

Wright, L.D., Coleman, J.M. and Thom, B.G. (1973). Geomorphic coastal variability, northwestern Australia. *Coastal Studies Bulletin* 7, 35-64.

Wright, L.D., Chappell, J., Thom, B.G., Bradshaw, M.P. and Cowell, P. (1979). Morphodynamics of reflective and dissipative beach and inshore systems: southeastern Australia. *Marine Geology* 32, 105-140.

Wright, L.D., Guza, R.T. and Short, A.D. (1982a). Dynamics of a high-energy dissipative surf zone. *Marine Geology* 45, 41-62.

Wright, L.D., Nielsen, P., Short, A.D. and Green, M.O. (1982b). Morphodynamics of a macrotidal beach: Broome, Western Australia. *Coastal Studies Unit Technical Report* No. 82/1, 53 pp.

2

Evolution of the Australian Coast and Continental Margin

J. J. JENKIN

Soil Conservation Authority
Melbourne, Vic.

Introduction

The coastal zone is the product of interaction between marine and terrestrial processes, set in a morphotectonic framework. These factors change through time and to view coastlines of the past simply in terms of present conditions is highly misleading. Seafloor spreading and consequent plate movements, accompanied by passive marginal rifting and active subduction, have resulted in both horizontal and vertical changes in oceanic dimensions. Such changes lead inevitably to the modification of marine circulation, swell regimes and the position of continents relative to climatic zones. Freeboard, the relative elevation of continents with respect to sea level (Wise, 1974), is likely to have varied through time, possibly as a direct result of plate movement and marginal tectonics, but also involving the tilting of continental plates as a whole. Differential earth movements within plates are often expressed as broad, shallow depositional basins, many being inundated by the sea at some stage in their history. Some tectonic effects, however, are more restricted in extent and individually can have noteworthy local results, while together they largely control the distribution of major coastal segments.

Past coasts as such are rarely seen, particularly those older than Pleistocene. However, their position and sometimes their form may often be deduced from the composition and structure of sedimentary deposits, as well as being inferred

COASTAL GEOMORPHOLOGY
IN AUSTRALIA
ISBN 0 12 687880 3

from the geometry of the continental margin or epicontinental basins. Using this kind of information, an attempt is made to trace in broad outline the development of the Australian continental margin, the evolution of the contiguous oceans, the relative sea level fluctuations since the breakup of Gondwanaland, and the role that these changes have played in the establishment of the present coast.

Development of the Australian Continental Margin

Gondwanaland Fragmentation

Eastern Gondwanaland began to break up in the late Jurassic (c. 150 Ma), with a distinct rupture which separated the incipient Indian and Australian plates (Johnson, *et al.*, 1976). This was accompanied by marine inundation, an event forshadowed as far back as the Late Carboniferous (185 Ma) by rifting, terrestrial deposition, alkaline volcanism and periodic marine incursion (Veevers & Evans, 1975; Veevers & Cotterill, 1978). However, northwestern Australian had continually faced the ocean as far back as the Ordovician (c. 450 Ma). In eastern, south-western, southern and south-eastern Australia, deposition during the Jurassic was entirely fluviatile, largely in troughs which later were to be successively inundated by the sea (Falvey & Mutter, 1981; Veevers & Evans, 1975). In the west, from Mid Jurassic into the Early Cretaceous, extensive marginal marine sedimentation commenced offshore in the rift zone extending from the Browse Basin in the north to the Perth Basin in the south (Fig. 1). This was followed during the later Early Cretaceous to the early Late Cretaceous by the expansion of marine conditions into the southwest of Western Australia, eastwards to the Eyre Peninsula in South Australia, with marginal marine extensions entering the Victorian Otway Basin. Towards the end of the Cretaceous, oceanic conditions became firmly established in western and northwestern Australia, accompanied by extensive carbonate sedimentation (Veevers & Evans, 1975).

Shallow marine inundation in the north began towards the end of the Early Cretaceous and covered a large part of east-central Australia, although this event did not coincide with the final tectonic isolation of the continent (see Figs. 1 and 6). The epicratonic sea of the Cretaceous is difficult to explain directly in terms of rifting and seafloor spreading. However, at this time, pre-rift swells were developing along the eastern and southern continental margins (Ollier, 1979; 1981). This could have been accompanied by complementary intra-continental downwarping, or even by a more general NNW tilting resulting in freeboard changes, allowing inundation of the Canning and Great Artesian Basins (see Veevers, 1980). The subsequent Late Cretaceous retreat of the sea in the north and northwest, exposing the Papuan shelf, was accompanied by inundations in the southern marginal basins, not all of which seem to be associated with rifting. This suggests the possibility of a reversal in continental tilt.

In the Cretaceous, extensive rifting took place off eastern Australia (Hayes & Ringis, 1973) producing a succession of arcuate splinters, now represented by New Zealand, the Lord Howe Rise, and parts of New Guinea (Carey, 1976), which converged on the southern Pacific Plate. It is unlikely that any remnant of the original east coast of Cretaceous age remains.

Isolation Completed (Early Tertiary)

The separation of Australia from Antarctica by the Southern Ocean, and from the Lord Howe Rise by the Tasman Sea,

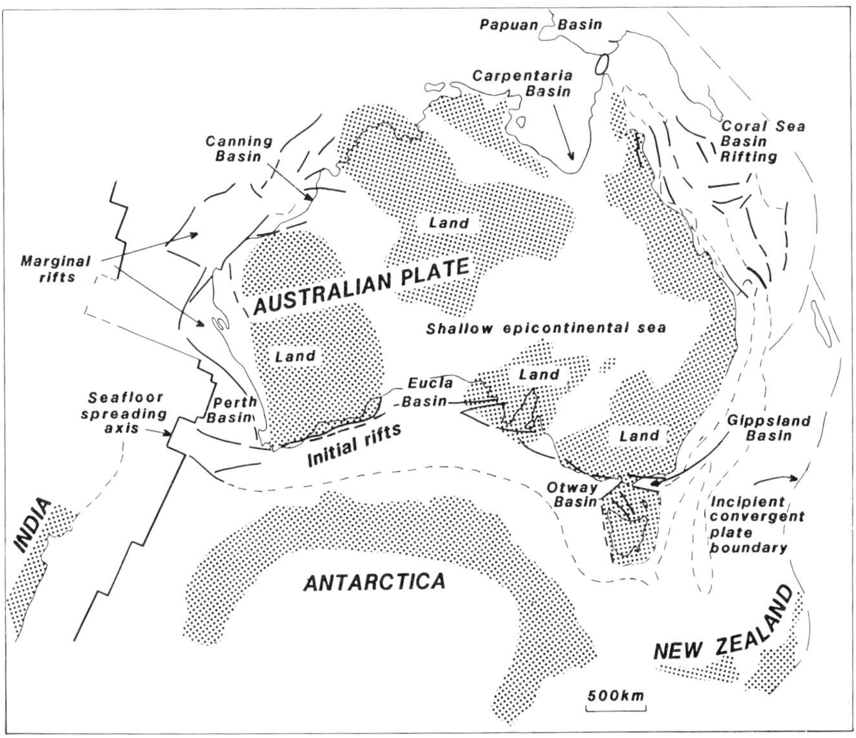

FIGURE 1. Australia in the Early Cretaceous, c. 110 Ma.

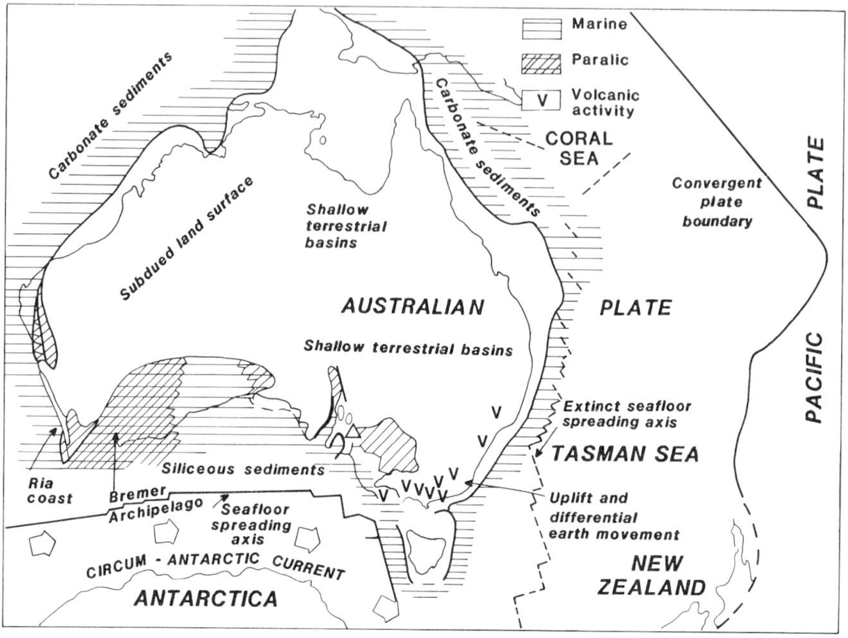

FIGURE 2. Australia in the Early Tertiary, c. 55 Ma.

combined with the Early Tertiary development of the Coral Sea, finally established the Australian continent in roughly its present form. However, the landmass still included Torres Strait and the Fly Platform area of Papua. The predominance of carbonate deposition at this time (Veevers & Evans, 1975) is consistent with the movement of the Australian plate into progressively more tropical regions (Fig. 2).

The eastern Australian continental margin has several unusual characteristics, including an exceptionally narrow shelf and coastal lowlands backed by the Great Escarpment. The now-dead rifted margin (Falvey & Mutter, 1981) has been likened by Ollier (1982) to the Western Rift of East Africa, with the Great Escarpment being due to the erosional retreat of the original rift scarp.

The broad outline of the southern half of the continent was very much as it is today, except for marginal marine incursions in the Eucla, St. Vincent, Murray, Otway and Gippsland Basins. The stratigraphy of these basins is reviewed in Playford *et al.*, (1975), Glaessner and Wade (1958), Ludbrook (1980), Wopfner & Douglas, eds. (1971) and Abele, *et al.*, (1976). These inundations occurred at about the same time (Mid Eocene) although they were foreshadowed in the Late Cretaceous by local marine episodes in the Eucla and Otway Basins.

In the Mid to Late Oligocene, with the final separation of Antarctica from the south Tasman Rise, the Circum-Antarctic Current became clearly established, producing an oceanic circulation system which was essentially the same as that operating at present (Kennett, *et al.*, 1974). In southern Australia, the markedly cool Oligocene, accompanied by low sea levels, could have been due partly to the development of an ice sheet in Antarctica, resulting in a cold current impringing on the outer shelf (see Frakes, 1979).

Maximum Marginal Inundations (Mid to Late Tertiary)

Between the Paleocene and Late Oligocene the pattern of marine advance and retreat is confused and may have been partly controlled by differential earth movement. There is undoubted direct evidence of such movements in eastern Australia (Jenkin, 1968; Ollier, 1981). However, beginning in the Late Oligocene and extending to a relatively sharp cutoff in the Late Miocene, there was a general advance of the sea in all marginal basins, indicated now by prominent limestone exposures in coastal and river cliffs, particularly in southern Australia (Fig. 3). Carbonate sedimentation became well-established in these shallow seas and terrestrial components were fine-grained, mainly fine sands and muds. In narrow zones where the coast impinged on bedrock or coarse sediments, there were areas of coarser deposition including shelly conglomerates, gritty limestones and calcareous sands, for example in the Eucla Basin (Playford, *et al.*, 1975) and in Western Port and East Gippsland (Jenkin 1962, 1981). The dominance of carbonates may be related to higher temperatures, hinterland aridity or general low relief, or a combination of all three.

Retreat of the Late Tertiary Sea

The subsequent retreat of the sea from the Tertiary basins occurred in the Late Miocene and Pliocene. Some local advances have been recorded during this period in the St. Vincent and Murray Basins (Glaessner & Parkin, eds., 1958), in the Otway Basin (Abele, *et al.*, 1976), and in the Gippsland Basin (Hocking, *et al.*, 1976). The retreat in the Murray Basin is marked by a succession of straight to arcuate sand ridges (Figs. 4 and 5). These represent stranded shoreline features (Blackburn 1962; Bowler and Magee, 1978; Jenkin, 1976, 1981). Similar shoreline features, also Pliocene in age, have recently been recognized in southern Victoria (Tuddenham, pers. comm.; Jenkin 1981, 1982).

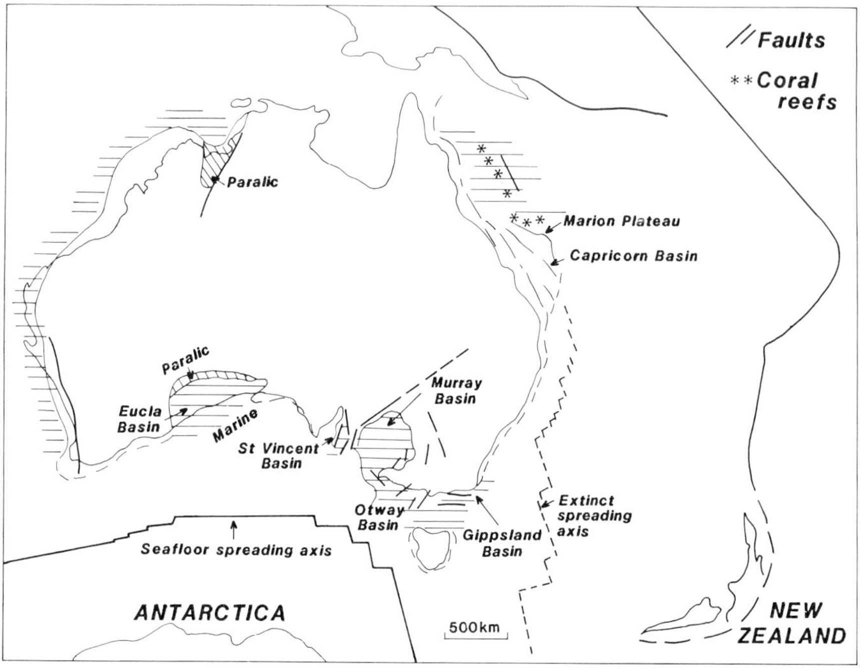

FIGURE 3. Australia in the Mid Miocene, c. 15 Ma.

Quaternary Shoreline Changes

Around most of Australia the landward Quaternary shoreline has transgressed inland very little beyond the present coast. However, there are three major exceptions. In the western part of the Otway Basin, Pleistocene coastal forms and deposits occur 100 km or more inland (Figs. 4 and 5). The inner limit is uncertain, but in Victoria, it would probably have been at the foot of the Kanawinka and contiguous scarps (Kenley, 1971, 1976). This assumes that the upper part of the Whalers Bluff formation is Early Pleistocene at its inner edge and that diachronism is not involved. Extending from there, and perhaps seaward from the present coast, is a series of straight to arcuate 'dune ranges' interpreted as foredunes resting on a sandy barrier or beach ridge and separated by estuarine, lagoonal and beach ridge deposits. An overall relative fall in sea level in western Victoria of at least 65 metres is indicated, but there were also spasmodic rises during the period which resulted in breaching and truncation of older dunes (Kenley, 1971, 1976). In South Australia, Hossfeld (1950) and Sprigg (1952) attempted to relate the dunes to the then current Quaternary stratigraphic system. Subsequently,

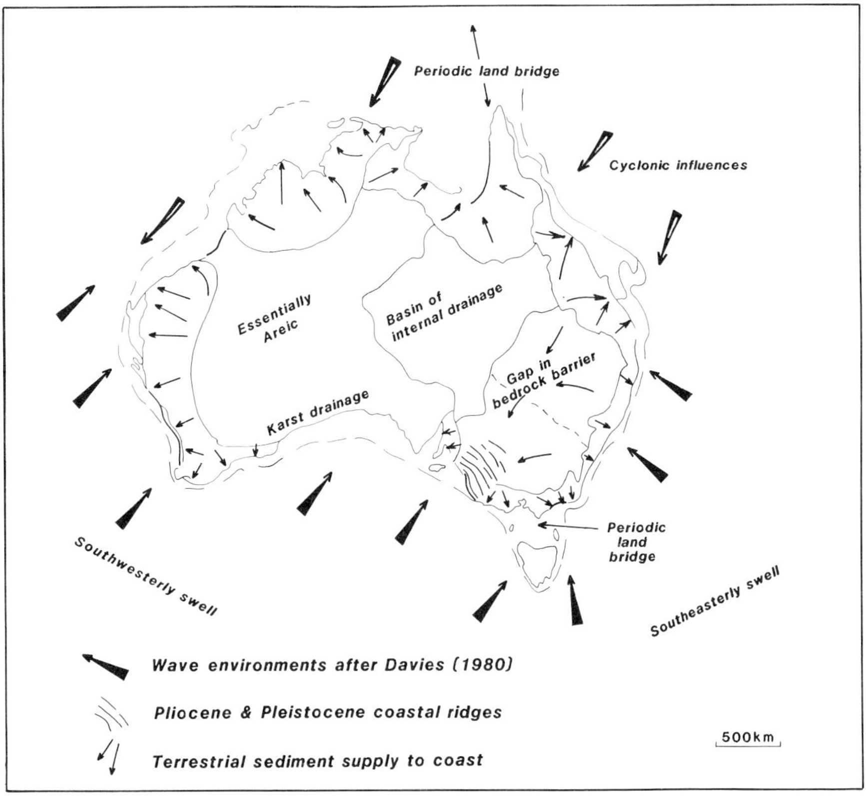

FIGURE 4. Australia in the late Cainozoic.

Cook *et al.*, (1977) interpreted the data in terms of at least twenty major high sea level stands over the last 690,000 years (see also Schwebel, this volume).

In the Carnarvon Basin of Western Australia, Quaternary marine deposits are widespread and extend for at least 50 km inland in the Rough Range - Lake McLeod area (Playford, *et al.*, 1975). Except at Shark Bay (Logan, *et al.*, 1970), the Carnarvon Basin Quaternary has not been studied in detail. In the third area, the Carpentaria Basin in Queensland, Pleistocene beach ridges, believed to date from c.110,000 yr. B.P., have been recorded up to 30 km inland. Holocene beach ridges up to 20 km inland are reported (Grimes and Doutch, 1978).

In the northeastern part of Spencer Gulf in South Australia, it appears that the peak of the Holocene transgression occurred at 6000-4000 yr. B.P., represented by a shingle ridge lying at +3 to +4 m. This was followed immediately by an apparent fall in sea level of about 1 m., then by an apparent fall of 0.5 m

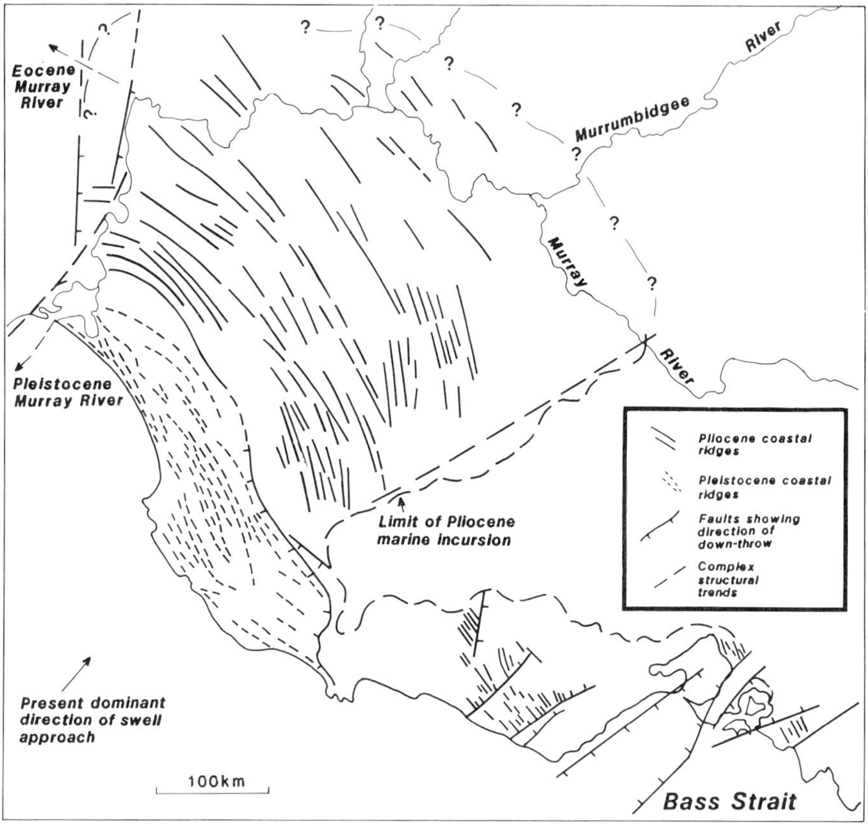

FIGURE 5. Pliocene and Pleistocene stranded coastal ridges, southern Australia. Sources: Blackburn (1962), Bowler and Magee (1978), Jenkin (1981) and ERTS imagery.

between 4000 and 3000 yr. B.P. During the succeeding 1000 years a 1 m fall is recorded, and a 0.5 to 1 m fall is suggested for the last 2000 years. High seismicity in the neighbouring Flinders Ranges accompanied by continuing tectonic uplift, is regarded as a contributing factor (Burne, 1982). This work illustrates the difficulty of determining past absolute sea levels. The area is not only tectonically active, but is at the head of an embayment which narrows inland. Because of this, any major change in oceanic, wind or estuarine circulation regimes would have an exaggerated effect at the head of the Gulf. Consequently, this evidence does not prove that any Holocene sea level has exceeded that of the present.

The position of the shoreline seaward of the present coast has fluctuated widely in response to glacio-eustatism in the Quaternary. At times, virtually the whole of the continental shelf has been exposed, establishing land bridges which connected Tasmania and Papua-New Guinea to the mainland (Fig. 4). At the end of the Pleistocene, during the 'last Glacial' maximum, when the sea was more than 100 metres below its present level, the Australian landmass was one third larger than it is at present (Gill, 1971).

Supposed submerged coastal features on the continental shelf have been mentioned by various authors. Gill (1971) listed coral and algal reefs, beach rock and shallow water shellbeds, sand ridges, aeolianite dunes, terraces and low sea cliffs. A submerged cliff off the Otway coast of Victoria has been interpreted as a wave-modified fault scarp. Calcarenites at the cliff top and at its base (-45 m) were regarded as relict beaches (Gill, *et al.*, 1980). Further, linear terraces and ridges on the shelf, lying more or less parallel to the present east coast, have been regarded as fossil strandlines (Davies, 1979; Jones & Davies, 1979). Roy and Thom (1981), however, pointed out that some of the low ridges off the northern part of this coast probably result mainly from present-day strong ocean currents, while accepting that remnants of early barrier structures have possibly survived on the N.S.W. shelf.

Factors Influencing the Evolution of the Continental Margin

The two basic factors influencing coastal change through geological time are tectonics, including rifting and continental earth movement, and climate, both local and distant.

Role of Continental Rifting

The general trend of most major segments of the Australian coast is a direct expression of passive continental rifting. The exception is in the north where the plate boundary is actively convergent (Fig. 3). The eastern edge of the Australian Plate is also convergent although the present mainland margin is the result of rifting. The form of this east coast rifting, like that of the Western Rift in East Africa, explains the position of the main drainage divide many kilometres west of the rift margin, the production of the erosional Great Escarpment and the dismembering and reversal of streams (Ollier, 1982). Rifting, also analogous to that of East Africa, has been proposed for the submerged western margin of Australia (Veevers and Cotterill, 1978). The complex rift systems of the eastern and western continental margins are in marked contrast to the relatively simple southern rifts (Falvey and Mutter, 1981).

The widening of oceans also increased the available fetch significantly. This is shown in the Perth Basin where exposure of soft shelf sediments to increasingly strong marine erosion resulted in the production, from the Eocene onwards, of wide marine abasion platforms backed by cliffs (Finkl and Fairbridge, 1979).

Continental Earth Movements

Broad warping as well as more localised movements within the Australian Plate have produced shallow epeiric seas and marginal embayments and gulfs. The Mid Cretaceous epicontental sea, which flooded much of the Australian craton at the same time as world sea level was low, was associated with rapid convergence of Australia's eastern margin. The Late Cretaceous retreat however, occurred when sea level was at a maximum and was related to the decay of local convergence. This is believed to have generated uplift of the continental interior (Veevers, 1980).

Although many intra-cratonic movements are directly related to events at plate margins, other forms of tectonic activity are also operative (Ollier, 1978, 1981). Post-Palaeozoic faulting within the craton is usually regarded as normal (Jenkin, 1968; Kenley, 1971; Ludbrook, 1981; Playford, *et al.*, 1975). In south-eastern Australia, Gibson (1982) has suggested that compressive stress has produced alternate zones of tectonically positive and negative expression, a view which is not consistent with that of Ollier (1981). The marginal basins which extend into the craton usually have fault or monoclinal margins, principally the Carnarvon, Perth, St. Vincent, Otway and Gippsland Basins (Figs. 2 and 3). The Canning and the Eucla Basins on the other hand are broad downwarps of the craton margin with little if any faulting involved (Ludbrook, 1980; Playford, *et al.*, 1975).

The intensity of earth movements and the trend of the structural lineaments relative to the coast are important factors in controlling the form of major coastal segments. Where the structural grain lies at a marked angle to the continental margin there is a tendency towards elongate gulfs or bays occupying at least part of the sunkland. These alternate with prominent cliffed sections in resistant rock as in St. Vincent Gulf in South Australia (see Glaessner and Wade, 1958). However, where the structural grain is parallel to the continental margin, as in the Perth Basin, long, straight cliffed coasts are formed (Finkl and Fairbridge, 1979).

The Murray - Otway Basin system of south-eastern Australia, although fault controlled, does not fit this pattern. Here, the Late Miocene - Pliocene withdrawal of the sea coincided with the later phase of Cainozoic tectonic activity, which appears to have been oscillatory in nature. Thus, in addition to the effects of eustatic retreat, uplift resulted in some former shorelines being higher than they would otherwise have been. The retreat is also marked by a sequence of barrier-lagoon systems developed at successively lower levels (Blackburn, 1962; Bowler and Magee, 1978; Jenkin, 1981, 1982). Further, the succession of coastal ridges continued through the Quaternary and has been tilted by earth movements which probably spanned at least the later Pliocene and the Pleistocene (Boutakoff, 1963; Kenley, 1971; Sprigg, 1952).

The coastal segments defined by particular structural lineaments do not necessarily have the same characteristics throughout their history. In some cases, there has been an alternation of truncation, cliffing and barrier formation. However, most rocky coasts probably retained their essential characteristics, except when the sea retreated to the edge of the marginal rifts, exposing large areas of continental shelf; the Archipelago of the Recherche is a possible example of the latter (Playford, *et al.*, 1975).

Localised movements which influenced coastal form have been detected in some areas and are probably more widespread than generally realised. These include synclinal, anticlinal and domal neotectonic forms, and faults or monoclinal structures with distinct surface expression (Jenkin, 1981; van de Geer, *et al.*, 1979). There is no doubt that both regional and local earth movements would have affected antecedent coasts, but with different intensities at different times. In southern Australia, for example, it is expected that earth movement, and structural control of the coast, would have intensified on several separate occasions. Such occasions include the time of west to east continental rifting which occurred between the Mid Cretaceous and the Paleocene, several times in the Early Tertiary and, as far as the present coast is concerned, in the Late Pliocene. However, significent movements continued into the Late Pleistocene.

Tectonics and Coastal Classification

The close correspondence between plate tectonics and the regional characteristics of coasts has been pointed out by

Inman and Nordstrom (1971). However, the distribution of coastal types suggested by these authors is difficult to reconcile with actual Australian conditions. The east coast of Australia is mapped as a collision coast, but this is not consistent with the tectonic environment envisaged by Ollier (1981, 1982) and others. The coastal uplands are not mountains in the strict sense but rather a plateau upwarped at the rifted margin, subsequently retreating along the erosional Great Escarpment. This is more akin to the Afro-trailing-edge coastal type. The alternative morphological designation of Inman and Nordstrom (1971) is also not consistent with this tectonic reconstruction as most of the Australian east coast is shown as 'mountainous'; Davies (1980) interpreted the east coast as a 'marginal sea coast', extending from northern Australia to Tasmania. In addition, the remainder of the Australian coast is regarded by Inman and Nordstrom (1971) as an Amero-trailing-edge coast and, alternatively, as a wide shelf, plains coast. However, these parts of Australia are highly varied and, although some of its features would fit certain of the Inman-Nordstrom criteria, there are many anomalies. These include the lengthy stretches of cliffed, indented and rocky coasts, as well as the requirement that the hinterland function as a major source of sediment which extensively modifies the coast by active deposition. Consequently, this section extending south from the Kimberleys around the continental margin to Tasmania in the east is more appropriately designated an Afro-trailing-edge coast. Thus, except for the marginal sea coast of the far north, the Australian coast is best regarded as being an Afro-trailing-edge type.

Tectonics, Climate and Sediment Supply

Drift of the Australian continent through geological time, from colder to warmer climates, had a considerable influence on the coast. This is shown by the establishment of carbonate sedimentation in northern Australia, starting in the Late Cretaceous, consolidating in the Early Tertiary and extending to the southern basins in the Miocene (Veevers and Evans, 1975). This calcareous development includes the construction of coral/algal bioherms in the northeast starting in the Miocene, in north-western Australia in the Eocene, and continuing to the present (Veevers, 1971).

Earth movement in the hinterland has had an important effect on the coast through regulating the supply of sediment. Uplift of the Eastern Highlands, for example, resulted in deep dissection and retreat of the Great Escarpment which supplied siliceous sediment by means of short, steep-gradient streams to the eastern and southeastern seaboard. The relationship between Quaternary history and sediment distribution in eastern Australia is discussed by Roy and Thom (1981). Also, sediments from the Eastern Highlands were supplied to the low-gradient Murray-Darling system of south-central Australia. In the Eocene, the Murray River flowed due west in the valley of the present Broughton River and produced the 100 km wide Broughton Delta near the head of Spencer Gulf. The Morgan Fault, associated with the Flinders-Mt.Lofty uplift (post Eocene), then diverted the Murray southwards (Williams and Goode, 1978), thus confining the sediment supply to the Murray-Otway Basin system. In addition, river captures and diversions in the Eastern Highlands (Ollier, 1981) affected sediment supply to the coast.

In the more stable western part of the Australian craton, there is an extensive system of paleorivers most of which can be related to Early Tertiary shorelines (van de Graaff, *et al.*, 1977). However, the oldest of these valleys developed before the Early Cretaceous and presumably originated in what is now Antarctica, to be beheaded in the Eocene by rifting in southern Australia (Falvey, 1974; Ollier 1977, 1981). This left only low gradient channels which soon silted up, aided by the Mid Tertiary change from a humid to an arid climate (Finkl and Fairbridge 1979). These paleorivers ceased to be a source of sediment to the coast in the early Miocene (van de Graaff *et al.*, 1977).

Global Sea-Level Changes

Far greater attention has been given to Quaternary eustatic changes in Australia than to earlier changes in sea levels. Longer term global sea-level curves, such as that of Vail *et al.*, (1977), have been used locally to assist in the interpretation of plate tectonic phenomena. However, Veevers (1980) developed an Australian curve. He suggested that the retreat of the Mid Cretaceous sea, allowing the spread of terrestrial conditions onto the continental shelf, can only be partly accounted for eustatically. Transgressions are also recorded in the mid-Late Eocene, to be followed by a major withdrawal in the Oligocene (Fig. 6 and 7).

The subsequent Mid to Late Miocene transgression penetrated far inland in places. It has been further suggested (Jenkin, 1981) that, if a maximum Mid Miocene eustatic rise of 300 m (Fleming and Roberts, 1973) is accepted, then all Australian marine deposits of this age can be accounted for without the need to invoke uplift. Moreover, Miocene marine sediments should occur at higher levels than those at which they are actually found. During the subsequent regression the sea

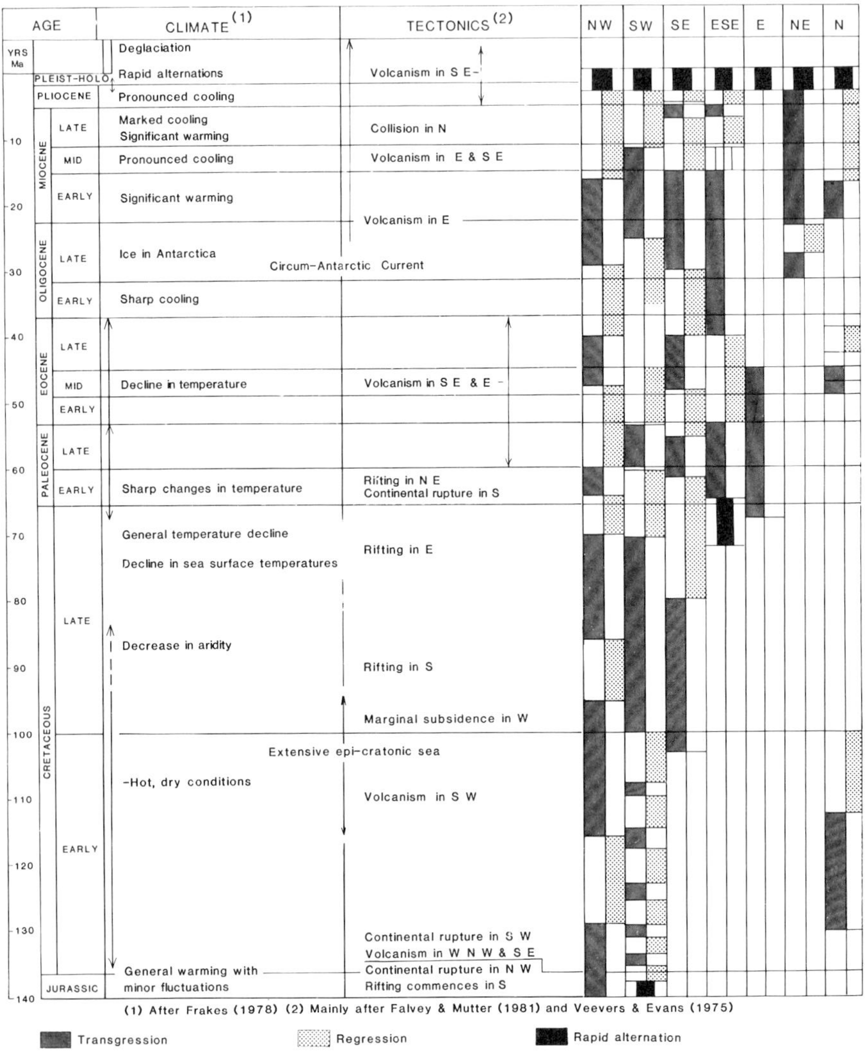

FIGURE 6. Transgression and regression in relation to tectonics and climate.

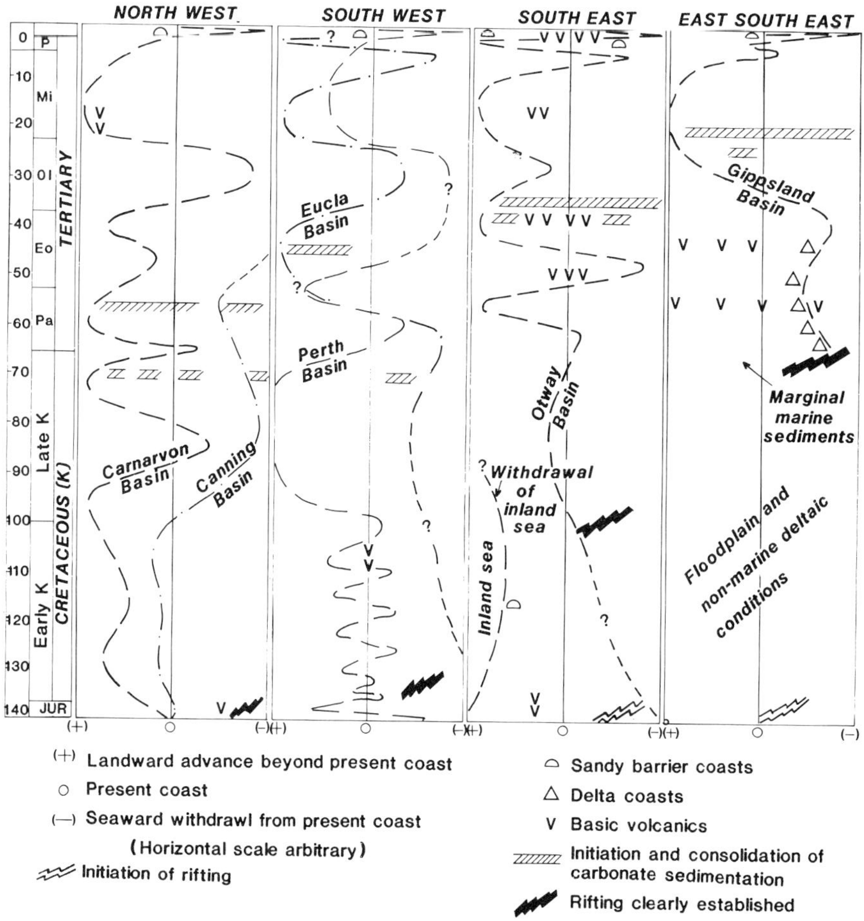

FIGURE 7. Generalised coast shifts in the main peripheral sedimentary basins. Sources: various.

level, according to Fleming and Roberts (1973), fell 800 to 1000 m exposing the shelves and upper continental slopes. This is far from fully established, although there is some evidence of an appreciable late Mid Miocene fall in sea level in Victoria (Abele, *et al.*, 1976; Mallett, 1978) and in the Perth Basin (Playford *et al.*, 1975). Notwithstanding, Ollier (1981) considers that glacioeustatism could account for the later Tertiary sea level changes, but not the early Tertiary and Mesozoic changes. These are believed to be tectonic in origin.

The principal climatic and tectonic events influencing the development of the Australian continental margin are shown in Figure 6, while Figure 7 gives a generalised view of coastal shifts in the peripheral sedimentary basins.

Conclusions

Analysis of coastal shifts, indicated by transgressions and regressions and detectable coastal forms shows that:

(i) One factor may be dominant at a particular time, climate on one occasion, regional tectonics on another.

(ii) Localised tectonic events have considerable influence in defining the form and extent of sites in which coastal processes operate.

(iii) The topography of the hinterland and its climate significantly affect coastal phenomena, even in an essentially flat, low-relief country such as Australia.

There are still many unresolved aspects of Australian coastal evolution over the last 150 Ma. Nevertheless, a rapidly growing literature on the structure and sedimentology of Australian continental margins now provides the basis for more detailed reconstructions of coastal forms of the past. Although the evidence is often fragmentary and indirect, an historical approach is seen as the only means of providing the factual control necessary for the accurate prediction of long-term coastal change.

Australia has certain unique features which have considerably influenced the evolution of its coast. Its flatness and lack of high mountains, the effect of this on climate and deprivation of the coasts of sediment from the hinterland are fundamental. Further, rifting and spreading within the plate itself has produced an eastern seaboard which is an eroded rift margin, the eastern (collision) edge of the Australian plate now lying far to the east. The northern plate boundary also is separated from the mainland by shallow seas. Thus, no mainland coast is of collison type.

Australia is essentially a plateau rifted along most of its margins, which has suffered broad warping, more localised differential tectonism, vertical uplift in restricted areas, and possibly temporary overall tilt in different directions at different times depending on the varying rates of northerly drift. Inevitably, there have been wide swings in coastal position through geological time accompanied by a succession of coasts which did not necessarily retain the same form as they migrated back and forth across the shelf. Individual coastal segments rarely reproduced their predecessors precisely, or even approximately in many cases. Strict cyclicity or uniformity in Australian coastal evolution therefore cannot be sustained. The coastal complex has been changing constantly on both local and continental scales, with influences from both the landward and seaward side varying in response to the tectonic, climatic and hydrodynamic conditions current at the time.

References

Abele, C., Kenely, P.R., Holdgate, G. and Ripper, D. (1976). Tertiary, Otway Basin. *In* Douglas, J.G. and Ferguson, J.A. (eds.) *Geology of Victoria, Geol. Soc. Aust., Spec. Publ.* 5, 198-229.

Blackburn, G. (1962). Stranded coastal dunes in north-western Victoria. *Aust. J. Sci.* 24, No. 9, 388-389.

Boutakoff, N. (1963). The geology and geomorphology of the Portland area. *Geol. Surv. Vic. Mem.* 22.

Bowler, J.M. and Magee, J.W. (1978). Geomorphology of the Mallee region in semi-arid northern Victoria and western New South Wales. *Proc. Roy. Soc. Vic.* 90, No's 1 & 2, 5-25.

Burne, R.V. (1982). Relative fall of Holocene sea level and coastal progradation, northeastern Spencer Gulf, South Australia. *B.M.R. J. Aust. Geol. Geophys.* 7, No. 1, 35-45.

Carey, S.W. (1976). The Expanding Earth. Elsevier, Amsterdam.

Cook, P.J., Colwell, J.B., Firman, J.B., Lindsay, J.M., Schwebel, D.A. and von der Borch, C.C. (1977). The late Cainozoic sequence in southeast South Australia and Pleistocene sea-level changes. *B.M.R. J. Aust. Geol. Geophys.* 2, 81-88.

Davies, J.L. (1980). *Geographical Variation in Coastal Development.* 2nd ed. Longman, London.

Fairbridge, R.W. and Finkl, C.W. (1978). Geomorphic analysis of the rifted cratonic margins of Western Australia. *Zeit. Geomorph.* 2, No. 4, 369-389.

Falvey, D.A. (1974). The development of continental margins in plate tectonics theory. *The APEA Journal* 14, No. 1, 95-106.

Falvey, D.A. and Mutter, J.C. (1981). Regional plate tectonics and the evolution of Australia's passive continental margins. *B.M.R. J. Aust. Geol. Geophys.* 6, 1-29.

Finkle, C.W. and Fairbridge, R.W. (1979). Paleogeographic evolution of a rifted cratonic margin: S.W. Australia. *Palaeogeog., Palaeoclim., Palaeoecol.* 26, 221-252.

Flemming, N.C. and Roberts, D.G. (1973). Tectono-eustatic changes in sea level and seafloor spreading. *Nature, 243, 19-22.*

Frakes, L.A. (1979). *Climates Throughout Geologic Time.* Elsevier, Amsterdam.

Gibson, G. (1982). Earthquakes. *In* Duncan, J.S.W. (ed.) *Atlas of Victoria,* Victorian Government Printing Office, Melbourne, 17-19.

Gill, E.D. (1971). The far-reaching effects of Quaternary sea level changes on the flat continent of Australia. *Proc. Roy. Soc. Vic.* 84, 189-205.

Gill, E.D., Segnit, E.R. and Hunt, I. (1980). Pleistocene submerged cliff off the Otway coast of Victoria, Australia. *Proc. Roy. Soc. Vict.* 91, 43-51.

Glaessner, M.F. and Parkin, L.W. (eds.) (1958). The Geology of South Australia. *J. Geol. Soc. Aust.* 5, No. 2.

Glaessner, M.F. and Wade, M. (1958). The St. Vincent Basin. *In* Glaessner, M.F. and Parkin, L.W. (eds.). *The Geology of South Australia, J. Geol. Soc. Aust.* 5, No. 2, 115-126.

Grimes, K.G. and Doutch, H.F. (1978). The late Cainozoic evolution of the Carpentaria Plains, North Queensland. *B.M.R. J. Geol. Geophys.,* 3, 101-112.

Hayes, E.D. and Ringis, J. (1973). Seafloor spreading in the Tasman Sea. *Nature* 243, 454-8.

Hocking, J.B., Gloe, C.S. and Threlfall, W.F. (1976). Tertiary, Gippsland Basin. *In* Douglas, J.G. and Ferguson, J.A. (eds.) *Geology of Victoria, Geol. Soc. Aust., Spec. Publ.* 5, 248-273.

Hossfeld, P.S. (1950). The late Cainozoic history of the south-east of South Australia. *Trans. Roy. Soc. S. Aust.* 73, 232-279.

Inman, D.L. and Nordstrom, C.E. (1971). On the tectonic and morphologic classification of coasts. *J. Geol.* 79, No. 1 1-21.

Jenkin, J.J. (1962). The Geology and Hydrogeology of the Western Port Area. *Geol. Surv. Vic. Underground Water Invest. Rept.* 5.

Jenkin, J.J. (1968). The geomorphology and Upper Cainozoic geology of South-East Gippsland, Victoria. *Mem. Geol. Surv. Vict.* 27.

Jenkin, J.J. (1981). Evolution of the Victorian coastline. *Proc. Roy. Soc. Vict.* 92, 37-54.

Jenkin, J.J. (1982). Physiography. *In* Duncan, J.S. (ed.) *Atlas of Victoria,* Victorian Government Printing Office, Melbourne, 20-25.

Johnson, B.D., Powell, C.McA. and Veevers, J.J. (1976). Spreading history of the eastern Indian Ocean and India's northward flight from Antarctica. *Geol. Soc. Amer. Bull.* 87, 1960-1966.

Jones, J.G. and Veevers, J.J. (1979). A Cainozoic history of Australia's Southeast Highlands. *J. Geol. Soc. Aust.* 29, 1-12.

Kenley, P.R. (1971). Cainozoic geology of the eastern part of the Gambier embayment, south-western Victoria. *In* Wopfner, H. and Douglas, J.G. (eds.) *The Otway Basin of Southeastern Australia, Spec. Bull. Geol. Survs. S. Aust. Vic.*, 89-153.

Kenley, P.R. (1976). Southwestern Victoria. *In* Douglas, J.G. and Ferguson, J.A. (eds.) *Geology of Victoria, Geol. Soc. Aust. Spec. Publ.* 5, 290-298.

Kennett, J.P., Houtz, R.E., Andrews, P.B., Edwards, P.B., Gostin, V.A., Hajos, M.A., Jenkins, D.G., Margolis, S.V. Ovenshine, A.T. and Perch-Nielson, K. (1974). Development of the Circum-Antarctic current. *Science* 186, 14-47.

Logan, B.W., Davies, G.R., Read, J.F. and Cebulski, D.E. (1970). Carbonate sedimentation and environments, Shark Bay. Western Australia. *Amer. Assoc. Pet. Geol. Mem.* 13.

Ludbrook. N.H. (1980). *A Guide to the Geology and Mineral Resources of South Australia.* Dept. of Mines and Energy. Government Printer, South Australia.

Mallett, C.W. (1978). Sea level changes in the Neogene of southern Victoria. *The APEA Journal*, 18, 64-69.

Ollier, C.D. (1977). Early landform evolution. *In* D. Jeans (ed.), *Australia, a geography.* Sydney University Press, Sydney, 85-98.

Ollier, C.D. (1978). Tectonics and geomorphology of the eastern highlands. *In* J.L. Davies and M.A.J. Williams (eds.). *Landform Evolution in Australasia.* ANU Press, Canberra, 5-47.

Ollier, C.D. (1979). Evolutionary geomorphology of Australia and Papua New Guinea. *Trans. Inst. Brit. Geogr.* 4, 516-539.

Ollier, C.D. (1981). *Tectonics and Landforms.* Longman, London.

Ollier, C.D. (1982). The Great Escarpment of eastern Australia: tectonic and geomorphic significance. *J. Geol. Soc. Aust.* 29, No. 1, 13-23.

Playford, P.E., Cope, R.N. and Cockbain, A.E. (1975). Palaeogeographic development of Western Australia, Phanerozoic. *West. Australia Geol. Surv. Mem.* 2, 451-460.

Sprigg, R.C. (1952). The geology of the south-east province, South Australia, with special reference to Quaternary coastline migrations and modern beach development. *Bull. Geol. Surv. S. Aust.* 29.

Roy, P.S. and Thom, B.G. (1981). Late Quaternary marine deposition in New South Wales and southern Queensland - an evolutionary model. *J. Geol. Soc. Aust.* 28, No. 4, 471-489.

Vail, P.R., Mitchum, R.M. and Thompson, S. (1977). Global cycles of relative changes of sea level. *Amer. Assoc. Petrol. Geol. Mem.* 26, 83-97.

van der Geer, G., Colhoun, E.A. and Bowden, A. (1979). Evidence and problems of interglacial marine deposits in Tasmania. *Geol. en Minjb.* 58, No. 1, 29-32.

van de Graaff, R.W.A., Bunting, J.A. and Jackson, M.J. (1977). Relict Early Cainozoic drainages in arid Western Australia. *Zeit. Geomorph.* 21, No. 4, 379-400.

Veevers, J.J. (1971). Phanerozoic history of Western Australia related to continental drift. *J. Geol. Soc. Aust.* 18, 87-96.

Veevers, J.J. (1980). Basins of the Australian craton and margin. *In* Bally, A.W., Bender, P.L., McGetchin, T.R. and Walcott, R.I. (eds.) *Dynamics of Plate Interirors.* American Geophysical Union, Geodynamics Series, Washington, 1, 73-80.

Veevers, J.J. and Evans, P.R. (1975). Late Palaeozoic and Mesozoic history of Australia. *3rd Internat. Gondwana Sympos.*, A.N.U. Press, Canberra, 579-607.

Veevers, J.J. and Cotterill, D. (1978). Western margin of Australia: evolution of a rifted arch system. *Geol. Soc. Amer. Bull.* 89, 337-355.

Williams, G.E. and Goode, A.D.T. (1978). Possible western outlet for an ancient Murray River in South Australia. *Search* 9, 443-7.

Wise, D.U. (1974). Continental margins, freeboard, and volumes of continents through time. *In* Burke, C.A. and Drake, C.L. (eds.), *The Geology of Continental Margins.* Springer-Verlag, Berlin, 45-58.

Wopfner, H. and Douglas, J.G. (eds.) (1971). The Otway Basin of southeastern Australia. *Spec. Bull. Geol. Survs. S. Aust. Vict.*

3

Morphodynamics of High Energy Beaches: An Australian Perspective

A. D. SHORT

L. D. WRIGHT[1]
Coastal Studies Unit
Department of Geography
University of Sydney, NSW

Introduction

The Australian continent lies between 11 and 44° south latitude exposing the coast to a wide range of zonal wind and wave conditions. Modal deep-water wave heights range from low (<1m) through moderate (1-2.5m) to high (>2.5m). Waves are predominantly low across monsoon-trade wind northern Australia and in the dunerock reef protected areas of West Australia. Moderate to occasionally high waves dominate the east coast from south Queensland to Tasmania. High waves prevail in the west coast swell environments across the south of Australia (Davies, 1980). The relative levels of wave heights referred to here are for Australian - southern hemisphere conditions, as lower values are often used in northern hemisphere (see e.g. Hayes, 1979).

[1] *Present address: Department of Geological Oceanography, Virginia Institute of Marine Science, Gloucester Point, Virginia.*

COASTAL GEOMORPHOLOGY
IN AUSTRALIA
ISBN 0 12 687880 3

This chapter is concerned with the sandy beach systems located in the moderate to high wave environments of southern Australia, that is below 25°S. More specifically it will present a model of beach-surfzone morphodynamics based on field investigations by the Coastal Studies Unit between 1977 and 1983. Experiments were conducted on 28 separate beach systems the details of which are given in Wright *et al.* (1982c) and Figure 1. In addition representative surveys of beach profiles, morphology and sediments covered most of southern Australia. The morphodynamic approach (see Wright and Thom, 1977) employed in these studies considers both the dynamic regimes (waves, tides, wind, etc.) that drive the surfzone and nearshore processes, and the morphology or form of these environments (beach, bar, channels etc.) that represent the morphological response to the dynamic environment. The assessment of beach morphodynamics therefore entails an examination of both the energy sources (dynamics) and the terrigenous response (morphology).

The framework within which the beaches are examined is based on the beach-state model of Wright and Short (in press). This model is out-growth of two earlier complementary beach models, one based on beach morphological sequences (Short, 1979b) and the other on morphodynamic types (Wright *et al.*, 1979). The beach-state model consists of three beach types dissipative, intermediate and reflective. The intermediate type has four states, making a total of six beach states (Fig. 2). The degree of wave reflection versus dissipation can be defined using the surf scaling parameter, ε,

$$\varepsilon = a_i \omega^2_t / g \tan^2 \beta$$

where a_i = incident wave amplitude; $\omega_t = 2\pi/T$; T = wave period; g = acceleration of gravity; and β = beach/surfzone slope (Guza and Inman, 1975). When $\varepsilon < 2$ wave reflection dominates, whereas $\varepsilon > 20$ favours wave dissipation (Table 1). Actual beach state is defined by morphological configuration and basic surfzone circulation as illustrated in Figure 2, and by wave-sediment thresholds. Using Dean's (1973) parameter Ω,

$$\Omega = Hb/w_s T$$

where Hb is breaker height, w_s is mean sediment fall velocity and T is wave period, beaches in southern Australia have been defined as being dissipative when $\Omega > 6$, intermediate when Ω is 1 to 6, and reflective when $\Omega < 1$ (Wright *et al.*, 1982c). Basically high waves and fine sand favour dissipative beaches,

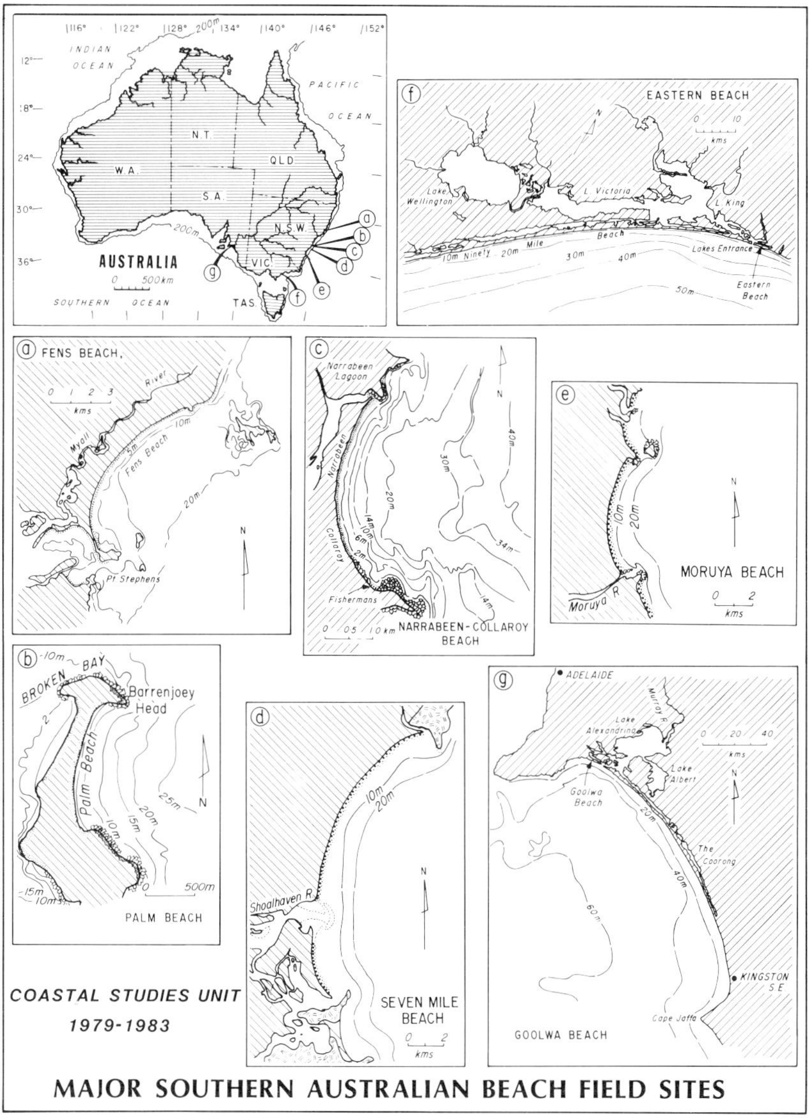

FIGURE 1. Location and regional morphology of some of the major field sites discussed in text.

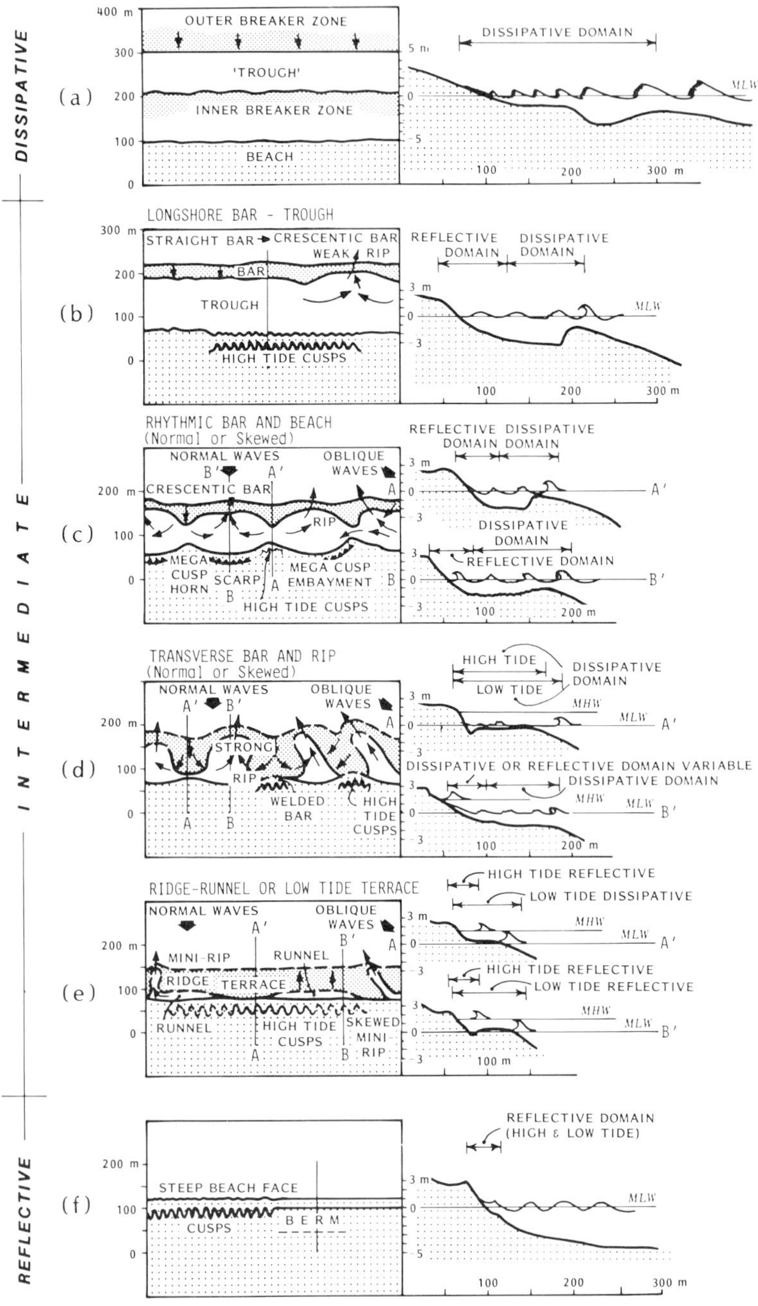

FIGURE 2. Plan and profile configuration and basic surfzone circulation of the six beach states.

TABLE 1 SOME CLASSIFICATIONS OF BEACH TYPE

Beach Type[1]	Beach State[2]	Surf Scaling[3]	Hb[4]	Hb/T_bW_s[5]	Sasaki[6]
Dissipative	Dissipative	> 20	> 2.5m	> 6	Infra-gravity
Intermediate	Bar Trough	20	2-2.5	6	Insta-bility
↓	Rhythmic Bar & beach	↓	1.5-2	↓	↓
	Transverse bar & rip		~1.5		
	Ridge & Runnel Low-Tide Terrace	2.5	1-1.5	1	
Reflective	Reflective	< 2.5	< 1	< 1	Edge Wave

1. for medium-fine sand beaches only; 2. see Figure 2; 3. from Guza and Inman, 1975; 4. for fine-medium sand beaches from Short and Hesp, 1982; 5. from Wright et al.1982b; 6. see Sasaki and Horikawa, 1979.

moderate waves and medium to fine sand intermediate beaches, and low waves and coarser sediment reflective beaches (see Table 1).

This chapter will only deal with moderate to high energy beaches, that is transverse bar and rip, rhythmic bar and beach, bar-trough and dissipative (Fig. 2). In southern Australia these systems, which are characterised by surfzone circulation, are usually driven by waves greater than 1m. Lower energy ridge and runnel and reflective beaches are not discussed. The discussion is based on field experiments, long term beach profiling, and sediment and morphological analysis conducted on representative type beaches around southern Australia (Fig. 1). The morphodynamic characteristics of each beach state, the environmental conditions required to produce the beach state, and the mode and scale of shoreface mobility and erosion for each state is presented.

Morphodynamic Characteristics

The Dissipative Beach State

The combination of high waves and fine sand result in high energy dissipative beaches characterised by low beach and surf-zone gradients, and high surf scaling parameters (i.e. $\varepsilon > 20$) and high values of Ω (i.e. > 6). Dissipative beaches are characterised by a wide low gradient beach face, which extends from the foot of the foredune to a lower gradient lower beach face and surfzone (see Fig. 2a). Depending on breaker height, the surfzone width may extend 200-500m seaward of the shore-line (Fig. 2a and 3). Multiple bars or breaker zones (usually 2 but up to 5) may be present across the surfzone. They are found in regions exposed to high breakers where fine sand is abundant. Such waves occur year round in the storm wave and west coast swell environments of southern Australia. The dis-sipative extreme is analogous to the 'storm' or 'winter' profile of seasonally varying beaches. They occur seasonally on the US west coast (see Aubrey, 1979), and periodically in areas where high waves persist for periods of at least several days to weeks, particularly where fine sand is abundant (see Bowman and Goldsmith, 1983).

Two beaches - north Seven Mile Beach on the N.S.W. south coast and Goolwa Beach in South Australia (Fig. 1) provide examples of fully dissipative beaches. Both beaches are sinks for fine sands and exhibit very low gradients ($\tan \beta$ = 0.01 - 0.03) and wide, double barred surfzones (Fig. 2a).

Longshore rhythms or significant irregularities are rarely present. During field experiments modal wave heights and periods were 3m and 12 seconds respectively at Goolwa (Ω = 10.4) and 2m and 10 seconds at Seven Mile Beach (Ω = 6).

High incident waves begin spilling across the outer breaker zone, and decay progressively across the surfzone. This mode of wave breaking causes a dissipation of the incident wave and a shift in energy to infragravity frequencies (period >30 seconds). This shift results in low frequency oscillations in water level at the shoreline called 'surfbeat'. These oscillations take the form of standing waves that rise and fall in the inner surfzone. Water accumulates in the inner surfzone as set-up, followed by a seaward return flow during set-down. This provides a mecha-

FIGURE 3. A high energy dissipative surfzone at Goolwa Beach (Fig. 1). Outer breakers are 3m high. Note the outer and inner breaker zones (OB and IB) and final bore and swash (B & S) dissipation zone.

nism for storing water from incident bores in the surfzone and then returning it seaward across the breaker zone as a single infragravity wave (Wright *et al.*, 1982a). Circulation is onshore at incident frequency in the outer surfzone shifting to infragravity nearer the beachface.

Power spectra of surface elevation η, and horizontal nearbottom currents, u and v from Seven Mile Beach and Goolwa Beach have been described in detail by Wright (1982) and Wright *et al.* (1982a), respectively. Figure 4 shows typical u spectra from Goolwa. The important feature of these spectra, and of η, u and v spectra from high energy, fully dissipative surfzones is the dominance of wide-banded infragravity energy in the region of 100-200 seconds period. Phase angles of $\pi/2$ between η and u for infragravity frequencies on Seven Mile Beach and Goolwa Beach indicate the oscillations at those frequencies are consistently standing (Wright, 1982; Wright *et al.*, 1982a). At Goolwa the velocity amplitudes of u corresponding to the infragravity surfbeat were on the order of 1.2 to 1.5m sec^{-1}, accompanying surfbeat heights of roughly 1m on the beach (Wright *et al.*, 1982a). Standing oscillations at subharmonic frequencies are consistently absent from dissipative surfzones.

The net circulation at Goolwa was weak relative to the surfbeat motions. In the surfzone flow is predominantly shoreward towards the surface and offshore towards the bed (Wright *et al.*, 1982a), a characteristic also observed by Sallanger *et al.* (1983). Under shore normal waves longshore and rip circulation is weak or non-existent although migratory rips may exist during erosive conditions. It should be pointed out, however, that on strongly embayed and headland bounded beaches such as those near Sydney, very strong, large scale rips frequently accompany the dissipative conditions which occur during storms (Short, 1979b; Wright *et al.*, 1979).

Figure 5 describes the 'process signature' observed in the Goolwa surfzone. The bar graphs indicate the *relative* amplitudes of near-bottom flows associated with incident waves, surfbeat, and net circulation. The bars are expressed in dimensionless form relative to the observed incident wave velocities, which are assigned a value of one. The pronounced shoreward growth of flows due to infragravity oscillations and the strong dominance of those flows in the inner surfzone are the most important features of the graph and are typical of dissipative beaches in general.

Intermediate Beach States

The most complex process signature of different beach states are associated with the intermediate states because

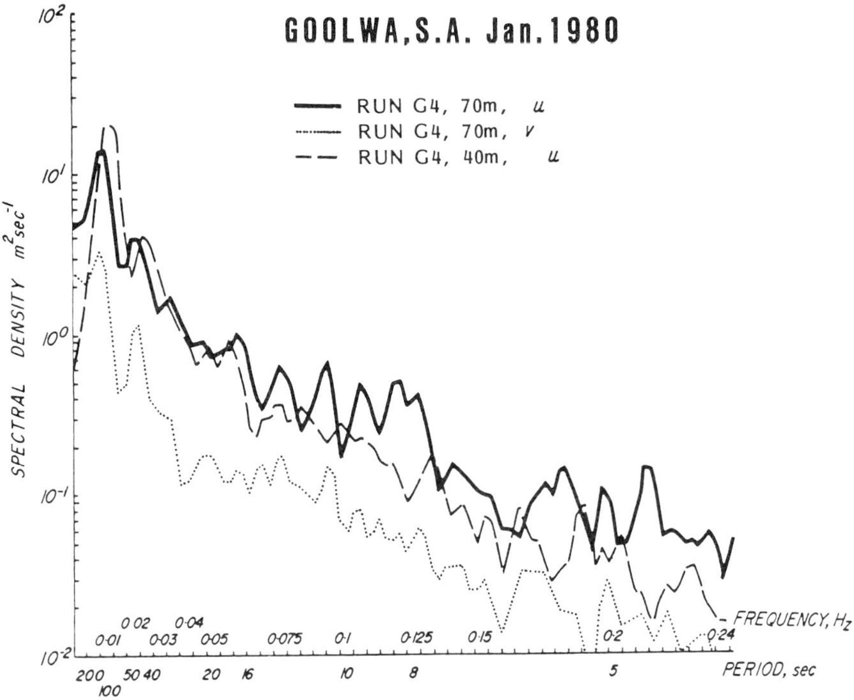

FIGURE 4. Power spectra of shore normal currents, u, and shore parallel currents, v, from mid (70m, Fig. 5) and inner (40m) surfzone regions of a high energy dissipative surfzone, Goolwa, S.A. Note the dominance of low frequency infragravity energy (after Wright, et al., 1982a).

they possess both dissipative and reflective elements. Figure 2 shows the four intermediate beach types (bar-trough , rhythmic bar-beach, transverse bar-rip and ridge and runnel). They represent a transition from dissipative to reflective, from high wave conditions when beach sediment is stored in and seaward of the surfzone to low wave conditions when it is stored on the subaerial beach (Short, 1979b). Consequently, mobile sediment usually resides in the surfzone in the form of bars. Prerequisite for intermediate types are fine to medium sand and a highly variable moderate to high (1-2.5m) wave environment (Short, 1980). The value of the surf scaling parameter, ε, lies between 2.5 and 20, and Ω between 1 and 6 (Table 1). On medium sand beaches in the Sydney Region, Short (1980) found that the higher energy bar-trough system is maintained when modal waves exceed 2m, the rhythmic bar-beach when waves are between 1.5-2m, transverse bar and rip with

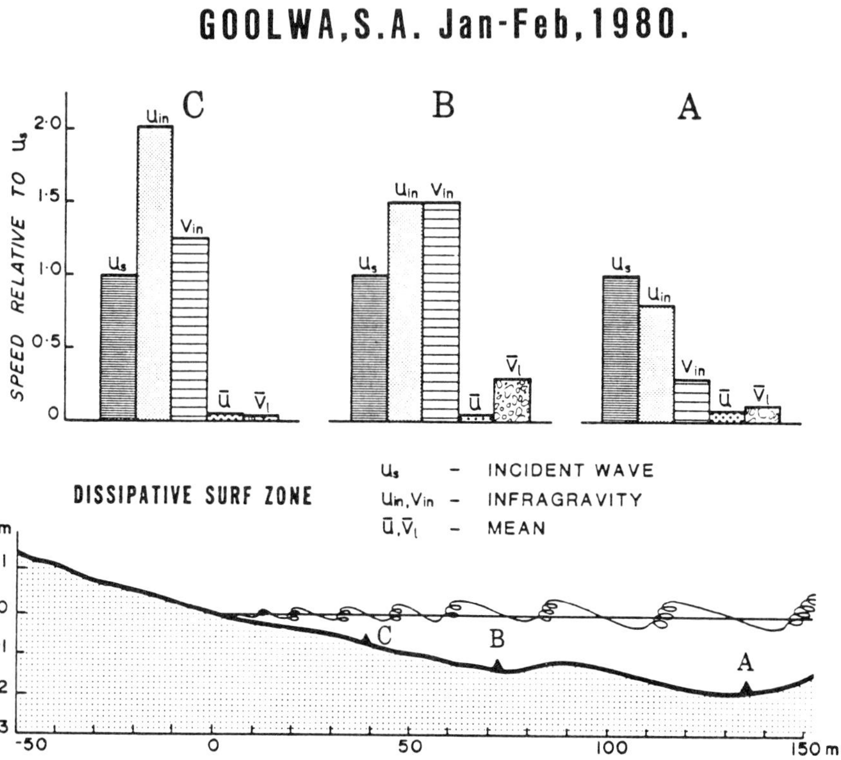

FIGURE 5. Process signature across the inner 150m of a 500m wide high energy dissipative surfzone, Goolwa, S.A. (see Fig. 3). The bar graphs indicate the relative magnitudes, scaled relative to the significant velocity amplitude u_s of the incident waves. Other parameters are the velocity amplitudes of shore normal and shore parallel flows at infragravity frequencies (U_{in}, V_{in}); and $\bar{u}$, $\bar{v}$ net shore normal and alongshore circulation, (from Wright and Short, in press).

waves 1-1.5m, and ridge and runnel with waves <1m. Because waves are rarely stable at the above levels beaches move from one state to another in response to varying wave conditions (Short, 1979a; Wright *et al.*, 1979). Intermediate beaches commonly occur on coasts exposed to high deepwater waves, but where some energy is lost to wave shoaling and/or refraction. They also persist on moderate energy east coast swell and trade wind and monsoonal coasts (Short, 1979a).

Morphologically intermediate states are characterised by pronounced longshore variability caused by alternating rip and bar topography. Rip circulation dominates and is best

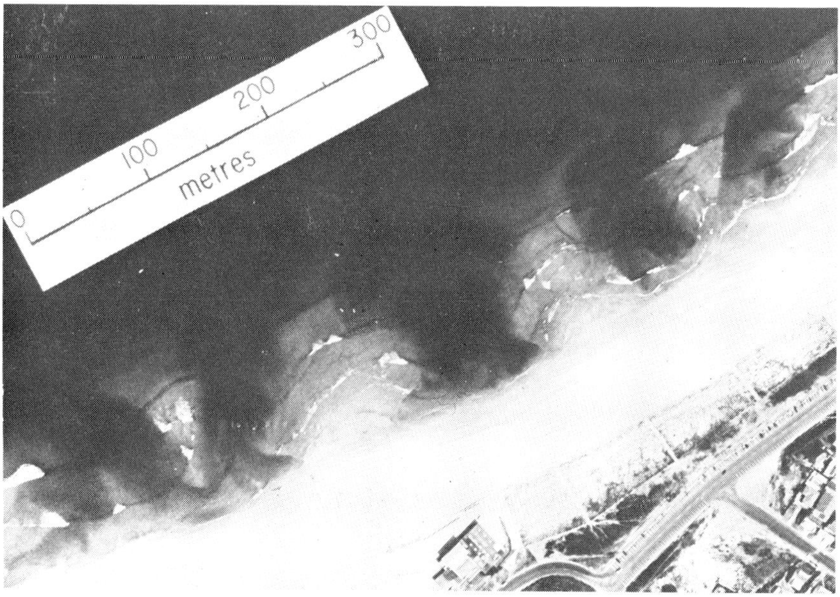

FIGURE 6. Intermediate surfzones are characterised by rip circulation as illustrated by this transverse bar and rip morphology at Cronulla Beach, N.S.W.

developed in the rhythmic bar and beach and transverse bar and rip states (Figs. 2c and d, and 6).

Longshore bar-trough. The bar-trough system is distinguished by a reflective beach face, a relatively wide, deep, shore-parallel trough and a pronounced shore-parallel bar (Fig. 2b). Processes associated with the longshore bar-trough and rhythmic bar and beach states are illustrated by results from Moruya Beach, some of which have been discussed by Wright *et al.* (1979) and Wright (1982), and by more recent and more conclusive results from Eastern Beach in eastern Bass Strait (Fig. 1; Wright *et al.*, 1982b). Figure 7 shows spectra of η and u from the trough and bar regions of pronounced rhythmic bar and beach topography on Eastern Beach, Gippsland, Victoria. The spectra shown in Figure 7, like all spectra from bar-trough and crescentic bar and beach topography on Eastern Beach and on Moruya Beach show dominance of energy at swell frequency, but with appreciable standing wave energy over a band which includes both subharmonic and infragravity frequencies. The infragravity standing waves occur at much higher frequencies than in fully dissipative surf-

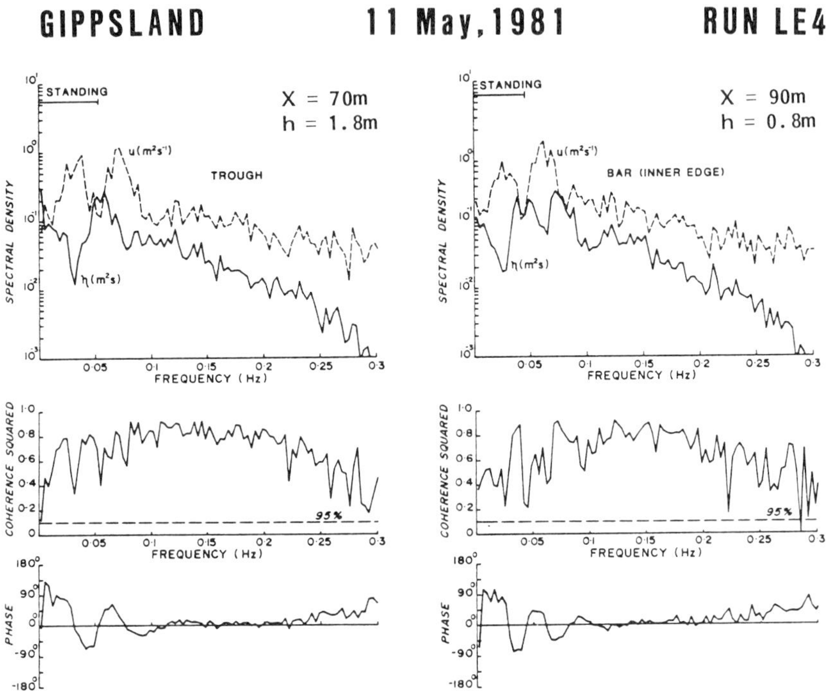

FIGURE 7. Spectra and cross spectra of η and u from the bar and trough regions associated with a beach state transitional between the longshore bar-trough and rhythmic bar and beach states (Eastern Beach, Gippsland, 11 May 1981). The broad peak at incident wave frequency is dominant; however, there is also considerable standing wave energy at relatively short infragravity periods (e.g. around 0.025hz or 40 seconds) (from Wright, et al., 1982b).

zones; typical periods of the dominant infragravity oscillations are on the order of 35 to 50 seconds. Chappell and Wright (1979), Short (1979b) and Wright (1982) have discussed these intermediate frequency standing waves and believe that they may be edge waves which initiate weak spatially controlled rip circulation. The circulation consists of seaward return flow alternating with onshore flow, the latter leading to the initiation of onshore migrating sand lobes called crescentic bars and the former to the maintenance and/or excavation of rip channels.

The net mean currents in the presence of bar-trough and crescentic bar and beach topographies are characterized by moderate strength, widely spaced (~100-150m) rips with longshore rip feeder currents within the troughs. On straight bar-trough topographies the rips are relatively weak and may initially migrate alongshore (Short, 1979b). Their positions become topographically arrested by the development of crescentic bars which lead to an intensification of the rip currents. When waves break oblique to the shore, it is the longshore bar-trough and to a lesser degree the crescentic bar and beach topographies which exhibit the strongest longshore currents, with longshore velocity maxima concentrated in the trough. A typical vertical profile of a longshore current, measured in the deep trough of Eastern Beach is shown in Figure 8.

Rhythmic bar and beach. This state possesses the most highly rhythmic bar and beach features (Fig. 2c) associated with the onshore migration of crescentic bars (Short, 1979b).

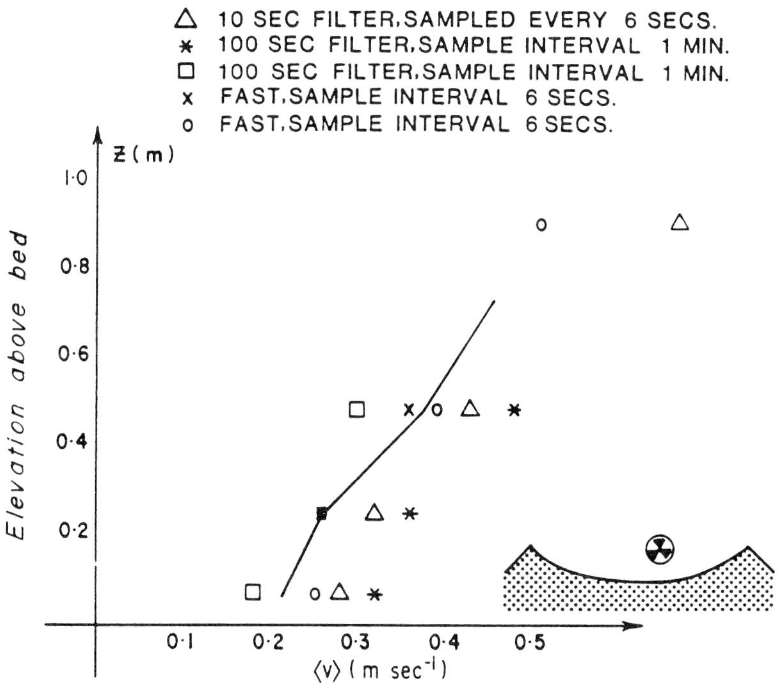

FIGURE 8. Vertical velocity profile of the longshore current as measured within the trough of Eastern Beach, 13 May, 1981. Note the high current speeds on the order of 25cm sec^{-1} *which are present less than 10cm above the bed (from Wright, et al., 1982c).*

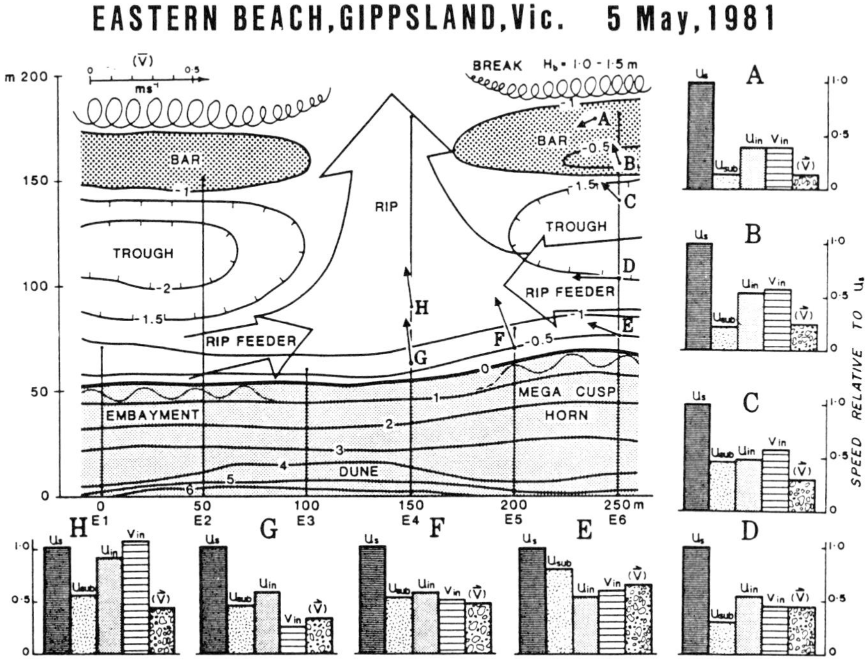

FIGURE 9. Net circulation pattern and process signature observed in the presence of rhythmic bar and beach topography, Eastern Beach, Gippsland (Fig. 1). The broad arrows indicate general circulation based on dye releases. Fine arrows indicate current speeds and directions measured at specific points. As in Figure 5, the bar graphs indicate the magnitude relative to incident wave orbital velocities of different modes of horizontal current motion: U_{sub} shore normal currents at subharmonic frequency; U_{in}, V_{in} shore normal and shore parallel currents at infragravity frequencies $\bar{V}$ = net time averaged velocities.

Experiments were conducted at Eastern Beach (Fig. 1) under highly rhythmic conditions. The process histograms from Eastern Beach (Fig. 9) indicate the magnitude, relative to the incident wave velocity amplitude, u_s, of near-bed velocity. related subharmonic oscillations (u_{sub}) , infragravity oscillations (u_{in},v_{in}), and net time averaged longshore and rip currents ($\bar{V}$) in different regions of the surfzone. Incident waves tend to dominate everywhere, but are overwhelmingly dominant over the bar. Waves break more heavily over the crescentic bars, with water flowing across and off the bars into the channel. The flow converges longshore to return seaward as a rip current. The current pulses at infragravity frequency (see H, Fig 9) indicating wave dissipation is also contributing to the circulation. At the beach face the waves alternate from surging in the lee of bars leading to growth of subharmonic edge waves (Fig. 9, E) and cusp formation on the megacusp horns, to plunging in the lee of the rip channels (Fig. 9, G) where less dissipation has occurred (Fig. 2c). This usually leads to scour in the rip embayment and scarping of the adjacent beach face even when the adjacent megacusp horn is accreting (see Fig. 2c) (Short, 1979a; Short and Wright, 1981; Wright 1980).

Transverse bar and rip. Continued onshore migration of the crescentic bar causes it to weld to the megacusp horn. Alternately, a period of partial bar erosion produces the transverse bar-rip state (Short, 1979b). In both cases, the state is characterised by alternating transverse bars and rip channels (Figs. 2d and 6). The result is the most pronounced rhythmic longshore variations in dissipation versus reflectivity. It is under these conditions that the strongest rip circulations occur for any given breaker height and may reach velocities exceeding 1m sec^{-1}, even under relatively low waves. Figure 10 shows the process signature associated with transverse bar and rip topography as exemplified by experiments conducted on the moderately embayed Palm Beach (Fig. 1). The important features are extremely strong rip currents and rip "feeder" currents (Fig. 10, C and B respectively). Although these currents remain persistently strong and unidirectional, they alternately strengthen and slacken at infragravity frequencies. On this type of topography, we observe rip current speeds to be typically on the same order as the breaking incident wave orbital velocities. This is the typical beach condition on many south east Australian beaches during summer and causes many problems and fatalities for inexperienced swimmers. The bars are relatively shallow, particularly at low tide, with relatively deep rips cutting transversely across them, the amount of skewing depending on wave obliquity (Fig. 2d).

Nature and Scale of Shoreline Mobility and Erosion

The dissipative extreme represents beach response to persistent high waves. It is at the erosional end of the spectrum of beach types, the end toward which lower energy beaches will strive during periods of high waves. Consequently, they are relatively stable beaches so long as waves remain

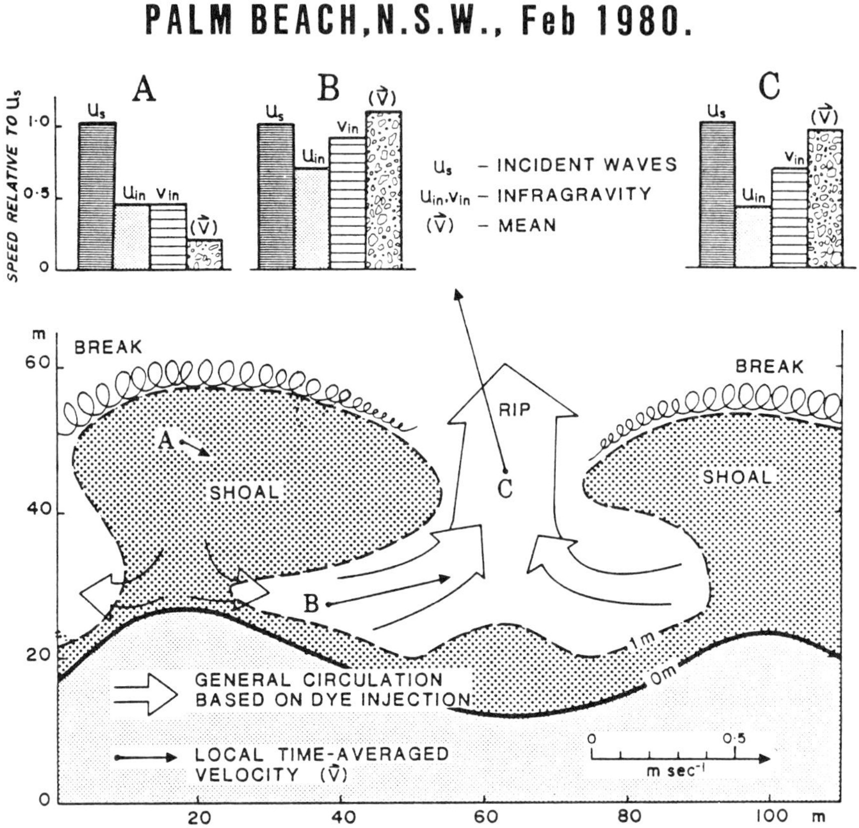

FIGURE 10. Process signature associated with transverse bar and rip topography, Palm Beach, N.S.W., 29 February, 1980. Note the strong rip and rip feeder currents.

high. Most potentially active sand is stored in and seaward of the surfzone and will only move shoreward when waves fall below modal wave height. Unless this occurs over a prolonged period, only slow and minor beach accretion will occur (see Fig 11, North Seven Mile). An increase in wave height will produce a relatively minor planing down of the subaerial beach and seaward movement of the breaker zones on the order of 10's to 100's m. Most severe erosion occurs in the backbeach-foredune as a result of an increase in standing wave amplitude which permits large infragravity bores (1-2m high) to flow right across the beach and penetrate the backbeach where they produce continuous, parallel foredune scarping and/or overwash at low points (Short and Hesp, 1982).

TABLE 2:

MORPHOMETRIC PARAMETERS OF SELECTED BEACH PROFILE LOCATIONS

LOCATION	GOOLWA	NORTH SEVEN MILE	MID SEVEN MILE	NARRABEEN	COLLAROY	FISHERMANS
BEACH TYPE	DISSIPATIVE		INTERMEDIATE			REFLECTIVE
n	4	12	12	100	100	32
Months	63	33	33	84	84	36
Hb(m)	3	1.6	1.6	1-1.5	.5-1	.3
T	12	10	10	10	10	10
Gd(mm)	.2	.25	.27	.3	.3	.35
$\bar{W}s$	.02	.032	.036	.04	.036	.05
Ω	9.6	5.10	4.4	3.75	2.1	.60
Gradient	1:33	1:39	1:37	1:15	1:12	1:9
B-S	6	5-6	5	3-4	2-3	1
$\bar{y}b$	90	112	112	42	36	27
$\sigma\bar{y}b$	10	16	15	14	9	5
CV	.10	.14	.13	.33	.25	.18
$\bar{V}$	120	153	177	180	65	40
$\sigma\bar{V}$	low	23	30	90	45	12
$\bar{V}/\bar{y}b$	1.3	1.37	1.58	4.3	1.8	1.5

n-sample size;
months-length of survey;
Hb-modal wave height;
T-wave period;
Gd-mean grain size;
$\bar{W}s$-mean fall velocity (m/sec);
Ω=Hb/T $\bar{W}s$;
Gradient-subaerial beach slope;
B-S-modal beach state;
$\bar{y}b$-mean beach width;
$\sigma\bar{y}b$-standard deviation of $\bar{y}b$ (shoreline mobility index);
CV-coefficient of variation of $\bar{y}b$ (backshore mobility index);
$\bar{V}$-mean subaerial beach volume and standard variation $\sigma\bar{V}$.

(Modified from Short & Hesp 1982).

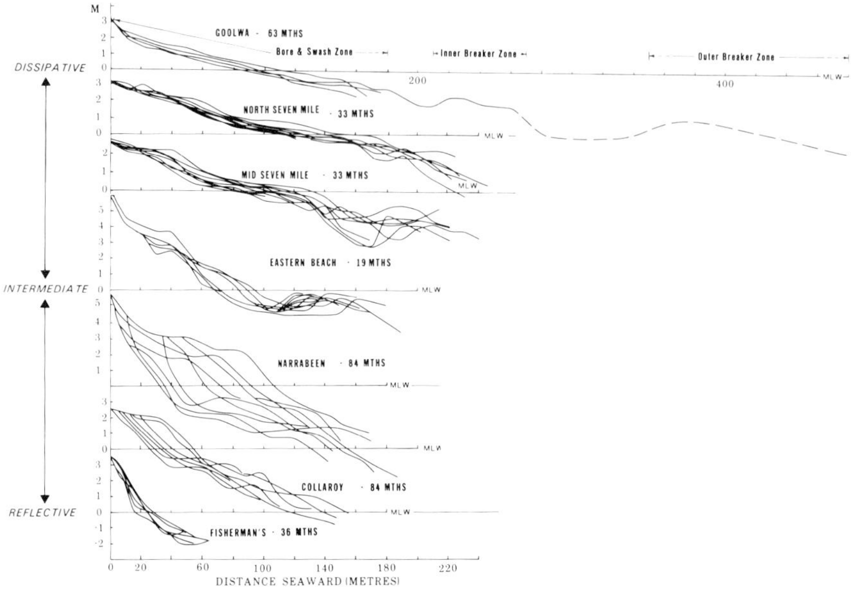

FIGURE 11. Sequential beach-surfzone profiles from long term survey sites chosen to be representative of the three beach types. The profiles illustrate the beach 'sweep' zone and morphological variability of each representative type. Months indicate duration of surveys, actual number of surveys given by n in Table 2. Note the steepening of beach gradients between dissipative and reflective, and the peak in shoreline mobility and sediment exchange associated with intermediate beach (modified from Short and Hesp, 1982). All surveyed by pole except outer (dashed) Goolwa by boat and echosounder.

Modally dissipative beaches are relatively stable both spatially, because they are uniform longshore (see Fig. 2a), and temporally because of relatively minor adjustments to the persistent high wave conditions, (see Fig. 11, Goolwa and North Seven Mile). This stability is reflected in the low shoreline (σyb) and backshore (CV) mobility indices (see Table 2, Goolwa).

Intermediate States

Intermediate beaches are the most mobile in terms of sediment exchange between the subaerial beach, surfzone and nearshore (Short, 1981). Long-term, multiple surveys of representative beaches in N.S.W. illustrate the active sweep zone on a variety of intermediate beaches (Fig. 11). In all cases they exceed the sweep zone of both the higher energy dissipative and lower energy reflective beaches.

The longshore bar-trough state is the most stable of the intermediate types with maximum sediment mobility associated with exchange between the beach face and low tide terrace, and with bar migration as it moves landward and seaward in response to falling and rising waves. Actual sediment exchange between the bar and the beach is relatively minor (Wright *et al.*, 1982c).

Rhythmic bar and beach and transverse bar and rip are the most mobile beach forms as crescentic bars rapidly migrate shoreward and rip channels infill under low wave conditions, while higher waves produce seaward bar migration and channel excavation. This highly dynamic state when combined with the inherent highly variable morphology between adjacent bars, rips and megacusps (see Fig. 2c and d) produces the greatest spatial and temporal variability in beach morphology and dynamics. This is reflected in the high shoreline ($\sigma_{\bar{y}b}$) and backshore (CV) mobility indices seen in Table 2.

As intermediate beaches lie between the dissipative and reflective extremes they possess characteristics of both the dissipative and reflective erosion modes (see Wright, 1980), as well as the distinctive rip embayment erosion which characterises intermediate systems. Rip embayment erosion is produced by three factors: (i) greater set-up of water as rip feeders converge toward the rip channel allowing higher swash excursion; (ii) greater wave attack due to less energy dissipation across the deeper rip channel; and (iii) close proximity of the eroding beach to the rip channel to convey eroded sediment seaward. When the rip is stationary as in an accreting rhythmic bar-beach state, the erosion produces embayment scarping (Fig. 2c). However, under rising wave conditions rip spacing increases leading to lateral shifts or longshore displacement in the location of rip embayments (McKenzie, 1958). This leads to a planing down of the beach and more subdued embayments as the rips stabilize (Short, 1979b). A typical sequence of intermediate beach erosion is reflective erosion and removal of the beach face, followed by rip planing of the megacusps and beach, and ultimately large widely spaced rip systems (500-1000m) interacting with wave dissipation to pulse at infragravity frequency. Under extreme conditions, particularly in topographically controlled systems (e.g. embayed beaches) very large 'megarips' or 'super rips' may drain an entire system and flow 2-3km seaward (Short, 1979b; Wright, 1980).

Higher energy intermediate beaches are morphologically and dynamically the most variable in spatial (see Figs. 2b, c, d, e) and temporal (see Fig. 11) terms. Spatial variability is an inherent characteristic of the bar-rip topography, while temporal variability is a response to varying wave conditions and particularly its effect on relocation and/or migration of rip channels (Wright and Short, in press). The degree of temporal variability is expressed diagrammatically in Figure 11 and quantitatively by the shoreline (σyb) and backshore (CV) mobility indices in Table 2.

In addition to the morphodynamic and mobility characteristics, Short (in press a and b) conducted sediment sampling, box coring and scuba observations across the beach, surfzone and nearshore of seven dissipative, intermediate and reflective beaches. He found the extent of the beach subenvironments (berm, step, bar, trough, nearshore, etc.) and their associated sediments, bedforms and structures vary significantly and systematically between beach states.

Environmental Determinants and Modal Beach State

The morphology of a beach at any particular time is a function of its sediment characteristics, prevailing and antecedent wave, tide and wind conditions and the antecedent beach state. However, over the long term a given beach will tend to exhibit a *modal* or most frequently recurrent state which depends on environment. The forcing wave climate can, in turn, be described in terms of the modal or most frequently occurring wave conditions around which a spread of higher or lower waves will occur. The beach in response to the wave climate consists of a modal beach type determined by the modal wave conditions, and a range of beach morphologies dependent on the range of the wave conditions. Associated with the temporal variations of beach state around the modal state, is corresponding variation in shoreline position, profile shape and morphology. The range of this variability defines the mobility of a beach. Because the same environmental factors which determine the modal state of a beach also determine its variability, there is a close association between modal beach state and beach mobility. Modal beach state can express the modal value of Ω, whereas the temporal variability of beach state should express the temporal variability of Ω. Plotting the value of Ω for numerous wave-sediment conditions in southern Australia has given values of $\Omega = 1$ as the threshold between reflective and intermediate beach types. The intermediate-dissipative threshold lies at $\Omega = 6$, though dissipative beaches may range in value from 6 to 30. On dissipative beaches the morphodynamic state does not change so long as Ω remains above 6. The surfzone width simply expands with increasing Ω and contracts with decreasing Ω. However, once Ω drops below about 6, the intermediate longshore bar-trough state replaces the fully dissipative state. Intermediate states result when 1 < Ω > 6. Within this relatively narrow band, the greatest amount of temporal variation occurs.

Figure 12 shows per cent frequency distributions of Ω for several beaches in southern Australia together with corresponding frequency histograms of beach states. Figure 13 shows plots of modal Ω against modal beach state and the average temporal ranges of Ω and state.

What these figures illustrate is that if modal wave and sediment conditions are known a beach can be typed; if the variability in breaker wave climate is known the temporal range of beach types can be calculated. As waves vary and thresholds of Ω are crossed so too will beach type change. In southern Australia, the high energy dissipative beaches such as Goolwa exposed to a persistent high wave environment tend to remain dissipative most of the time. Along the east coast and through Bass Strait where the wave energy is more moderate and more variable the temporal shifts in wave conditions result in greater temporal variability in beach

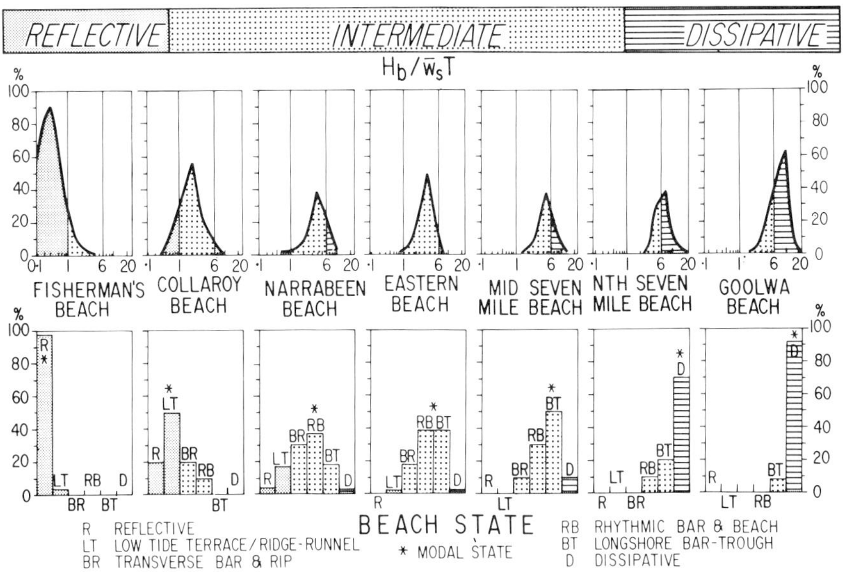

FIGURE 12. Frequency distribution of $\Omega = Hb/w_sT$ and corresponding percentage occurrence of beach states for the seven beaches illustrated in Figure 11. (From Wright and Short, in press).

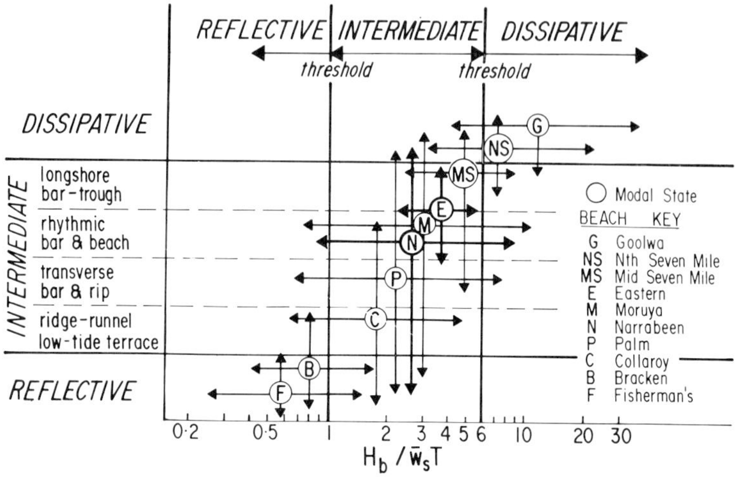

Figure 13. Modal beach state and temporal variability of beach state as functions of the modal values and temporal variability of Hb/w_sT. Horizontal arrows indicate average temporal range of Hb/w_sT; vertical arrows indicate average temporal range of beach state (from Wright and Short, in press).

type (Figs. 12 and 13). Associated with the degree of temporal variability is the auxiliary nature and scale of shoreline mobility and beach erosion discussed in the previous section.

A simple method of determining beach type is given in Figure 14. If modal Hb, T and W_s (or mean grain diameter) are known, the associated beach type can be determined for east coast (modal T = 10 sec; Fig. 14a) and south and west coast beaches (modal T = 12 sec; Fig. 14b). The figure illustrates the importance of Hb, W_s and to a lesser degree T in controlling beach type. High steep waves and fine sand favour higher beach types, whereas low, long waves and coarser sand produce lower beach types. Just as Short (1980) found that on medium-fine grained size Sydney beaches there are thresholds in Hb for each beach state (exemplified by the Narrabeen data, Fig. 14a), so too are there thresholds in W_s. Under modal east coast wave conditions (Hb = 1.5m, T = 10 sec) W_s must exceed 0.15m sec^{-1} (0.6ϕ) to remain reflective or be less than 0.026m sec^{-1} (2.8ϕ) to be dissipative. Consequently, dissipative beaches only occur modally on very fine sand east coast beaches (e.g. North Seven Mile, Fig. 14a), or periodically when Hb > 1.5m (e.g Narrabeen 14a). Bracken Beach

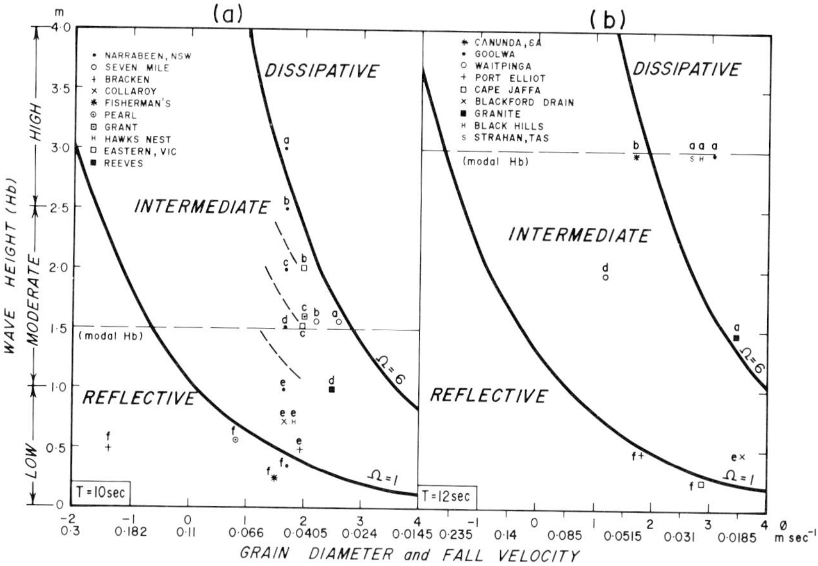

FIGURE 14. Graphs for determining beach type based on Hb, T and w_s, for a) East coast (modal T=12sec), and b) South and West coast (modal T=12sec). a to f refer to beach states in Figure 2. All sites refer to modal wave-sediment conditions and beach state, except Narrabeen and Eastern beaches which illustrate the effect of temporal variation in wave height, and Bracken which changed type because of a temporal variation in grain size.

(Fig. 14a) illustrates one beach that has changed types (reflective to intermediate) as a result of a decrease in grain size.

Figure 14 is designed for modal wave periods in southern Australia. In eastern Australia T can range from 6 to 14 sec. Consequently a family of curves is required to calculate the prevailing beach type using this method.

Within the intermediate type the threshold values of Ω required to discriminate the four states requires more field data. Long term observations at Narrabeen (Fig. 14) give an indication of both the wave height and Ω value required for a beach with medium-fine sand (1.7ϕ). Using the Narrabeen values the approximate level of Ω required is <1 for reflective, 1 to 3 for ridge and runnel, 3 to 4 for transverse bar and rip, 4 to 5 for rhythmic bar and beach, 5 to 6 for bar-trough, and >6 for dissipative. Data are being compiled for a large number of Australian and overseas beaches in an attempt to define the Ω thresholds required to change beach type and/or state over a wide range of wave-sediment conditions (Short, in preparation).

Conclusion

The coast of Australia offers an ideal natural laboratory for the examination of coastal morphodynamics. Extreme ranges in wave, tide and sediment conditions along with climate and biota permit the selection of field sites to suit prescribed environmental conditions. The authors have utilized this situation to study wave-dominated sandy beaches in southern Australia (Fig. 1), particularly microtidal beaches in low, moderate and high wave environments. Numerous field experiments and surveys have resulted in the identification of three basic beach types which consist of six individual beach states (Fig. 2). This chapter has examined the morphodynamics associated with the four highest energy beach states in southern Australia,those produced by waves exceeding 1m in height.

Each beach type is a response to environmental parameters. High steep waves and fine sand favour dissipative beaches, moderate to high waves and fine to medium sand result in intermediate beach types, and long low waves and coarser sediment favour reflective beaches (Fig. 14). For most beaches with a given sediment size, the modal breaker wave will determine the modal beach type while temporal variations in wave climate will affect a change in beach state and possibly beach type. Between the beach states is a systematic yet distinctive transition in beach-surfzone morphodynamics. Furthermore, the three beach types are characterised by three distinct modes of erosion and absolute scales of shoreline mobility.

High energy beach systems in southern Australia are highly distinctive in form and character. The combinations of waves, tide and sediment that produce the Australian beach systems are expected to maintain similar systems where they occur elsewhere on the globe.

Acknowledgements

This chapter summarises studies supported by the U.S. Office of Naval Research, Coastal Sciences Program, and the Australian Research Grants Committee. Figures were drafted by J. Roberts and J. de Roder.

References

Aubrey, D.G. (1979). Seasonal patterns of onshore/offshore sediment movement. *J. Geophys. Res.* 84, 6347-6354.

Bowman, D., and Goldsmith, V. (1983). Bar morphology of dissipative beaches: an empirical model. *Mar. Geol.* 51, 15-33.

Chappell, J., and Wright, L.D. (1979). Surf zone resonance and coupled morphology. *Proc. 16th Intl. Conf. Coastal Engr.*, Hamburg, 1359-1377.

Davies, J.L. (1980). *Geographical Variation in Coastal Development.* 2nd ed. Longman, London, 212 pp.

Dean, R.G. (1973). Heuristic models of sand transport in the surfzone. *Proc. Conf. Engr. Dynamics in the Surf Zone*, Sydney, 208-214.

Guza, R.T., and Inman, D.L. (1975). Edge waves and beach cusps. *J. Geophys. Res.*, 80, 2997-3012.

Hayes, M.O. (1979). Barrier island morphology as a function of tidal and wave regime. In Leatherman, S.P. (ed.) *Barrier Islands*, Academic Press, New York, 1-27.

McKenzie, P. (1958). Rip current systems. *J. Geol.*, 66, 103-113.

Sallanger, A.H. Jr., Howard, P.C., Fletcher, C.H., and Howd, P.A. (1983). A system for measuring bottom profile, waves and currents in the high energy nearshore environment. *Mar. Geol.* 51, 63-76.

Sasaki, T. and Horikawa, K. (1979). Observations of nearshore currents and edge waves. *Proc. 16th Intl. Conf. Coastal Engr.*, Hamburg, 791-809.

Short, A.D. (1979a). Wave power and beach stages: A global model. *Proc. 16th Intl. Conf. Coastal Engr.* Hamburg, 1145-1162.

Short, A.D. (1979b). Three dimensional beach stage model. *J. Geol.* 87, 553-571.

Short, A.D. (1981). Beach response to variations in breaker height. *Proc. 17th Intl. Conf. Coastal Engr.*, Sydney, 1016-1035.

Short, A.D. (in press, a). Sediments and structures in beach-nearshore environments, south east Australia. *In* McLachlan, A. and Erasmus, T. (eds.) *Sandy Beaches as Ecosystems*, Junk Publ. The Hague.

Short, A.D. (in press b). Beach and nearshore facies: south east Australia. *In* Greenwood, B. and Davis, R.A. (eds.) *Coastal Environments Dominated by Waves*, Elsevier, Amsterdam.

Short, A.D. (in preparation). A method for determining beach type.

Short, A.D., and Wright, L.D. (1981). Beach systems of the Sydney region. *Aust. Geographer* 15, 8-16.

Short, A.D. and Hesp, P.A. (1982). Wave, beach and dune interactions in southeastern Australia. *Mar. Geol.*, 48, 259-284.

Wright, L.D. (1981). Beach cut in relation to surf zone morphodynamics. *Proc. 17th Intl. Conf. Coastal Engr.*, Sydney, 978-996.

Wright, L.D. (1982). Field observations of long-period surf zone oscillations in relation to contrasting beach morphologies. *Aust. J. Mar. Freshwater Res.*, 33, 181-201.

Wright, L.D., Chappell, J., Thom, B.G., Bradshaw, M.P. and Cowell, P. (1979). Morphodynamics of reflective and dissipative beach and inshore systems: Southeastern Australia. *Mar. Geol.*, 32, 105-140.

Wright, L.D., Guza, R.T., and Short, A.D. (1982a). Dynamics of a high energy dissipative surfzone. *Mar. Geol.*, 45, 41-62.

Wright, L.D., Nielsen, P., Short, A.D., Coffey, F.C. and Green, M.O., (1982b). Nearshore and surfzone morphodynamics of a storm wave environment: Eastern Bass Strait, Australia. *Coastal Studies Unit, Tech. Rept.* No. 82/3, Coastal Studies Unit, University of Sydney, 154 pp.

Wright, L.D., Short, A.D. and Nielsen, P. (1982c). Morphodynamics of high energy beaches and surf zones: A brief synthesis. *Coastal Studies Unit. Tech. Rept.* No. 82/5, Coastal Studies Unit, University of Sydney, 64 pp.

Wright, L.D., and Short, A.D., (in press). Morphodynamics of beaches and surfzones in Australia. *In* Komar, P.D. (ed.) *Handbook of Coastal Process and Erosion,* CRC Press, Miami.

4

Foredune Formation in Southeast Australia

PATRICK A. HESP

Division of Resource Management
Department of Agriculture
Perth, WA

Introduction

Foredunes develop in the seawardmost vegetated sand surface behind active beaches (Fig. 1). They are generally shore-parallel, convex ridges. Concave swales (or slacks) commonly separate the ridges. In southeast Australia foredune-like sand ridges have been commonly termed beach ridges (Davies, 1957). Series of such ridges form plains in many coastal embayments, and constitute a significant feature within coastal embayments in New South Wales (Langford-Smith and Thom, 1969; Davies 1977; Thom *et al.*, 1981; Thompson and Bowman, this text). Whilst several modes of beach-ridge formation have been proposed in the international literature including accreted bars and berms (e.g. Hine, 1979), and storm-built berms and beach deposits (e.g. Psuty, 1966), with a few exceptions Australian hypotheses have centred on berm initiation. To date little evidence in the form of topographic or photographic surveys over time, or data on internal sedimentary structures have been presented to support these hypotheses. This paper presents new evidence on the formation of foredunes and beach ridges. The data are examined in the light of previous hypotheses of beach-ridge evolution.

Two basic types of foredunes are distinguished; incipient and established foredunes (Hesp, 1982). Figure 1 shows how these two types are recognized on the basis of their location with respect to the backshore zone of an active beach and their ecological status. 'Incipient foredunes', the initial or embryo foredunes, are formed by the trapping of sand within pioneer plant species. These dunes may eventually be colonised by an

COASTAL GEOMORPHOLOGY
IN AUSTRALIA
ISBN 0 12 687880 3

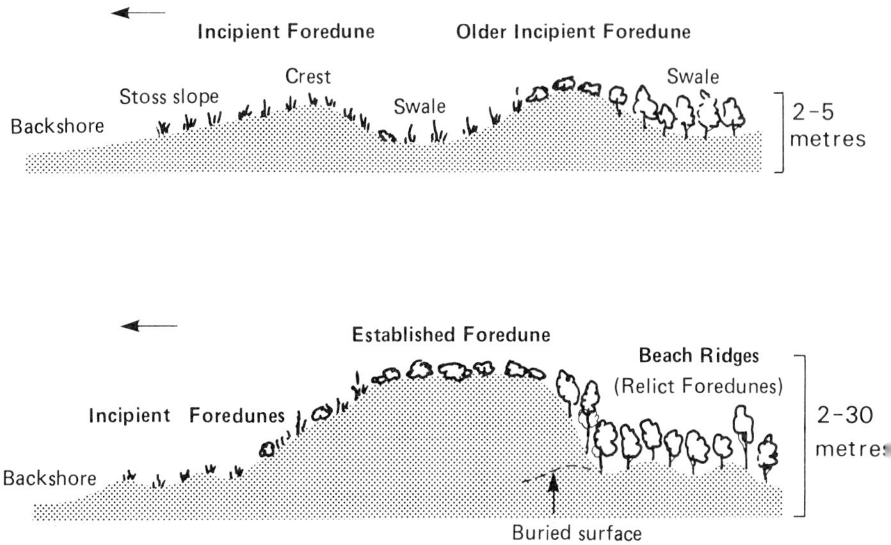

FIGURE 1. Terminology of foredunes as used in this paper.

intermediate group of 'woody', mat or tufted plants, often following the formation to seaward of a new incipient foredune (Fig. 1). Where semi-established foredunes are colonised by shrubs and trees they are called 'established foredunes'.

Foredune Formation: Some Observations

In the period 1976 to 1980, the author monitored several foredune-beach sites within the Myall Lakes National Park, N.S.W. Changes in surface elevation and plant density per vegetation species were surveyed in this region at six sites, at monthly intervals over a three year period (1977-1980) (Hesp, 1982). Occasional surveys have continued since 1980. In addition, incipient foredunes were photographically monitored at selected beaches extending along the N.S.W. coast between 1976 and 1980 (see Hesp, 1982, for details).

The Myall Lakes region comprises five coastal compartments within which wind and wave exposure increases northwards from low to high (Short and Hesp, 1982). Tides are micro-tidal with an average spring range of 1.2 m. Significant wave heights exceed 1.5 m, 2.5 m and 4 m for 50%, 10% and 1% of the time respectively (Lawson and Abernathy, 1975; Wright, 1976). The region experiences a bi-modal wind regime consisting of onshore summer sea-breezes (ESE to E) and offshore winter westerlies. These are underscored by occasional onshore southeasterly storm winds. Rainfall is fairly uniformly distributed with slight summer maxima (101 mm, median) and spring minima (50 mm). Figure 2 shows the location of the study area.

Hesp (1982) distinguished four modes of incipient foredune formation. Type 1a includes those which are initiated by aeolian sand deposition within and in the lee of *discrete* plants. Type 1b are those which are initiated by sand deposition within *discrete* zones of seedlings. Type 2 includes incipient foredunes formed by sand deposition within *laterally extensive* colonies of either seedlings (type 2a) or rhizomes (type 2b). Examples of the development of types 1a, 2a and 2b are given below. Although no actual example of the type 1b is given, its evolutionary pattern is similar in many respects to that of type 2a incipient foredunes (see Godfrey and Godfrey, 1976; and Woodhouse, 1978).

Type 1a Incipient Foredunes

Type 1a incipient foredunes are formed by aeolian accretion within discrete annual and perennial plants which have germinated either from wind dispersed seed, or from seed contained within swash-aligned drift material (Ridley, 1930; Salisbury, 1952; Bond, 1952; Olson, 1958a, b; Steers, 1964; Leatherman, 1979). Plant fragments, usually storm eroded vegetative material, can also act as the nucleii for type 1a dunes (Gemmell *et al.*, 1953; Laing, 1958). In the European literature, dunes formed on drift lines are commonly termed 'embryo' dunes, and are often initially dominated by the annual, *Cakile maritima* (Salisbury, 1952; Steers, 1964; Zenkovich, 1967; Quinn, 1977). Discrete aeolian bedforms formed in association with the plants have been variously termed 'hummocks', 'mounds', and 'tussock dunes' where the bedforms are formed within the plant body (Sokolow, 1894; Kuhlman, 1960; Steers, 1964; Quinn, 1977). The terms 'tongue dunes', 'lenticular heaps', 'miniature dunes' and 'hillocks' have been used where the bedforms are formed to the lee of the plant body (Hogbom, 1923; van Dieren, 1934; Pidgeon, 1940, 1942; Guilcher, 1958; Olson, 1958a; Kuhlman, 1960; Cooper, 1958, 1967; McBride and Hayes, 1962; Bigarella, 1972; Goldsmith, 1973). Many of these have similarities to coppice dunes (Melton, 1940). Such discrete bedforms are here termed 'shadow dunes', a term Bagnold (1954, p.188) used for 'deposits caused directly by fixed obstructions in the path of the sand-driving wind' (see Hesp, 1981). Examples are shown in Figure 3.

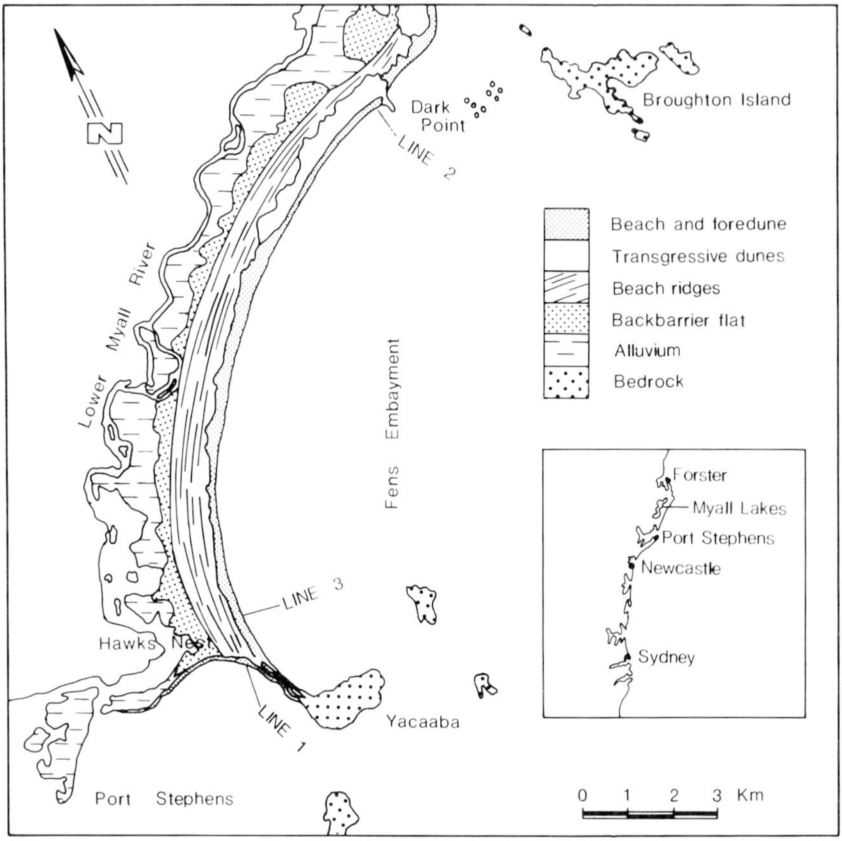

FIGURE 2. Morphology of Fens embayment, Myall Lakes, N.S.W., and location of survey sites.

Observations on the southeast Australian coast indicate that the main plant species involved in type la incipient foredune development are *Cakile edentula, Cakile maritima, Senecio lautus, Senecio spathulatus, Arctotheca nivea, Festuca littoralis, Sonchus megalocarpus, Stackhousia spathulata* and *Ammophila arenaria*. *Cakile* species are the most common plants involved. Pure stands of *Senecio* species have also been observed (e.g. Pidgeon, 1940). However, species other than *Cakile* are fewer in number, and rarely appear to initiate an incipient foredune. Shadow dunes may be formed behind almost any obstacle (e.g. single *Spinifex sericeus*[1] stems; rocks;

[1] The Eastern States *Spinifex hirsutus* has been renamed *Spinifex sericeus* (Craig, in press).

(a) 5.8.77

(b) 1.12.77

(c) 9.6.78

(d) 17.11.78

(e) 26.4.80

FIGURE 3. Photographic record of a type 1a incipient foredune at Line 1, Port Stephens.

seaweed clumps). Although these obstacles may aid in the formation of incipient foredunes, they have never been observed to be the sole initiating agent.

The evolution of a type la incipient foredune was monitored in the Fens/Port Stephens area (Line 1 in Fig. 2). The reflective beach in this location is typically narrow and alternates between a linear (no berm) form and a subtle berm. The type la incipient foredune was associated with the growth of *Cakile edentula* (Fig. 3a). *Cakile edentula* (Bigel.) Hook, spp. californica (Heller) Hult is a glabrous annual herb. It germinates in spring (October - November) in southeast Australia in almost any beach or dune location, where the seeds or fruit have been wave or wind transported (for further information on *Cakile* see Barbour, 1970a, b; Keddy, 1980).

A 'zone' comprising discrete *Cakile edentula* plants was formed in October, 1976, at the limit of spring tide swash (Fig. 4). By March, 1977, large *Cakile* plants had trapped sand to form shadow dunes. By June, 1977 (Fig. 4) the *Cakile* had died but the bedforms, held by the plants root material, remained (Fig. 3a). During the following spring the *Cakile* regrew *in situ* and was invaded by *Spinifex sericeus* rhizomes and stolons from a landward ridge (Fig. 3b). Although the incipient foredune was overwashed in June, 1978, by storm waves (Fig. 3c), *Spinifex* growth was re-established in the 1978 spring season (Fig. 3d). It continued to build very gradually and by 1980 had a complete cover of *Spinifex*

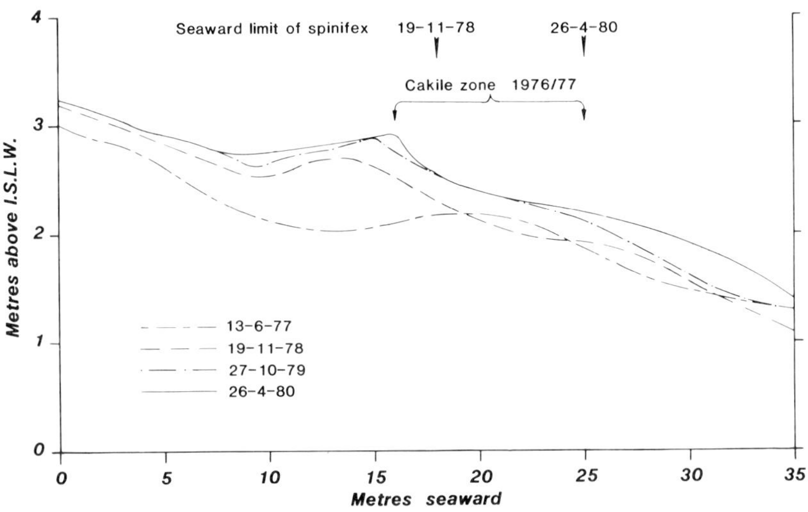

FIGURE 4. Morphologic survey of a type la incipient foredune at Port Stephens.

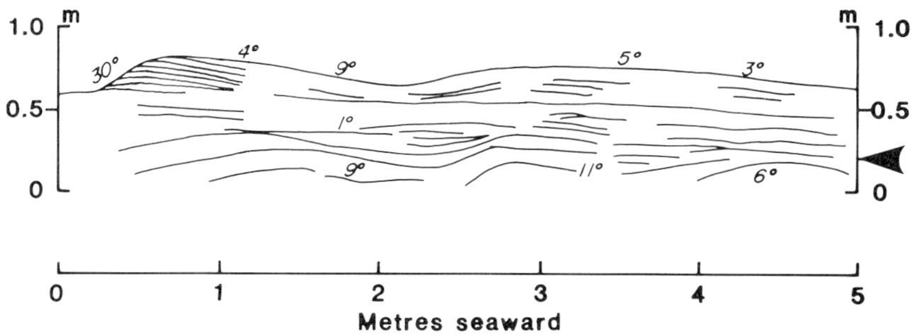

FIGURE 5. Internal sedimentary structure of part of the type la incipient foredune near Line 1, Port Stephens. The arrow on right indicates the approximate divide between basal shadow dune sets and Spinifex *dominated sets. Prevailing wind blows from the right to left.*

sericeus (Figs. 3e; 4). Some *Senecio lautus* and *Festuca littoralis* also colonised the ridge during 1979. The gross morphology of the type la incipient foredune monitored at this site was relatively simple due to the low rate of aeolian sand deposition (limited beach fetch and low modal wind and wave energy; see Short and Hesp, 1982). A low hummocky terrace gradually developed into a subtle asymmetric ridge over the 1976 to 1980 period. A swale formed in the limited to non-depositional zone in the lee of the ridge.

The record of type la dune development (as shown in Figure 4) is further illustrated by an analysis of the internal structure of the ridge. In Figure 5 the shadow dunes forming the initial 'terrace' may be seen. These form quasi-discrete sets of simple, convex, low angle, small to medium scale cross-laminae. Strata below these sets (not shown) were simple, seaward dipping, low angle cross-laminae typical of the back-shore zone. Cross-laminae above the sets formed by the shadow dunes at first drape the lower sets and infill the intervening lows at low angles. These indicate the initial invasion of *Spinifex*. Cross-laminae then become more continuous and medium-scale as the dune surface is approached, and simple, lenticular cross-laminae dominate.

The mode of type la incipient foredune formation outlined above parallels that of English examples except that *Ammophila arenaria* is generally the 'invading' species rather than *Spinifex* (Salisbury, 1952; Quinn, 1977). However, type la incipient foredunes as described here are fundamentally different from those described by Cooper (1958, 1967) and McBride and Hayes (1962). The proportion of incipient foredune consisting of shadow dune bedforms is quite small in the Australian case compared with some of the American examples.

In the latter, shadow dunes formed principally in and behind *Ammophila sp.* However, various other species may form major portions of the foredune (Cooper, 1958, 1967). Coalescence of large shadow dunes eventually produces a semi-continuous foredune ridge (e.g. Cooper, 1958, plate 14). In the Australian case, although discrete annuals may initiate type la incipient foredunes, the main bulk of the foredune is formed by laterally continuous accretion within *Spinifex sericeus*.

Type lb Incipient Foredunes

Although no type lb incipient foredunes were monitored in this study, it is relevant to briefly discuss their evolution here. Incipient foredunes are formed by aeolian accretion within discrete zones of seedlings which germinate on the backshore within storm or high tide debris lines (Land, 1964; Bigarella *et al.*, 1969; Bigarella, 1972; Goldsmith, 1973; Godfrey and Godfrey, 1976; Woodhouse, 1978). The morphology of type lb foredunes is typically a discrete, semi-circular, convex dome. Adjacent areas are usually unvegetated (e.g. Woodhouse, 1978, Fig. 3). Bigarella *et al.*, (1969) and Bigarella (1972) term these discrete incipient dunes 'pro-dunes' or 'dome-dunes' (see also McKee, 1979). Here they are termed incipient dunelets following Godfrey and Godfrey (1976) and Woodhouse (1978).

Type lb incipient foredunes develop into a continuous alongshore ridge either by (i) laterally extensive invasion of the so-called 'embryo' drift line zone by perennial seedlings, or (ii) gradual coalescence of discrete dunelets. The latter usually occurs by shoot and rhizome colonisation of adjacent nonvegetated areas. In most cases hummocky or undulating ridges result (Hogbom, 1923; van Dieren, 1934; Cooper, 1958, 1967; Goldsmith, 1975; Carter, 1977).

Type 2a Incipient Foredunes

Type 2a incipient foredunes are formed by aeolian accretion in laterally continuous (alongshore) zones colonised by perennial seedlings (Cowles, 1899; Laing, 1958; Olson, 1958a, 1958b; Cooper, 1958; Willis *et al*, 1959; Ranwell, 1972; King, 1972; Godfrey and Godfrey, 1976; Quinn, 1977; Woodhouse, 1978; Godfrey *et al.*, 1979). The seedlings usually germinate within litter or drift material deposited at the limit of high tide swash (Salisbury, 1952; Leatherman, 1979).

A type 2a incipient foredune was monitored at Dark Point, Fens embayment (line 2 in Fig. 2). The moderate energy, intermediate beach at this location is typically a wide, low gradient berm. The type 2a incipient foredune was initiated by the growth of *Spinifex sericeus*. *Spinifex sericeus* is a perennial native grass commonly about 25 cm high, with stout rhizomes and stolons which can extend for several metres. Seeds (usually caryopses in spikelets) are blown from landward source areas or may be wave transported to the beach and backshore. In N.S.W. westerly offshore winds which occur throughout autumn and winter appear primarily responsible for the transport of summer produced seeds onto the beach. Here they are incorporated into the beach sediments by the action of swash and onshore aeolian sand transport. Maximum germination occurs in spring, particularly early October through November (Hesp, 1982).

At the initiation of the survey the monitored zone consisted of a low (5.5 m) foredune with an accreted scarp, an unvegetated swale and a low terrace-shaped incipient foredune (see Fig. 6). The older foredune displayed a vegetation cover of *Spinifex sericeus, Scaevola calendulacea, Carpobrotus glaucescens, Hydrocotyle bonariensis*, and *Acidanthera brevicolis*. The intervening swale was colonised by a few *Spinifex* plants, and the incipient foredune was dominated by *Spinifex sericeus* with some rare *Arctotheca nivea* plants (Fig. 7a). The stoss face (accreted scarp) of the older foredune was produced by storms in 1974, and the incipient fore-

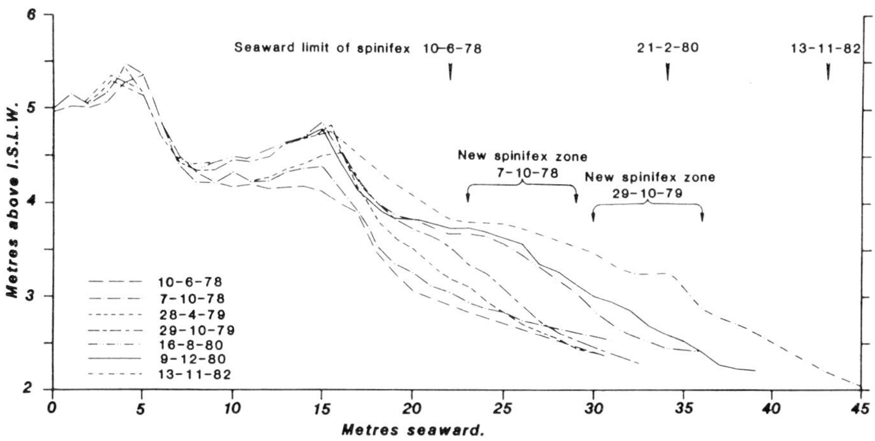

FIGURE 6. Morphologic surveys of the type 2a incipient foredune at Line 2, Fens embayment.

(a) 10.6.78 (b) 17.11.78 (c) 24.3.79

(d) 24.9.79 (e) 24.9.79 (f) 16.8.80

FIGURE 7. Photographic record of a type 2a incipient foredunes at line 2, Fens embayment: (a) first foredune; (b) seedling zone of second incipient foredune; (c) second foredune; (d) high velocity event on first and (e) second incipient foredune.

dune was initiated in the spring (October) of 1977. The lack of *Spinifex* rhizomes in the swale (Fig. 7a) suggested that the incipient foredune was initiated by discrete seedling colonisation.

In the months following the first survey, the incipient foredune continued to accrete and *Spinifex* rhizomes colonised the backshore lying immediately to seaward of it. Figure 6 illustrates that most of the accretion until October, 1978 occurred on the crest of the incipient foredune. During October, 1978, *Spinifex sericeus* seedlings colonised a semi-discrete zone of the backshore, seaward of the first incipient foredune. Seedling distribution was variable, with a maximum of 232 seedlings per square metre recorded. Figure 7b illustrates a portion of this zone near line 2 whilst Figure 7c shows the first incipient foredune and new incipient zone in the following month. The *Spinifex* population of this second (most seaward) incipient foredune reached approximate adult size during February, 1979, and the overall numbers of *Spinifex* were at a seasonal maximum at this time (Fig. 6, 7d). During this five month period (October-February) the crestal area of the first incipient foredune tended to accrete at the expense of the second seaward incipient foredune (Fig. 6). This accretion of the first incipient foredune illustrates the development of a zone of maximum deposition on a foredune. For a range of windspeeds, there is a downwind (vegetated) distance over which most of the incoming saltating sand will be deposited. Since the rate of deposition increases downwind, an asymmetric deposit forms and eventually leads to the formation of a ridge. *Spinifex* growth is greatest in the zone of maximum deposition and thus provides a positive feedback to continued deposition in that zone. A swale is indirectly formed in the lee of the ridge as the dune accretes.

Figure 7e, f, illustrates the effect of a strong southerly storm on the foredune zone in September, 1979. Wind velocities of this storm ranged between 4 and 17 m sec^{-1} (measured at $z = 7$ m). Dune height decreased and width increased on both incipient foredunes.

By late October, 1979, new *Spinifex* seedlings again colonised a discrete zone 6-10 m seaward of the seaward edge of *Spinifex* vegetation. Initial species numbers per m^2 were much less than in the previous year, and survival was low due to both wind deflation and high tide swash erosion during the first month of growth.

From October, 1979, the degree of topographic change on the first incipient foredune was small as the second, seaward lying incipient foredune trapped most of the incoming aeolian sand (Fig. 6). As the second incipient foredune developed, a subtle swale was formed to the lee of it. The second dune was apparently extending seawards, predominantly by rhizome and shoot growth, at the last survey (13.11.82).

Figure 8 illustrates the internal structure of the two type 2a incipient foredunes trenched in the vicinity of line 2 in 1979. The small incipient ridge shown in Figure 8a is part of the second or seaward incipient ridge depicted in Figure 6. The lowermost set in this seaward ridge is bounded at the top by a line (arrowed) of drift material (seed and wood fragments). This material was deposited by high spring tide swash on the backshore, and the *Spinifex sericeus* growing above originated from seeds deposited within this litter. The lowermost set is a simple, tabular set comprising extremely low-angle (1-2°), parallel, swash-deposited laminae. The sets above are aeolian in origin, being simple lenticular sets dominated by low-angle cross-laminae. Maximum cross-strata dips were 13°. Set shape results from minimal backset deposition and maximum topset and foreset deposition. Further details are described in Hesp (1983).

The other ridge illustrated (Fig. 8b) forms part of the first incipient foredune of line 2. The bounding surfaces essentially mimic the time lines shown in the Figure 6 survey (10-20 m zone). Simple lenticular sets comprising gently convex, low-angle (2-15°) cross-laminae dominate this type 2a incipient foredune ridge.

Type 2b Incipient Foredunes

Incipient foredunes are formed by rhizome-stolon colonisation of the immediate seaward backshore zone (Gooding, 1947; Tague, 1947; Salisbury, 1952; Olson, 1958a, b; McKenzie, 1958; Hill, 1966; Thom , 1964, 1965; Wright, 1970; Quinn, 1977; Davies, 1980). The rhizomes usually grow from a landward source area (e.g. a foredune). In many cases the rhizome-stolon colonisation takes place in conjunction with shoot growth as shown in the type 2a example above.

In the case of *Spinifex sericeus,* field observations indicate that mature *Spinifex* plants are in the range of one to two years old before they produce rhizomes. It appears rhizome growth can take place throughout the year, although there is a maximum growth period in spring-summer (Oct.-Feb.). At intervals of 10-15 cm new nodes are formed on the rhizomes or stolons from which discrete plants emerge. Field measured rates of rhizome growth range from 0.4 m to 1.4 m per month (Hesp, 1982).

A type 2b incipient foredune was monitored near Hawks Nest, Fens embayment, a low-moderate energy site characterised by moderate gradient berms (line 3, Fig. 2). At this location

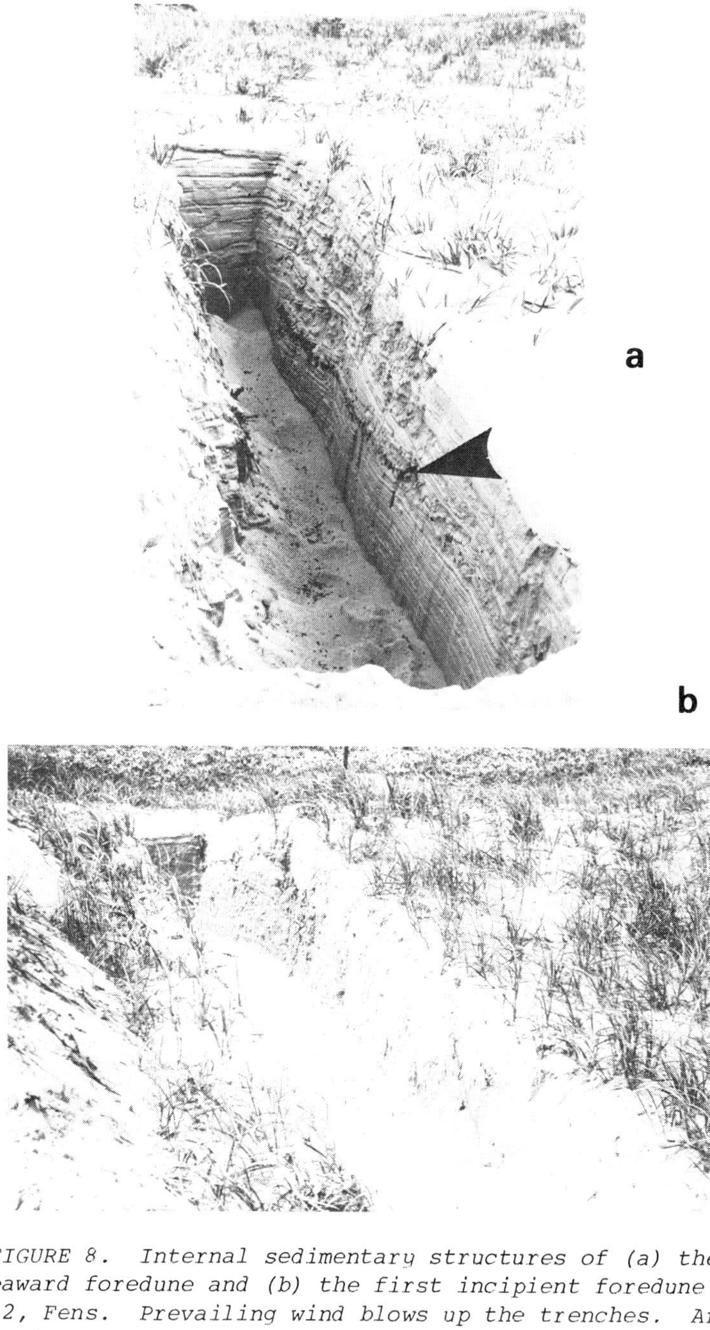

a

b

FIGURE 8. Internal sedimentary structures of (a) the second or seaward foredune and (b) the first incipient foredune at Line 2, Fens. Prevailing wind blows up the trenches. Arrow in (a) indicates swash deposit.

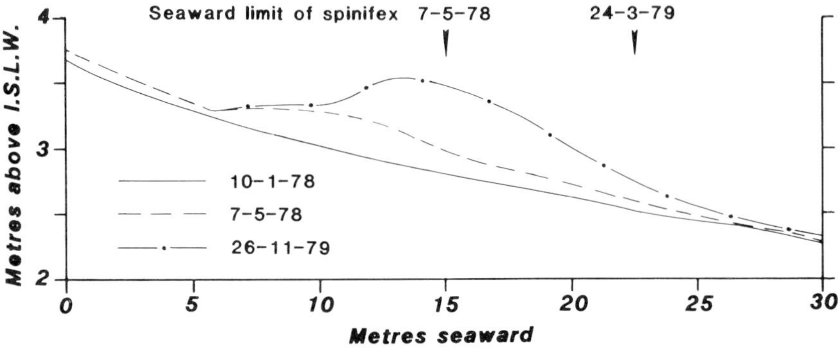

a

b

FIGURE 9. (a) Morphologic survey and (b) illustration of the type 2b incipient foredune at Line 3, Fens.

Spinifex rhizomes and stolons gradually colonised the seaward slope of the backshore from an older incipient foredune source area. This is similar to the process observed by McKenzie (1958) where an incipient ramp was initially formed as the foredune stoss face accreted. As rhizome and shoot density increased, and plant colonisation of the seaward beach took place, incoming aeolian sand was preferentially trapped in a discrete zone. A ridge and swale was gradually formed (Fig. 9a, b).

Sedimentary structures of type 2b incipient foredunes are similar in many respects to those described for type 2a dunes, except in the lowermost beds. In the region where rhizomes initially colonise the backshore, lenticular aeolian cross-laminae interdigitate with simple tabular and lenticular swash laminae. As aeolian deposition takes place within a continuously seaward migrating vegetation, overlapping wedge-shaped and lenticular sets of low-angle cross-laminae typify the initial internal structure.

In summary, observations in the Myall Lakes region indicate that the foredune sub-types discussed above are initiated, in association with plants, on the backshore zones of beaches. The most commonly occurring foredune sub-type is type 2a, although a combination of 2a and 2b (seedling and rhizome) is also frequently found. The type 1a foredunes (initiation by annuals) do not occur as frequently as the 2a and 2b types, and are most common on the lower energy, more protected beaches.

Previous Australian Literature on Beach Ridges

In order to examine the relationship between the mode of modern foredune development presented above, and that proposed for the development of sand beach ridges, it is necessary to review the Australian literature on beach-ridge formation. This review is restricted to the post-1950's Australian literature on beach-ridge formation, which has been principally concerned with the formation of berms and incipient foredunes; these are viewed by most authors as analogous to beach-ridge formation.

The post-1950's debate on the origin of beach ridges was initiated by Sprigg (1952). He proposed two modes of beach-ridge formation. One, that minor variations in sea level caused oscillatory migrations of the strandline producing a ridge and swale topography; and two, that periodic tidal range variations, possibly caused by cyclic changes in the orbit of the moon (18-61 year cycle), enabled sand to be transported to higher levels on a beach, and in larger quantities, forming a ridge. Sprigg favoured the latter explanation. He considered that preservation of the ridge resulted from colonisation by vegetation, and that vertical growth was due to deposition of aeolian sand.

In 1957, Davies proposed the 'cut and fill' hypothesis of sand beach-ridge formation. In this hypothesis a berm is built by calm weather waves (fill). If not subsequently eroded (cut), the berm, assisted by its convex cross profile and drift material pushed up above high water mark, becomes the nucleus for a beach-ridge. Plants colonise the ridge and trap aeolian sand. After a certain time a new berm may be built in front of the ridge and the process repeated. Davies (1957) also proposed that swales were formed by the new berm being initially lower relative to the older beach ridge. This new berm forms a 'distinct nucleus around which windblown sand can accumulate' (p 108).

According to Davies, variations in berm convexity, and therefore efficiency as a sand accumulating nucleus explained why some ridges are distinct and some abut the preceding ridge. Beach ridges owed their alignment to the disposition of the berm nucleus which parallels the beach outline in plan, in his view. Davies believed that 'frontal dunes are the evolutionary descendents of beach ridges' (1957, p 214). Jennings (1959), Hails (1969) and Wright (1970) agreed with Davies' (1957) mode of beach-ridge evolution.

McKenzie (1958) disagreed with Davies' (1957) hypothesis, stating that berms were not sufficiently stable to form the nuclei of beach ridges. In McKenzie's view the sand comprising the berms would be blown landwards. McKenzie (1958) stated that incipient foredunes (beach ridges) were formed where pioneer vegetation, particularly *Spinifex hirsutus,* or *S. sericeus* as it is now named, grows seaward from an established landward source, and traps aeolian sand. He described the temporal development of this process where a progression from ramp to terrace to ridge morphology occurred as the beach was colonised by rhizomes and aeolian sand deposition took place.

Bird (1960) took issue with McKenzie (1958), stating that McKenzie's mechanism did not explain the pattern and spacing of modern beach ridges in Gippsland. Bird (1960) described the occurrence of incipient foredunes (beach ridges) which paralleled the plan of the berm and the shoreline, and which were only vegetated on their crests (by *Ammophila arenaria* and *Festuca littoralis).* Intervening swales were unvegetated. Bird (p. 350) considered these may have been too saline for colonisation. He argued that the ridges were initiated upon berm crests, and thus agreed with Davies' hypothesis. Subsequently, Bird (1961) used the term 'parallel dunes' for beach ridges (see also Bird, 1963; 1965).

Thom (1964, 1965) presented a modified combination of Davies' (1957) and McKenzies' (1958) modes of beach-ridge formation. In Thom's view a berm was built and then colonised by seaward growing plants. The plants grew seawards until they reached a point frequently awash at high tide. Onshore winds deposited sand in the plant zone and formed a ridge. According to Thom (1965, p. 33) 'oscillation in the growth of the prograding beach controlling the seaward position reached by pioneer plants is suggested as the important factor in ridge and swale development'.

Jenkin (1968) agreed with Davies (1957) and Bird (1960) in that he considered beach ridges originate as berms. He disagreed on the role of vegetation as stressed by McKenzie (1958), stating that it did not play a significant part in the formation of 'very regular beach ridges' (p. 91). Rather, Jenkin believed that the preservation of ridges was dependent on rapid removal from the influence of wave and wind action, by the building of new ridges to seaward.

In 1969 (2nd edn., 1976), Bird modified the 'cut and fill' mode of ridge formation. In this publication, Bird described a scenario in which a foredune forming on top of a berm, is scarped by storm waves. Subsequently waves build a new berm in front of, and 'parallel to the trimmed foredune, separated from it by a low-lying trough or swale' (p. 156). Dune grasses colonise the berm first, leaving the swale unvegetated, and a new foredune is built up. Bird (1969) stated that it was the 'effect of cut' which was responsible for the separation of dunes into parallel ridges. In the absence of cut, a foredune became broadened as a coastal terrace. Bird (1976) provided a model diagram of this 'cut and fill' mode of beach-ridge formation (Fig. 54, p. 157).

Davies (1972; 2nd edn., 1980) has more recently placed sand beach ridges into his larger category of 'impeded primary dunes' or foredunes. He stated that the foredune's "... close approximation in plan to the beach is brought about partly by the way in which the back of the berm acts as a nucleus for sand accumulation, and partly by the way in which high waves periodically trim the front of the growing ridge" (p.160). This line of thinking is continued in Davies' (1977) review of Australia's coast, in which he views frontal dunes, beach ridges and foredunes as analogous dune forms.

Genesis of Sand Beach Ridges

The author believes that the modes of incipient foredune (and foredune) formation described above are identical to the processes involved in sand beach-ridge formation, i.e. beach ridges are in fact relict foredunes. A discussion follows in which the field evidence presented above is examined in the light of previous studies.

In regard to the berm initiation hypothesis proposed by Davies (1957), and supported by Bird (1960), in all incipient foredune cases that were monitored (see Figs. 3, 4, 6, 7, 9) the *crests* of berms were never successfully colonised by pioneer plant seedlings. This was the case even under conditions of rapid beach progradation. The zone of vegetation colonisation was restricted to the backshore zone (i.e. well landward of the berm crest). Overseas evidence also supports this conclusion (e.g. Salisbury, 1952; Olson, 1958a). In addition, it is evident that incipient foredunes are forming, and have formed, on modally dissipative beaches which *do not* have berms. The beach face is linear to slightly concave on most dissipative beaches (see e.g. Wright *et al.*, 1979; Short and Hesp, 1982). It appears that Davies (1980) in part recognises the lack of berm crest initiation when he states that the *back* of the berm acts as a nucleus for beach-ridge formation.

Davies (1957) stated that beach ridges owe their alignment to the disposition of the berm nucleus which parallels the beach outline in plan. Results of the present study show that it is not necessary to invoke a berm nucleus to explain the regular, parallel nature of beach ridges or foredunes. Where the foredune is initiated by seedling colonisation, the seedlings are most commonly deposited at the height of spring tide swash. They are thus aligned conformably with the beach outline and direction of prevailing swell. Where the foredune is initiated by rhizome/stolon or shoot growth, or where seeds are scattered by winds, the seaward extent of vegetative growth is controlled by spring tide swash and storm wave inundation (Thom, 1964; Hesp, 1982). The seaward edge of the incipient foredune again parallels the shoreline in plan by the action of these processes.

Bird's (1976) second 'cut and fill' mode of formation where wave scarping is responsible for the *separation* of dunes into beach ridges is not substantiated by the present study. Ridge morphologies may be formed where incipient terraces have been scarped, and where later localised scarp and crest accretion occurs, as Bird (1976) seems to suggest. However, this does not appear to be a common mode of formation of ridges. Bird's (1976) view that in the absence of 'cut' a foredune becomes broadened as a 'coastal terrace' was also not observed in this study. However, this is not to say that some beach ridges do not display 'terrace-type' morphologies. Where the rate of coastal progradation has been rapid, as for example in the recent case of Brunswick Heads, N.S.W., rhizome and shoot growth may maintain colonisation rates which preclude the formation of a ridge. Many incipient terraces eventually become ridges, however, as either the rate of seaward rhizome colonisation and shoot growth decreases, or the rate of beach progradation decreases. Under these conditions, a zone in which maximum aeolian sand depositon occurs usually develops on the seaward margin of the foredune.

According to Davies (1957), variations in berm convexity and therefore efficiency as a sand accumulating nucleus explains why some beach ridges are distinct, and some abut the preceed-ing ridge. Evidence collected in this study indicates that the distance between ridges is determined by one or a combination of the following: (i) the mode of initiation (seedling or rhizome); if seedling initiated, the new incipient zone may be located against, or any distance seaward of, the preceeding ridge (see Fig. 6), and if rhizome initiated the incipient zone will initially abut the preceeding ridge; (ii) dynamic interactions between vegetation density and wind velocity; and (iii) the rate of beach progradation (see also later discuss-ion).

Zones of discrete pioneer seedlings separated by unvegetated zones are common on the backshore zones of both stable and accreting beaches. The presence of such discrete zones of seedlings does not imply berm crest initiation, or that inter-vening non-vegetated zones are too saline for colonisation, as suggested by Bird, (1960). The present writer would argue that Bird's (1960) beach ridges are formed in the same manner as those described at line 2 (Fig. 6): that is, by rapid, succ-essive germination of discrete zones of pioneer plant species, the seeds of which were deposited and aligned by tidal processes (types 1a and 2a). Discrete plant zones then trapped most incoming sand, and swales were formed in limited-deposition zones to the lee of the ridges.

Evidence presented here (Figs 4, 6, 9) demonstrates that swales are commonly formed where a seaward vegetation zone traps most of the aeolian sand in transit. They are thus low to non-depositional zones. They become indirectly deeper as seaward ridges become higher. Invoking successive berm prog-radation and preferential accretion of the berm crest to explain swale formation as proposed by Davies (1957) and Bird (1976) seems unnecessary. However, these comments do not preclude the formation of swales by other mechanisms such as reversing vortices at the base of scarps (e.g. Zenkovich, 1967; Bowen and Lindley, 1977), or inheritance from former deflation basins (Ranwell, 1972; Godfrey and Godfrey, 1976).

Internal structures in the selected incipient foredunes provide further evidence against a berm crest initiation (see Fig. 8a for example). In all foredune excavations, beds forming the bounding base of the foredunes were low-angle ($<4^{\circ}$), parallel, large scale (i.e. >6 m), cross-laminae deposited as high tide backshore sediments. No berm crest type structures such as those illustrated by Panin (1967) were observed.

The internal structures of several beach ridges sited within beach-ridge plains were examined in various mining pits along the N.S.W. coast. Photographs of early mining works were also analysed. Figure 10 illustrates two examples of beach ridges at Tomago and Cudgen. In these ridges, as in all others observed, foreshore sediments comprise onlapping, simple, wedge-shaped sets of large-scale, low-angle cross-strata. These grade both vertically and horizontally into backshore cross-strata which are of similar shape and scale. The backshore facies tend towards lower angles as the aeolian facies are approached. No berm crest-type structures were observed. Aeolian cross-beds conformably intergrade and overlie the beach and backshore facies at low angles. The facies relationships and bedding structures are very similar to those found within the monitored incipient foredunes decribed above, and those observed elsewhere (e.g. Harms *et al.*, 1982).

Factors Influencing Foredune Morphology

Several authors have discussed variations in ridge and swale morphology and the following have been noted as significant factors influencing such variations: the rate of shoreline progradation (Davies, 1957, 1980; Thom, 1965; Bird, 1976); variations in berm convexity (Davies, 1957); the rate and magnitude of wave erosion (Bird, 1976; Wright, 1970); beach state (Wright, 1970; Short and Hesp, 1982); temporal factors, in particular the length of time a ridge remains in a seawardmost position (Davies, 1957, 1974, 1980; Thom, 1965; Wright, 1970; Shepherd, 1981); and variation in local and regional wind velocity (Shepherd, 1970, 1981). Whilst these factors are important in producing variations in ridge and swale form, the author believes that biologic processes and plant-aerodynamic interactions assume foremost importance in influencing initial morphologic variation on foredunes.

Hesp (1983) has shown that the near-surface aerodynamic flow structure within *Spinifex* varies according to variations in plant density/distribution. Near-surface flow velocity and plant canopy penetration decreases as plant density increases. For instance, when wind velocity profiles are measured simultaneously downwind across surfaces on which plant density is laterally variable, marked variation in flow velocity is observed. In such a case, near-surface velocities rapidly increase in areas of low plant density and decrease in areas of high plant density. Aeolian sand transport and deposition is directly influenced by these lateral variations in flow structure. For a given wind speed, saltating sand grains are trapped sooner and over less distance in high density plant canopies compared to low density canopies. In addition,

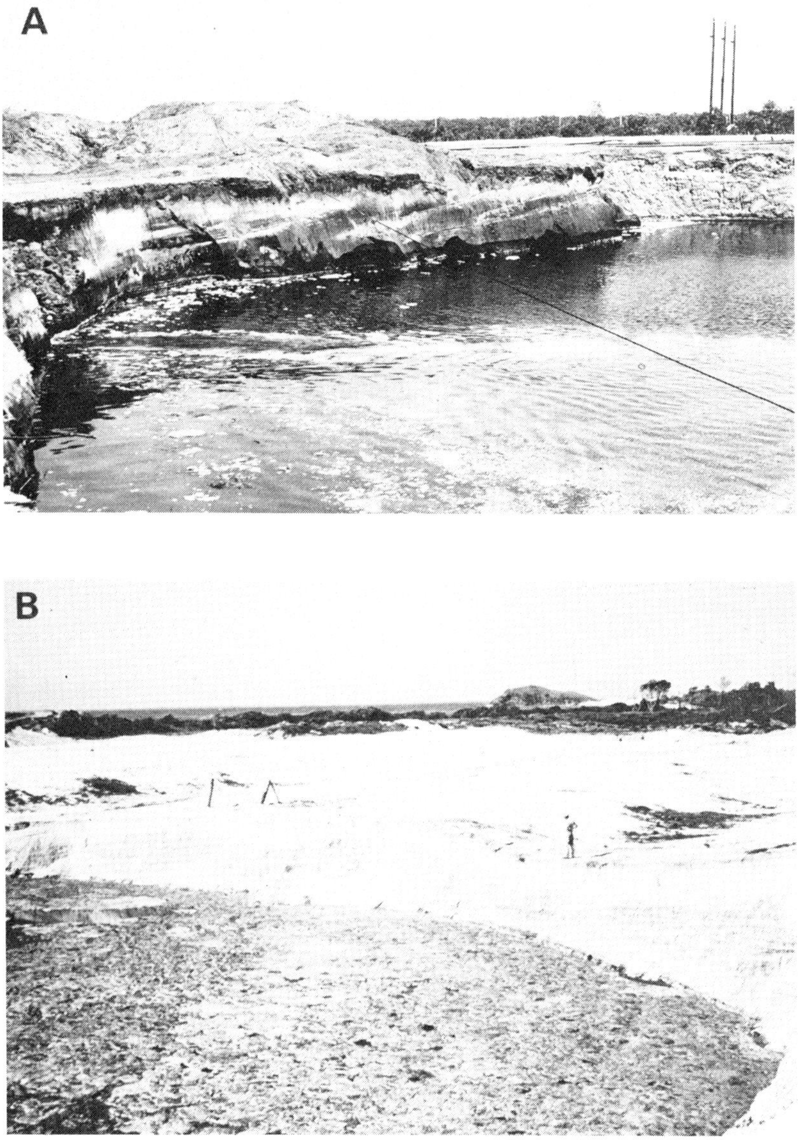

FIGURE 10. Internal sedimentary structures of sand beach ridges: (a) seaward dipping foreshore and backshore beach ridge facies at Tomago (aeolian facies have been stockpiled); (b) backshore and aeolian beach ridge facies at Cudgen. Sea is to left; mining ponds are visible in foreground.

saltation declines rapidly downwind in dense canopies, whilst it may be maintained, at least in part, in low density canopies. Since plant density and distribution commonly varies in both the alongshore and landward directions, these variations are accompanied by variations in dune morphology. Foredune height increases and basal width decreases as plant density increases. Foredunes will be locally higher, and of shorter width where plant density is high, and lower and wider where plant density is low. Longshore variation in the crestal and lee slope morphology of individual ridges is primarily a product of variations in plant density and distribution.

In addition to plant density, variations in wind velocity also produce lateral variation in dune morphology. For a given *Spinifex* density, sand transport and deposition will take place over a longer distance during high wind velocities than during low to moderate wind velocities (see Fig. 6). Thus, as wind velocity increases, dune height tends to decrease and dune basal width increases (Hesp, 1983). Swale depth and width is also influenced by the plant density and wind velocity variables. Where plant density is low, or wind velocities are high, swales have less chance of forming, or will be substantially shallower, than where plant density is high or wind velocities are low. This is because the downwind distance over which sand deposition occurs increases as plant density decreases or wind velocity increases.

The operation of ecological processes introduces a qualifying factor into the two variables above. Plant density or ground cover may change with time. Firstly, plant density varies seasonally (Hesp, 1982). This may induce variation in the location of the zone of maximum sand deposition. Secondly, there is a positive feedback between plant growth and active deposition. This acts to reinforce deposition within the highest density zone if there is little or no seaward vegetation growth. Thirdly, seaward growth of vegetation may gradually relocate the zone of maximum sand deposition further seaward, inducing progressive morphologic change on the incipient foredune, as discussed by McKenzie (1958) and shown in Figure 6.

Conclusion

It is suggested that data on the ecology, morphology and sedimentary structures of foredunes presented above provide evidence for an aeolian origin for the southeast Australian sand beach ridges. Figure 11 presents a summary of this evidence in the form of a synthesized evolutionary sequence of the

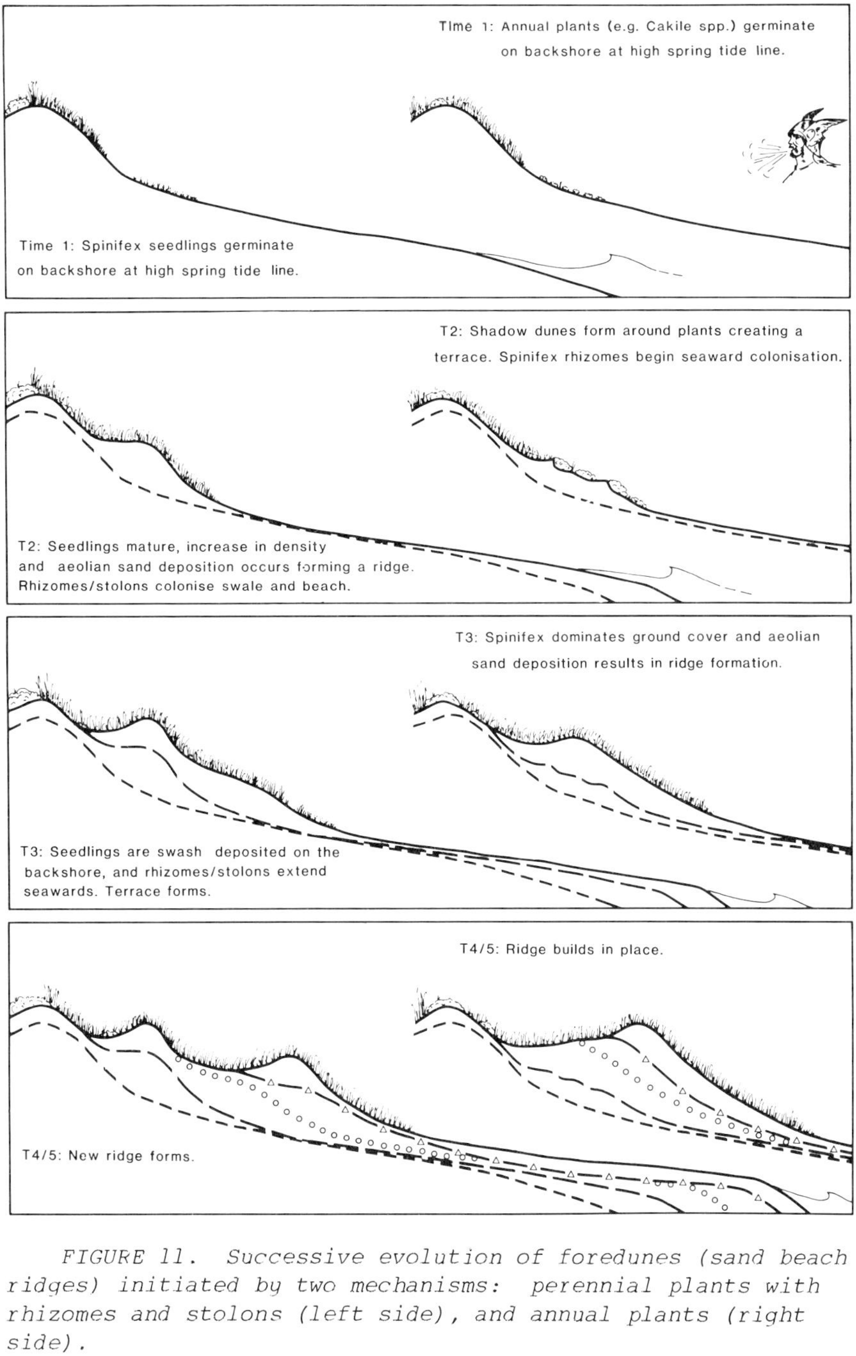

FIGURE 11. Successive evolution of foredunes (sand beach ridges) initiated by two mechanisms: perennial plants with rhizomes and stolons (left side), and annual plants (right side).

two major modes of foredune initiation and development detailed above. The sequences are based on actual examples. The left hand portion illustrates the initiation and development of an incipient foredune by perennial seedling germination, the right hand portion, initiation by annual plants. The time lines of each provide an indication of the various evolutionary paths the dunes may take depending on the operation of the various factors outlined above. Note that these paths are not mutually exclusive between types, and that at any point in time a particular morphology (ramp, terrace or ridge) may be preserved by the formation of a new incipient zone to seawards.

It is clear that not all sand beach ridges are formed as aeolian sand dunes, and that a case for multiple causality should be forwarded. For example, the storm-built ridges (Psuty, 1966), and emerged bars (Komar, 1976) are genetically different from the aeolian variety described above. In order to reduce the degree of terminological confusion, the author suggests that aeolian sand beach ridges of southeast Australia be more correctly termed relict foredunes.

Acknowledgements

The work presented here formed part of a doctoral research programme funded by the Department of Geography, University of Sydney, and the Soil Conservation Service of N.S.W. I thank Alex Colvin, Virginia Hesp, Linda Huzzey and Carol Luddington for surveying assistance; Bruce Thom, Andrew Short, Michael Shepherd and Brian Lees for very useful discussions on the contents of this paper; Peter Myerscough and Roger Carolin for plant identifications; and Gillian Craig for allowing me to refer to her work on *Spinifex*. Glennis Farrow and Janet Thompson kindly typed drafts of this manuscript, and Lloyd Morey prepared the figures.

References

Barbour, M.G. (1970a). Germination and early growth of the strand plant *Cakile maritima*. *Bull. Torrey Bot. Club, 97* (1), 13-22.

Barbour, M.G. (1970b). Seedling ecology of *Cakile maritima* along the California coast. *Bull. Torrey Bot. Club, 97* (5), 280-289.

Bigarella, J.J. (1972). Eolian environments: their characteristics, recognition and importance. *Society Econ. Paleontol. Mineral. Spec. Publication*, No. 16, 12-62.

Bigarella, J.J., Becker, R.N. and Duarte, G.M. (1969). Coastal dune structures from Parana (Brazil). *Marine Geol.*, *7*, 5-55.

Bird, E.C.F. (1960). The formation of sand beach ridges. *Aust. J. Sci.*, *22*, 349-350

Bird, E.C.F. (1961). The coastal barriers of East Gippsland, Australia. *Geogr. J.*, *127*, 460-468.

Bird, E.C.F. (1963). The physiography of the Gippsland Lakes, Australia. *Zeit. fur Geom.*, *7*, 233-245.

Bird, E.C.F. (1965). *A Geomorphological Study of the Gippsland Lakes*. Dept. of Geography Publication G/1, A.N.U., Canberra.

Bird, E.C.F. (1976). *Coasts*. A.N.U. Press, Canberra, (2nd edn.) 282 pp.

Bond, T.E.T. (1952). *Elymus arenarius*. Biological Flora of the British Isles. *J. Ecol.*, *40*, 217-227.

Bowen, A.J. and Lindley, D. (1977). A wind-tunnel investigation of the wind speed and turbulence characteristics close to the ground over various escarpment shapes. *Boundary-layer Met.*, 12, 259-271.

Carter, R.W.G. (1977). The rate and pattern of sediment interchange between beach and dune. *In* Tanner, W.F. (ed.), *Coastal Sedimentology*. Geol. Dept Florida State Univ., Tallahassee, Fla., 3-34.

Cooper, W.S. (1958). Coastal sand dunes of Oregon and Washington. *Geol. Soc. Amer. Mem.*, *72*, 168 pp.

Cooper, W.S. (1967). Coastal dunes of California. *Geol. Soc. Amer. Mem.*, 104, 131 pp.

Cowles, H.C. (1899). The ecological relations of the vegetation on the sand dunes of Lake Michigan. *Bot. Gaz.*, *27*, 95-117; 167-202; 281-308; 361-391.

Craig, G.F. (in press). Reinstatement of *Spinifex sericeus* R.Br., and hybrid status of *S. alterniflorus* Nees (Poaceae). To appear in *Nuytsia*.

Davies, J.L. (1957). The importance of cut and fill in the development of sand beach ridges. *Aust. J. Sci.*, *20*, 105-111.

Davies, J.L. (1974). The coastal sediment compartment. *Aust. Geogr. Studies*, *12*, 139-151.

Davies, J.L. (1977). The Coast. *In* Jeans, D.N. (ed.), *Australia. A Geography*, Sydney Univ. Press, Sydney, 134-151.

Davies, J.L. (1980). *Geographical Variation in Coastal Development*, Longman, London, (2nd edn.) 212 pp.

Gemmell, A.R., Greig-Smith, P. and Gimmingham, C.H. (1953). A note on the behaviour of *Ammophila arenaria* (L.) Link in relation to sand dune formation. *Trans. Bot. Sci. Edinb., 36,* 132-136.

Goldsmith, V. (1973). Internal geometry and origin of vegetated coastal sand dunes. *J. Sed. Pet., 43* (4), 1128-1142.

Goldsmith, V. (1975). *Eolian Sedimentation in Coastal Areas.* Virginia Inst. Marine Sci., Contrib. No. 533, 138 pp.

Godfrey, P.J. and Godfrey, M.M. (1976). *Barrier Island Ecology of Cape Lookout National Seashore and Vicinity, North Carolina.* Nat. Park Ser. Sci. Monograph Ser. No. 9, 160 pp.

Godfrey, P.J., Leatherman, S.P. and Zaremba, R. (1979). A geobotanical approach to classification of barrier beach systems. *In* Leatherman, S.P. (ed.), *Barrier Islands from the Gulf of St. Lawrence to the Gulf of Mexico.* Academic Press, New York, 99-126.

Gooding, E.G.B. (1974). Observations on the sand dunes of Barbados, W. Indies. *J. Ecol., 34,* 111-125.

Guilcher, A. (1958). *Coastal and Submarine Morphology.* Methuen & Co. Ltd, London, 274 pp.

Hails, J.R. (1969). The origin and development of the Umina-Woy Woy beach ridge system, Broken Bay N.S.W. *Aust. Geogr., XI* (1), 1-12.

Harms, J.C., Southard, J.B. and Walker, R.G. (1982). *Structures and Sequences in Clastic Rocks.* Society Econ. Paleotol. Mineral., Short Course No. 9

Hesp, P.A. (1981). The formation of shadow dunes. *J. Sed. Pet., 51* (1), 101-111.

Hesp, P.A. (1982). *Dynamics and Morphology of Foredunes in South East Australia.* Ph.D. Thesis, Univ. Sydney.

Hesp, P.A. (1983). Morphodynamics of incipient foredunes in N.S.W., Australia. *In.* Brookfield M.E. and Ahlbrandt, T.S. (eds.), *Eolian Sediments and Processes,* Elsevier, 325-342.

Hill, R.D. (1966). Changes in beach form at Sri Pantal, Northeast Johore, Malaysia. *J. Tropr. Geogr., 23,* 19-27.

Hine, A.C. (1979). Mechanisms of berm development and resulting beach growth along a barrier spit complex. *Sedimentology, 26,* 333-351.

Hogbom, I. (1923). Ancient inland dunes of Northern and Middle Europe. *Geogr. Annaler, 5,* 113-242.

Jenkin, J.J. (1968). The geomorphology and Upper Cainozoic geology of S.E. Gippsland, Victoria. *Geol. Survey, Vict., Mem.,* 27, 147 pp.

Jennings, J.N. (1959). The coastal geomorphology of King Island, Bass Strait, in relation to changes in the relative level of land and sea. *Rec. Queen Vict. Museum, Launceston, 11,* 1-39.

Keddy, P.A. (1980). Population ecology in an environmental mosaic: *Cakile edentula* on a gravel bar. *Can. J. Bot., 58*, 1095-1100.

Komar, P.D. (1976). *Beach Processes and Sedimentation.* Prenctice-Hall, New York, 429 pp.

Kuhlman, H. (1960). The terminology of the geo-aeolian environment, especially in relation to Danish landscapes. *Geograf. Tidsskrift, 59*, 70-88.

Laing, C.C. (1958). Studies in the ecology of *Ammophila breviligulata.* I. Seedling survival and its relation to population increase and dispersal. *Bot. Gazette, 119*, 208.

Land, L.S. (1964). Eolian cross-bedding in the beach dune environment, Sapelo Island, Georgia. *J. Sed. Pet., 34*, (2), 389-394.

Langford-Smith, T. and Thom, B.G. (1969). Coastal morphology of New South Wales. *J. Geol. Soc. Aust., 16*, 572-580.

Lawson, N.V. and Abernathy, C.L. (1975). Long term wave statistics off Botany Bay. *Proc. 2nd Aust. Conf. Coastal Engr.*, Inst. Engr. Aust., 167-176.

Leatherman, S.P. (1979). *Barrier Island Handbook.* Nat. Park. Serv. Coop. Res. Unit. Univ. Mass., Amherst, 101 pp.

McBride, E.F. and Hayes, M.O. (1962). Dune cross-bedding on Mustang Island, Texas. *Bull. Amer. Assoc. Petrol. Geol., 46*, 546-551.

McKee, E.D. (1979). Sedimentary structures in dunes. *In* McKee, E.D. (ed.), *A Study of Global Sand Seas.* Geol. Surv. Prof. Paper 1052; Chpt. E, 83-134.

McKenzie, P. (1958). The development of sand beach ridges. *Aust. J. Sci., 20*, 213-214.

Melton, F.A. (1940). A tentative classification of sand dunes: its application to dune history in the Southern High Plains. *J. Geol., 48*, 113-174.

Olson, J.S. (1958a). Lake Michigan dune development. 2. Plants as agents and tools in geomorphology. *J. Geol.*, 66, 345-351.

Olson, J.S. (1958b). Rates of succession and soil changes on southern Lake Michigan sand dunes. *Bot. Gazette*, 119 (3), 125-170.

Oosting, H.J. (1954). Ecological processes and vegetation of the maritime strand in the United States. *Bot. Rev., 20*, 226-262.

Panin, N. (1967). Structure de depôts de plage sur la côte de la mer noire. *Marine Geol., 5*, 207-19.

Pidgeon, I.M. (1940). The ecology of the Central Coastal area of N.S.W. III. Types of primary succession. *Proc. Linn. Soc. N.S.W.*, 65, 221-249.

Pidgeon, I.M. (1942). *Ecological Studies in N.S.W.* D.Sc. Thesis, Univ. of Sydney (unpub.).

Psuty, N.P. (1966). The geomorphology of beach ridges in Tabasco, Mexico. *Tech. Rept. 30.* Coastal Studies Inst., L.S.U., Baton Rouge, Louisiana, 51 pp.

Quinn, C.M. (1977). *Sand Dunes. Formation, Erosion and Management.* An Foras Forbartha, Dublin, 92 pp.

Ranwell, D.S. (1972). *Ecology of Salt Marshes and Sand Dunes.* Chapman and Hall, London, 258 pp.

Reineck, H.E. and Singh, I.B. (1975). *Depositional Sedimentary Environments.* Springer-Verlag, N.Y. 439 pp.

Ridley, H.N. (1930). *The Dispersal of Plants throughout the World.* L. Reeve and Co. Ltd., England, 744 pp.

Salisbury, E. (1952). *Downs and Dunes. Their Plant Life and Environment.* G. Bell and Sons Ltd. London.

Shepherd, M.J. (1970). *Coastal Geomorphology of the Myall Lakes Area, New South Wales.* Ph.D. Thesis, Uni. of Sydney (unpubl.).

Shepherd, M.J. (1981). The Rockingham coastal barrier system of Western Australia. *Western Geographer 5* (1-2), 67-81.

Short, A.D. and Hesp, P.A. (1982). Wave, beach and dune interactions in southeastern Australia. *Marine Geol., 48,* 259-284.

Sokolow, N.A. (1894). *Die Dünen; Bildung, Entwicklung und Innerer Bau.* (German transl. from Russian 4th ed.). Berlin, Springer, 298 pp.

Sprigg, R.C. (1952). The geology of the south-east province, South Australia with special reference to Quaternary coastline migrations and modern beach developments. *S.A. Dept. of Mines. Geol. Surv. of S.A. Bull.,* No. 29.

Steers, J.A. (1964). *The Coastline of England and Wales.* Cambridge Univ. Press, 750 pp.

Tague, G.C. (1947). The Post-Glacial geology of the Grand Marais Embayment in Berrien County, Michigan. In: *Occasional Papers for 1946 of the Geology of Michigan,* Pub. 45, Geol. Ser. 38, Pub. by Dept. Cons. and Geol. Surv. Div: Part 1, 1-82.

Thom, B.G. (1964). Origin of sand beach ridges. *Aust. J. Sci.,* 26, 351.

Thom, B.G. (1965). Late Quaternary coastal morphology of the Port Stephens-Myall Lakes area, N.S.W. *J. Roy. Soc. N.S.W.* 98, 23-36.

Thom, B.G. Bowman, G.M. and Roy, P.S. (1981). Late Quaternary evolution of coastal sand barriers, Port Stephens-Myall Lakes area, central N.S.W., Australia. *Quat. Res., 15,* 345-364.

van Dieren, J.W. (1934). *Organogene Dunenbildung.* Martins Nijhoff Haag, 304 pp.

Willis, A.J. Folkes, B.F., Hope-Simpson, J.F. and Yemm, E.W. (1959). Braunton Burrows: The dune system and its vegetation. I and II. *J. Ecol.*, *47*, 1-24 and 249-288.

Woodhouse, W.W.Jr. (1978). Dune Building and Stabilisation with Vegetation. *U.S. Army, Corps of Engrs., Spec. Rept.*, No. 3, 112 pp.

Wright, L.D. (1970). The influence of sediment availability on patterns of beach ridge development in the vicinity of the Shoalhaven River Delta, N.S.W. *Aust. Geogr.*, *11*, 327-335.

Wright, L.D. (1976). Nearshore wave power dissipation and the coastal energy regime of the Sydney-Jervis Bay region, N.S.W. : a comparison. *Aust. J. Mar. Freshwater Res.*, *27*, 633-640.

Wright, L.D., Chappell, J., Thom, B.G., Bradshaw, M.P. and Cowell, P. (1979). Morphodynamics of reflective and dissipative beach and inshore systems: Southeastern Australia. *Marine Geol.*, *32*, 105-140.

Zenkovich, V.P. (1967). *Processes of Coastal Development*. Oliver and Boyd, Edinburgh 738 pp.

5

New South Wales Estuaries: Their Origin and Evolution

P. S. ROY

Geological Survey of New South Wales
Department of Mineral Resources
Sydney, NSW

Introduction

Estuaries are environments with dynamic characteristics that continually change over geological time spans. According to Pritchard (1967) 'An estuary is a semi-enclosed coastal water body which has a free connection to the open sea and within which sea water is measurably diluted with freshwater derived from land drainage'. The transitional nature of estuaries between terrestrial and marine environments is reflected in the character and distribution of sediment, biota and water. Knowledge of the mechanisms producing change in estuarine properties is important for two major reasons:

1. The ecology of an estuary depends upon the geological stage it has reached in its evolutionary progression.
2. The rate and direction of natural change in an estuary provides a yardstick to assess impacts induced by man.

New South Wales has an embayed bedrock coast with a narrow discontinuous coastal plain; it is subject to moderately high energy ocean waves and a small (< 2m) tidal range (Jennings and Bird, 1967; Roy *et al.*, 1980; Roy and Thom, 1981). Virtually all estuaries occupy valleys excavated into bedrock by coastal rivers during past periods of lowered sea level. Estuarine conditions occur behind sand deposits located in the valley mouths. Seawards of these are embayments dominated by open ocean marine conditions; bays such as Botany Bay, Bate Bay, Jervis Bay, Batemans Bay etc. are estuaries only in a limited sense. At the other extreme are

COASTAL GEOMORPHOLOGY
IN AUSTRALIA
ISBN 0 12 687880 3

freshwater coastal lagoons. Some of these have evolved through estuarine stages of development (see below), while others, (e.g. perched dune lakes) have always been lacustrine.

The bedrock framework in which estuaries occur evolved over tens of millions of years (Roy and Thom, 1981) and has undergone multiple phases of excavation and infilling. During times of low sea level (glacial periods) estuaries are displaced out onto the continental shelf and sediments at the present coast are eroded by rivers cutting to new base levels; during interglacial periods, such as at present, coastal valleys are drowned and begin to infill with sediment from the land and from the sea. This cycle has been repeated many times over the last million years with the result that most estuaries in N.S.W. contain remnants of previous interglacial deposits in addition to those deposited in the last 10,000 years (the Holocene period).

With the melting of the continental ice sheets starting approximately 17,000 years BP, sea level rose about 150 m reaching to near its present level 6000-7000 years ago (Thom and Chappell, 1975; Thom and Roy, 1983). This was the Post-glacial Marine Transgression (P.M.T.), and corresponds to a period of rapid shoreline translation across the continental shelf (Roy and Thom, 1981). As the present coast was inundated, estuaries formed behind marine sand bodies that migrated into the drowned valley mouths. During the last 6000-7000 years of stable sea level conditions (the Holocene stillstand), the estuaries have been infilling with sediment.

Estuary Classification

Factors Influencing Estuary Development

The diversity in size, shape, salinity regime, circulation patterns and biota in N.S.W. estuaries is due to the interaction of two classes of factors:

1. Inherited factors of a mainly geological nature (e.g. bedrock type, coastal morphology) that control the size and shape of the estuary basin and the nature of the sediment supplied to it.
2. Contemporary factors of a process nature (e.g. tidal currents, river discharge, ocean waves) that influence modes of sedimentation, estuary hydrodynamics and the biota of the estuary.

In very general terms, all estuaries follow similar evolutionary paths as they infill with sediment (Davies,

1974). Water area and depth decrease with time and, as a result, hydrological characteristics and biological communities change. Estuary configurations and rates of infilling, however, are quite variable, consequently estuaries at virtually all stages of evolution can be recognised along the N.S.W. coast. For this reason present-day estuary properties such as size and shape alone are not sufficient criteria for distinguishing basic differences in estuary type. Previous classifications of Australian estuaries have largely focused on existing conditions (Rochford, 1951; 1959). Alternatively a geological classification needs to embody concepts of slow evolutionary change, a record of which is preserved in the accumulated sediments (Roy *et al.*, 1980).

Do all N.S.W. estuaries belong to a single evolutionary continuum? Studies of estuarine stratigraphy, morphology and present-day hydrology suggest that this is not the case. A synthesis of these studies has led to the hypothesis that different types of estuaries arise from the initiation of characteristic entrance conditions at the time the estuaries formed. In turn, these entrance conditions control the style of subsequent sedimentation, the salinity regime and water circulation in the estuary and, ultimately, the nature of the estuarine biotic communities. Thus present-day estuaries are classified at two levels: firstly, according to their basic *type* and secondly, the *stage* they have reached in their evolutionary progression.

Estuary Entrances

The salinity regime of N.S.W. estuaries is influenced more by salt water inputs via the estuary mouths than by river inflow. Regional and seasonal differences in rainfall pattern along the N.S.W. coast are not strongly developed (Jennings and Bird, 1967) and all except the smallest streams usually have perennial flow although discharges are highly variable and erratic. Typically, normal river inflow is considerably less than tidal discharge and, for this reason, the salinity regimes of N.S.W. estuaries are dominated by entrance morphology that control tidal exchange. This contrasts with estuaries in south-western Australia which are characterised by dramatic seasonal variations in rainfall (Hesp, 1984).

Considerable debate surrounds the factors controlling estuary mouth conditions. River flow, estuary size and shape, tidal discharge and prism, entrance channel geometry, exposeure to waves, rate of longshore sand transport, availability and type of sediment are all potentially important. Water movements in estuary mouths are usually described in terms of cross sectional area, tidal prism, discharge and velocity. Tidal prism in the estuary basin is commonly seen as the forcing, or initiating, factor to which the entrance channel responds (O'Brien, 1931; 1969; Bruun and Gerritsen, 1958). In contrast, Nielsen and Gordon (1981), following ideas of Bruun (1977), emphasise the way in which littoral drift and shoal development on the open coast affect tidal flow at an estuary mouth. Their detailed study of Wallis Lake suggests that the tidal regime of this estuary is responding to (not forcing) changes initiated at the estuary mouth.

It is postulated here that marine processes, acting at estuary mouths over geological time-spans, largely predetermine entrance channel configurations. Three basic entrance conditions are recognised on the N.S.W. coast and provide a framework for devising a classification of major estuary types:

(i) deep, relatively wide entrances with full tidal exchange;

(ii) narrow entrance channels in which tides are attenuated; and

(iii) entrances that are closed under most conditions.

These entrance conditions arise from two primary controls that act in a hierarchical manner (Fig. 1). Firstly, inherited coastal topography controls the mode of marine sand deposition in the drowned coastal valleys; estuaries with deep, open entrances have sub-aqueous tidal deltas, while estuaries with narrow or mostly closed entrance channels occur behind coastal sand barriers that rise above sea level. Secondly, normal river discharge, probably in combination with marine processes acting at the estuary mouth, determine whether entrance channels through coastal sand barriers remain open or closed for long periods. In this scheme, present-day estuary size and shape are thought to have little direct bearing on estuary hydrodynamics which is mainly a response to tidal exchange through estuary mouths of different types.

Compared to the relative smoothness of most of the continental shelf, its inner-most part near the present coast has an irregular terrain characterised by rocky promontories separating valleys of varying width and depth. Drilling in the coastal barriers and estuaries shows that palaeofluvial valleys (of Last Glacial age) beneath the present coast were as deep as 90 m below present sea level, but in most cases were less than 30 m (Thom *et al.*, 1978; Roy *et al.*, 1980). As the Postglacial sea transgressed the inner shelf surface, marine sands migrated shorewards in the form of wave-dominated shoreface deposits (c. 15-25 m thick) comprising dune, washover and tidal channel facies (Roy and Thom, 1981). Upon encountering the more rugged topography at the present coast,

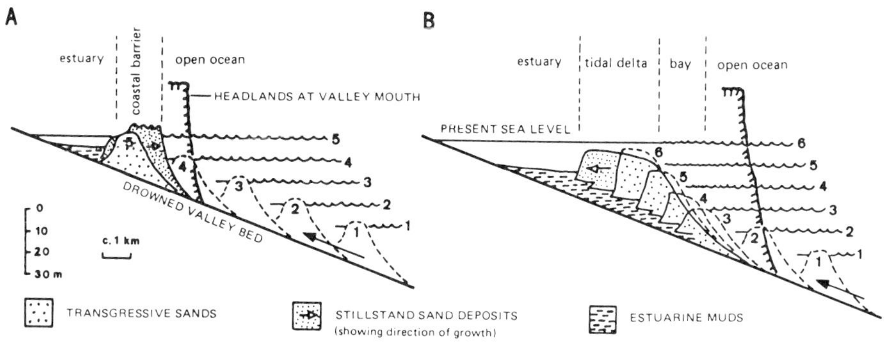

FIGURE 1. Contrasting modes of coastal sedimentation, (A) in broad, shallow valleys and (B) in deep narrow valleys. Numbers refer to sea level positions during the Postglacial Marine Transgression and correspond to successive positions of transgressive sand bodies as they migrate landwards; in (A) barriers retain a subaerial expression but in (B) they are drowned after S.L. 3 and form sub-aqueous tidal deltas (see below)

contrasting modes of marine sedimentation were initiated in shallow (<30m) and deep (>30m) drowned valleys (Fig. 1).

In broad, shallow embayments (Fig. 1A) landward sand migration virtually ceased and shoreface deposits stabilised inside the bay mouth when sea level stabilised. Continued additions of marine sand resulted mainly in barrier progradation (seaward growth). Estuaries that occur behind these bay barriers are connected to the sea by narrow channels that significantly restrict tidal exchange. Estuary mouth sand deposits were mainly emplace around the time sea level stabilised (Roy, 1984). Tidal channel and washover deposits predominate and in many cases lie directly on a Pleistocene substrate (e.g. Lake Macquarie, Tuggerah Lake, Lake Illawarra). In these estuaries, post-stillstand sand deposition appears to constitute a small proportion of the sand in the estuary mouth.

In deep, narrow-mouthed embayments (Fig. 1B) shoreface deposits were drowned before sea level stabilised. This is seen as a response to a reduction in wave energy inside the sheltered bay mouth. On the exposed coast, relatively high energy waves built subaerial barriers, but in more protected areas, tidal currents increasingly dominated sand transport to form subaqueous shoals. These shoals migrated landward as flood-tide deltas during the latter stages of the P.M.T. and the following stillstand. A prerequisite for this mode of sedimentation is a relatively deep estuary basin upstream of the tidal delta front into which the delta builds by slip-face accretion. In the Hawkesbury River mouth (Broken Bay) and Port Hacking, these transgressive sand deposits attain thicknesses of 30-40m (Nielsen and Roy, 1982; Roy, 1984). With the progressive upstream transfer of marine sand from the seaward face of the delta, valley mouths deepen and become wave-dominated, semi-enclosed bay environments (e.g. Broken Bay). Landward of these are estuaries with relatively open mouths that allow full tidal exchange (e.g. Sydney Harbour).

Fluvial discharge also plays an important role in determining estuary type. Estuaries with small streams have entrances that are closed for long periods, while those with large rivers have entrances that are permanently or mostly open. No comprehensive measurements have been carried out correlating entrance conditions with catchment area or stream discharge; certainly estuary size seems unrelated to entrance state (Smiths Lake is large with a mostly closed entrance while Narrabeen Lake is small with a mostly open entrance). Nor are there adequate data on frequency of entrance breaching to establish whether or not a continuum exists between permanently open and closed entrances.

Other factors undoubtedly affect the efficiency, in terms of tidal exchange, of open estuary mouths. The work of Nielsen and Gordon (1981) suggests that important factors are those controlling shoal development at the seaward end of the entrance channel (i.e. littoral drift and wave action). From this it follows that, where shoals are weakly developed, entrance channels are hydrodynamically efficient, tides are only weakly attenuated and, in the estuary, their ranges are relatively large. Channel efficiency probably also controls sediment transport and tidal delta growth. Here a feed-back relationship may develop: in efficient entrances, tidal delta growth that causes channel elongation increases tidal attenuation and thus reduces hydrodynamic efficiency. Inter-relationships such as this have yet to be proven on geological time scales.

Estuary Types

It is possible to classify N.S.W. estuaries into three basic types corresponding to their entrance conditions. The terminology used here follows in part that proposed by Roy *et al.*, (1980); synonymous terms applied to N.S.W. estuaries by Jennings and Bird (1967) are shown in brackets.

(1) Drowned river valley estuaries (ria-like, dendritic drowned valleys).

(2) Barrier estuaries (estuarine lagoons).
(3) Saline coastal lakes.

Drowned river valley estuaries have open mouths (semi-enclosed bay environments) with sub-aqueous tidal deltas and full tide ranges throughout. Barrier estuaries and saline coastal lakes occur behind coastal barriers. The former have permanently, or mostly, open entrance channels in which tides are attenuated. The latter are usually smaller with ephemeral inlets and are normally non-tidal. Each estuary type is described below in terms of its early, or 'youthful' development stage before extensive infilling takes place.

Examples of *drowned river valley estuaries* (Fig. 2A) occur mainly in central N.S.W. : the Hawkesbury River-Broken Bay system, Sydney Harbour, Georges River and Port Hacking (in the Sydney region). The Clyde River (Batemans Bay) on the south coast also belongs in this category as probably does Port Stephens, north of Newcastle. This latter estuary occupies a valley incised into Pleistocene deposits (including coastal barriers) rather than bedrock and shows features of both drowned river valley and barrier estuary types. It contains a large subaqueous tidal delta sand body, an open mouth and is fully tidal. A Holocene barrier complex that encloses its northeastern side is more characteristic of barrier estuaries.

Palaeo-valleys occupied by drowned river valley estuaries are deep, usually narrow with steep rocky sides. In the Sydney region, this erosional topography has been developed in massive, thick bedded sandstones that crop out as cliffs at the coast. Large sub-aqueous tidal delta sand bodies extend from the open bay mouth into the lower estuary and form sills (or thresholds) behind which fine river sediments accumulate from suspension in deep (c. 20-30 m) mud basins (Roy *et al.*, 1980). Tidal delta sands are typically fine grained and shelly and reach thicknesses of 45 m (Roy and Stephens, 1981; Nielsen and Roy, 1982; Roy, 1984). Delta surfaces are reworked by waves on their seaward face and by currents, especially in tidal channels. Steep depositional slopes (c. 20°- 30°) occur on active delta fronts where marine sands are deposited over estuarine muds. Upstream of the mud basins, fluvial deltas prograde into the headwaters of the estuary. These are mainly composed of muddy, organic-rich channel sands. In narrow valleys the sands reach thicknesses of about 10 m, but flood plain deposits are weakly developed in initial stages of in-filling. The entire estuary is fully tidal although steep valley sides restrict intertidal environments to deltaic areas in the upper estuary.

Barrier estuaries (Fig. 2B) range widely in size from large estuaries such as Wallis Lake and Lake Macquarie which

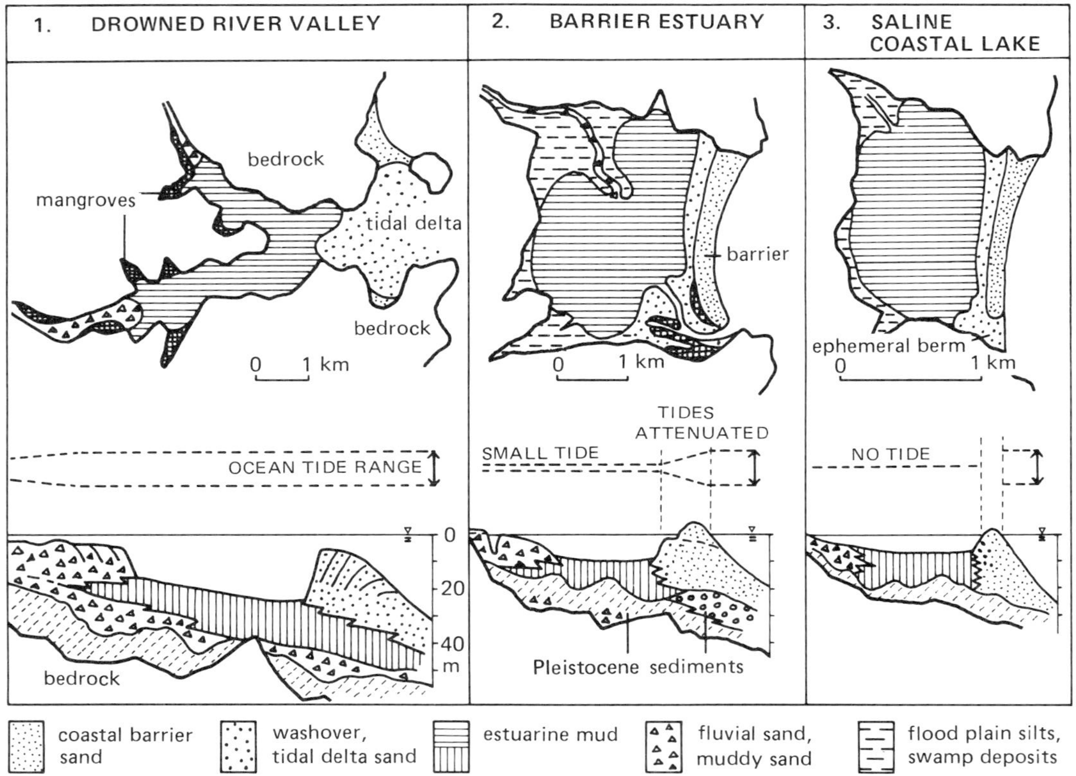

FIGURE 2. Three main estuary types in N.S.W. showing idealised sediment distributions in plan and in section (dimensions are only approximate). Tidal ranges in the estuaries are shown in relation to that in the open ocean.

exceed 100 km^2 in area to small estuaries on the south coast (Wallaga, Merrimbula and Pambula Lakes) less than 10 km^2 in area. In the geological past even larger barrier estuaries existed in the valleys of the north coast rivers (e.g. Clarence, Richmond, Macleay), but these are now largely infilled.

Barrier estuaries are characterised by narrow, elongated entrance channels within broad tidal and backbarrier sand flats. These estuary-mouth sands rarely exceed thicknesses of 15-25 m and comprise fine grained, shelly as well as coarse, low-shell sand facies. Over large areas they rise slightly above sea level and presumably formed as storm ridges thrown up by lake waves. Contemporary (stillstand) delta growth occurs at the inner end of active entrance channels but, in most estuaries, accounts for only a minor part of the total deposits.

The question of contemporary delta growth is important in regard to entrance channel elongation and its possible effect on tidal hydrodynamics. In Wallis Lake, there is radiocarbon dating evidence to suggest that marine channel sands have built considerable distances into the estuary during the stillstand, but in Queens Lake and Watson Taylor Lake they occupy old fluvial channels cut through the Pleistocene barrier and show little sign of contemporary extension. Drilling in Lake Macquarie suggests that active tidal delta growth is confined to the landward end of the present entrance channel and constitutes only 5% of the estuary mouth sand deposits (Roy, 1984).

Away from active channels, delta surfaces are shallow, low energy environments often densely covered by sea grasses (e.g. Tuggerah Lake and Lake Macquarie). Delta fronts slope steeply into moderately deep (c. 5-15 m) mud basins. Fluvial deltas prograde at river mouths into the mud basins and, because of limited tidal and wave action, delta lobes are commonly elongated. Silts and sandy muds in delta margin and overbank deposits are the dominant fluvial lithologies; these are intersected by thin (c. 5 m thick), ribbon-like channel sands and levee deposits (e.g. Shoalhaven flood plain, Thom *et al.*, 1981; Roy, 1984).

Strong tidal flows occur in the entrance channel, but tidal ranges are progressively reduced by frictional effects and may be as small as 10% of the ocean tide in the estuary basin (PWD, 1983). This is a major constraint on the extent of intertidal environments which are best developed near the estuary mouth. In terms of water circulation, wind waves and wind-induced flows are more important than tidal currents in the body of the estuary. The influence of freshwater discharge is restricted to the fluvial delta except during floods when temporary stratification may develop in deep estuary basins (e.g. Lake Macquarie; MacIntyre, 1959).

Most *saline coastal lakes* (Fig. 2C) are small features located in small coastal valleys (e.g. Dee Why Lagoon); large examples such as Smiths and Coila Lakes are rare. In their early stages of development, they are geologically similar to barrier estuaries; shallow mud basins and muddy fluvial deposits are the dominant environments. Fluvial inputs are characteristically very small. River delta sediments are readily reworked by wind waves and delta lobes are weakly developed or absent (e.g. Smiths Lake). Marine sands are relict except for minor washover activity inside the estuary mouths.

Under most conditions lake waters are saline to brackish but non-tidal. Winds cause mixing and some mass water movements in larger lakes. Normal fresh water inputs are accommodated by evaporation and percolation through the porous sand barrier. Freshwater conditions accompany heavy rains at which time inlets may be breached by super-elevated lake waters (Gordon and Cook, 1977; PWD, 1983) and storm waves (Coila Lake in 1974; B. Thom pers. comm.). The lake then becomes saline and tidal for weeks to months until open ocean processes reform the beach berm across the inlet. Unlike coastal lagoons in semi-arid areas (e.g. South Australia), hypersaline conditions rarely occur in N.S.W.

Estuary Evolution

Under stable sea level conditions, estuary evolution primarily results from sediment infilling which creates a progression from youth to maturity. Estuaries infill from the sea with marine sand and from the land with fluvial sand and mud. Calcareous tests of animals and organic remains of plants are incorporated in the sediments during or after deposition (typical estuarine lithofacies and depositional environments are listed by Roy *et al.*, 1980). Infilling has occurred at differing rates depending on estuary type, sediment supply and the size of the estuary basin.

Initial size and shape of Holocene estuaries is due, not only to valley cutting during the preceding period of lowered sea level, but also to the inheritance of older depositional landforms. These latter include remnants of estuary basins that have been reoccupied by the Holocene sea (e.g. Queens and Watson Taylor Lakes). Supply of fluvial sediment is directly related to river discharge, although its composition (relative abundance of sand and mud) is profoundly affected by the petrology and weathering characteristics of lithologies in the catchment. River size thus strongly influences rates of infilling and estuary type. In contrast, the supply of marine sand is related to estuary type because of their characteristic entrance conditions. However, considerable variability occurs within each type (especially barrier estuaries) due to marine processes that influence sand movements at their mouths.

Progressive estuary infilling causes a reduction in water area and average depth. As deltaic flood plains expand, estuarine water bodies are increasingly channelised. This occurs in a spatial as well as a temporal sense and estuaries typically show a trend towards decreasing maturity in a downstream direction. Characteristic evolutionary progressions in each estuary type are expressed in terms of morphology, stratigraphy, hydrology and, it is postulated, estuarine biology. Physical and hydrological stages in these progressions are indicated in Figures 3 to 5 for each estuary type.

Physical Stages in Estuary Evolution

In their youthful stage of development (Fig. 3A), *drowned river valley estuaries* contain extensive and deep mud basins behind shallow tidal delta sills (e.g. Port Hacking). Where studied, sedimentation rates at the upstream delta fronts are rapid (about 16000 m^3/yr of sand over the last 7000 years has resulted in the Port Hacking delta advancing approximately 4 km; see Roy, 1984). Mud basins progressively decrease in size as both tidal and fluvial deltas build into the estuary. Vertical rates of mud accumulation depend on fluvial inputs: they range from 1-2 mm/yr in the Port Hacking estuary, but were probably an order to magnitude greater at an early stage in the Hawkesbury estuary which is fed by a much larger river system (Roy, 1984). As fluvial deltas infill the upper estuary, channel sands are progressively superseded by levee and floodplain deposits. Channel sediments become finer grained and more muddy in a downstream direction (e.g. Hawkesbury River). Shallow water and inter-tidal environments are increasingly well developed around the expanding delta front in the middle of the estuary (e.g. Georges River, Fig. 3B). Estuary water movements become increasingly affected by tidal flow in the lower and middle estuary and by river flow in the upper part. Salt wedge development is common under high flow conditions. Wind waves stir the muds in shallow basins (e.g. western Port Stephens) although fetches are often restricted.

1. DROWNED RIVER VALLEY ESTUARY

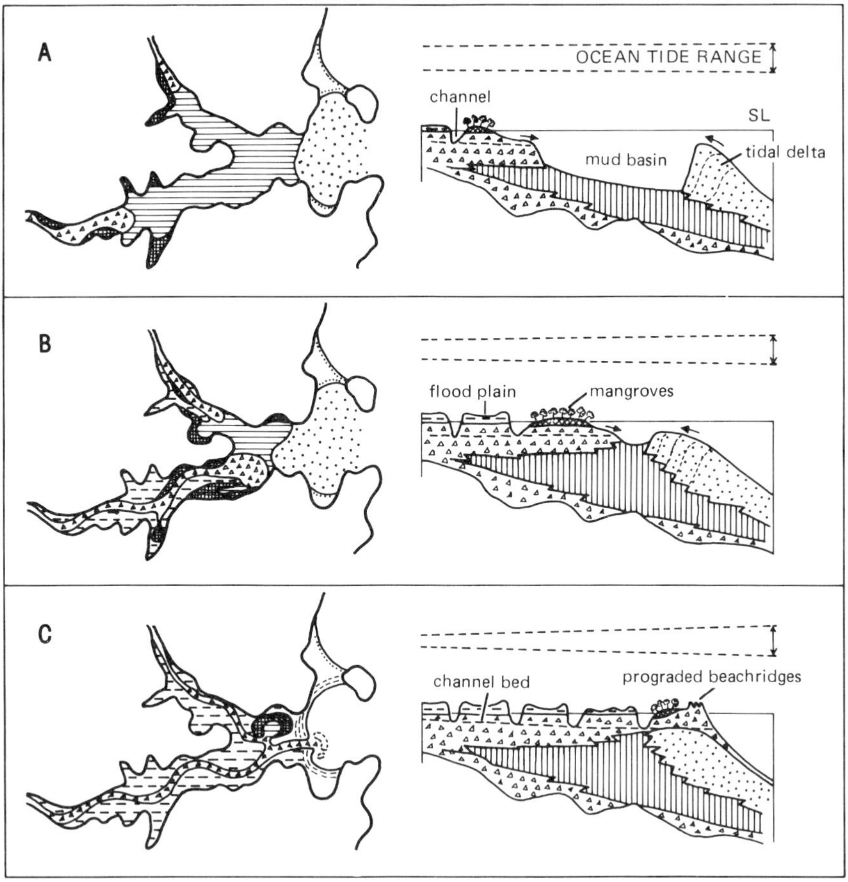

FIGURE 3. Stages of infilling in the evolution of a drowned river valley estuary. Arrows indicate direction of delta growth (sediment symbols, scales and tidal representations as for Figure 2).

Figure 3C depicts a more mature stage of development. Fluvial deposits infill the drowned valley and prograde seawards over the tidal delta. Channels are restricted by levee banks and intertidal environments are confined to cut off bays in the lower estuary. Fluvial sand, discharged at the river mouth during floods is reworked by waves into prograding shoreline deposits in the semi-enclosed bay (e.g. Batemans Bay). Under normal flow conditions 'river estuaries' remain strongly tidal for considerable distances inland although some attenuation occurs in their upper-most reaches.

The stratigraphy of drowned river valleys is characterised by sediment sequences up to 60 m thick (Roy, 1984). Flood plains form thin (c. 3 m thick) surficial deposits over thick fluvial channel sands and muddy sands. Estuarine muds and marine sands reach thicknesses of 20-45 m in the lower estuary. Marine sands typically show transgressive onlapping relationships with the underlying estuarine muds.

Barrier estuaries (Fig. 4) experience complex changes in shoreline morphology during infilling. In youthful stages their shorelines are often rocky and highly irregular (e.g. Lake Macquarie, Fig. 4A). As the estuary infills and shallows, smoother sedimentary shorelines initially develop (e.g. Tuggerah Lake, Fig. 4B), but later they become highly irregular as lobate deltas with bifurcating distributary channels and middle ground shoals prograde and cut off sub-embayments (Tuross estuary, see Roy and Peat, 1976; Fig. 4C). Final stages of infilling are characterised by sinuous channels with smooth levee banks (e.g. Shoalhaven delta, Fig. 4D).

Due to palaeo-valley geometry (see above), barrier estuaries occupy shallower valleys than drowned river valley estuaries and their sediment thicknesses are correspondingly less (c.< 30 m). Infilling is mainly with fine fluvially derived sediment. Two facies types are dominant: basin muds and delta-front sandy muds (Roy *et al.*, 1980). Rates of mud accretion range from 0.15-1.0 mm/yr in slow filling basins such as Lake Macquarie (Roy, 1981) to 5 mm/yr in the Shoalhaven proto-estuary (Roy, 1984). This (c.<5m thick) channel sands are confined to the upper part of the mud sequence (e.g. Shoalhaven delta, Thom *et al.*,1981; Figs 4C and D). Broad flood plains, with backswamps and cut-off bays, characterise mature estuaries and are particularly well developed in the deltaic flood plains of the north coast rivers (Tweed, Richmond, Clarence, Macleay). Tidal delta sand bodies show considerable diversity that seems unrelated to estuary size. Relatively large tidal deltas in small estuaries such as Wagonga Lake on the south coast may be associated with efficient entrance conditions; the reverse is presumably true in the case of larger estuaries with only weakly active deltas such as Lake Illawarra (PWD, 1983). However, further studies are needed to distinguish between tidal deltas facies that are relict (e.g. Narrabeen) and those that are still active.

In contrast to drowned river valley estuaries, youthful barrier estuaries are only weakly tidal. However, as they evolve their tidal ranges appear to increase until a strongly tidal regime is achieved in mature, channelised 'river estuaries'. The explanation proposed here emphasises the governing role of estuary mouth shoal development and channel

2. BARRIER ESTUARY

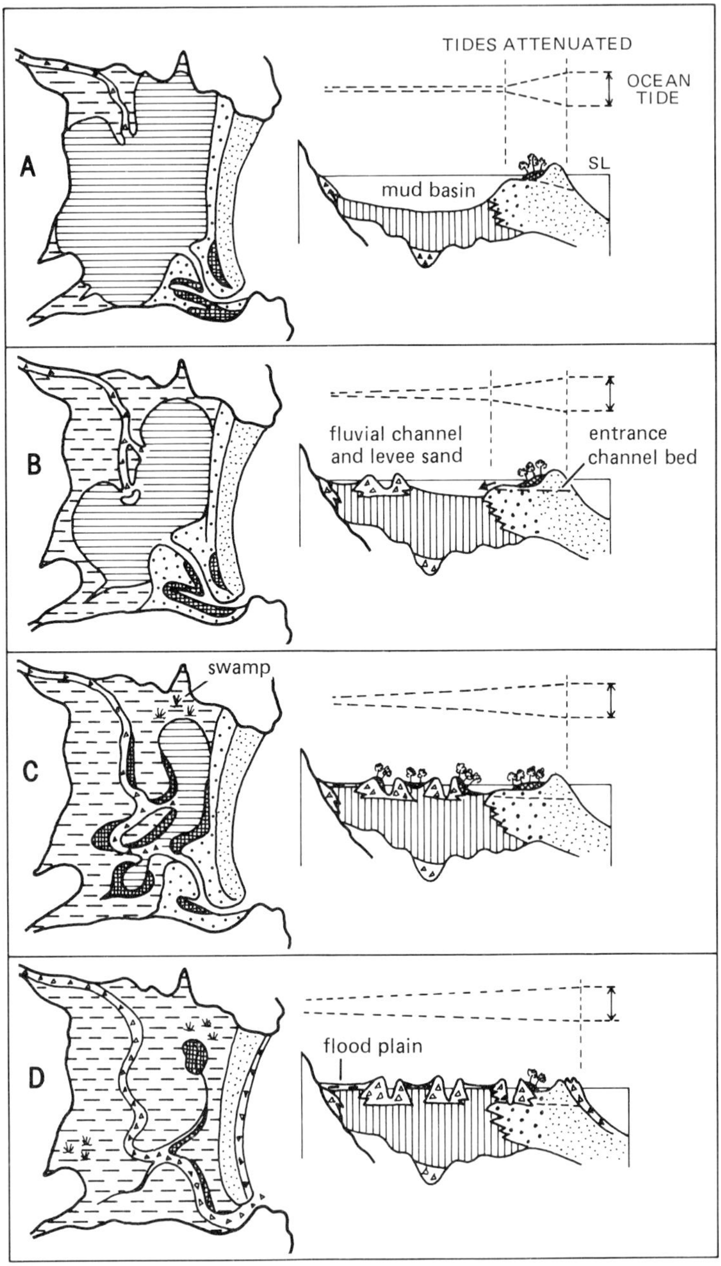

FIGURE 4. Stages of infilling in the evolution of a barrier estuary. (Sediment symbols, scales and tidal representations as for Figure 2).

configuration at the time estuaries form. Clearly the tidal prism (water area X tidal range) in the estuary basin is directly related to tidal exchange through the entrance channel. However, it has been suggested previously that the former is a passive response to the latter. Thus the reduction in water area that accompanies estuary infilling could cause the observed increase in tidal range without inducing significant alteration of the tidal discharge through the entrance channel (see also Boon and Byrne, 1981). A contributing factor that may further increase tidal ranges as the estuary evolves arises from a reduction in its freshwater storage capacity. As this diminishes, flood waters increasingly cause more severe scouring of the estuary mouth and channel as they are discharged to the sea. With more frequent over-deepending of the entrance channel, tidal exchange, and therefore tidal ranges in the estuary, could be expected to increase overall.

Limited tidal exchange is responsible for wide salinity fluctuations especially in the distal parts of broad, shallow estuaries (e.g. Tuggerah Lake). During youthful and most intermediate stages of infilling, slow water circulations are wind induced; waters are mostly well mixed and often turbid (PWD, 1983). Stratified conditions occur in deep, youthful basins (e.g. Lake Macquarie) and also in mature 'river estuaries' due to salt wedge development (e.g. Shoalhaven River; Wright, 1977). Intertidal environments are mainly limited to the tidal delta region in the early stages of infilling (Figs. 4A and B). They expand into the estuary as tides increase in amplitude and reach peak development during the stage of maximum shoreline complexity (Fig. 4C). Despite relatively high tidal ranges at mature stages of development, this trend is reversed as the estuary becomes channelised (Fig. 4D).

In their youthful, saline stages of evolution, _saline coastal lakes_ have similar configurations and sediment patterns to shallow barrier estuaries (Fig. 5). However, the water bodies are usually smaller and marine sand deposits are less well developed. In large coastal lakes with small fluvial inputs (e.g. Smiths Lake) saline conditions are maintained for long periods and muddy and shelly estuarine lithofacies accumulate in the low-energy basin and deltaic environments (Fig. 5A). Mangroves are absent but some salt marsh species colonise creek deltas and sea grasses fringe the estuary margin (e.g. Coila Lake; Australian Museum, 1980).

As the basins infill, salinity variations increase and fresh to brackish conditions are prolonged. Smaller water areas would experience more flushing. As a result, sea grasses

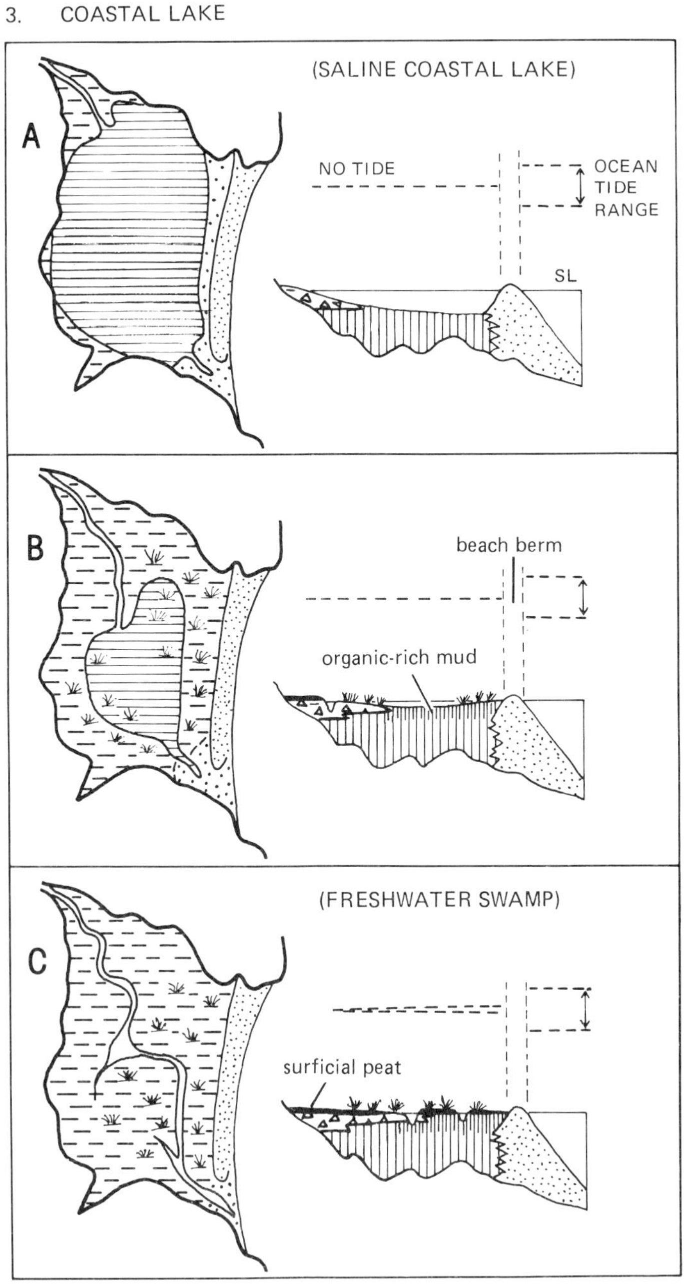

FIGURE 5. Stages of infilling in the evolution of a coastal lake (sediment symbols, scales and tidal representations as for Figure 2).

are progressively replaced by fresh-water tolerant species; extensive reed swamps develop and shelly estuarine sediments are superseded by organic-rich freshwater muds (Fig. 5B). Eventually the estuary is infilled by swampy flood plains intersected by narrow channels (Fig. 5C) While basins retain their flood storage capacity, entrance conditions remain unchanged (i.e. mostly closed), but with increasing channelisation, breaching possibly occurs more often.

Biological Changes Accompanying Estuary Evolution

The model presented here describes the evolution of basic estuary types in terms of changes in water depths, estuary configurations, sediment types and salinity regimes which can be expected to exert a major influence on estuarine biota. One general aspect of estuary evolution is a progressive reduction in water area. This affects the size, and possibly also the ecological viability of estuarine communities. Whether predictable changes in nutrient levels are also associated with the evolutionary model remains to be determined. Since tidal flushing with low-nutrient, open ocean water is generally more pronounced in drowned river valley estuaries, their nutrient levels may be lower than in more constricted barrier estuaries (and saline coastal lakes) at the same stage of infilling.

Figure 6 is a preliminary attempt to predict the way estuary biota might change as each estuary type evolves. These biological variations are envisaged to occur slowly over geological time-spans. However, in some estuaries they co-exist in a spatial sense. As estuaries infill from their upper reaches towards their mouth they show a trend of decreasing maturity in physical as well as biological parameters in the same direction (e.g. the Hawkesbury River - A. Jones, Aust. Museum, pers. comm.). In general terms, Figure 6 suggests that estuarine biota are best developed in drowned river valley and barrier estuaries during intermediate stages of infilling. However, no absolute comparisons between the two estuary types are implied in this figure. In coastal lakes the biota is truly estuarine only during initial stages of infilling.

During the early stages of infilling of drowned river valley and barrier estuaries mangrove colonisation is minimal due to restricted intertidal habitats around the steep estuary sides. In barrier estuaries and saline coastal lakes, the development of mangroves may also be influenced by the small tidal ranges. Sea grass distributions are also limited by

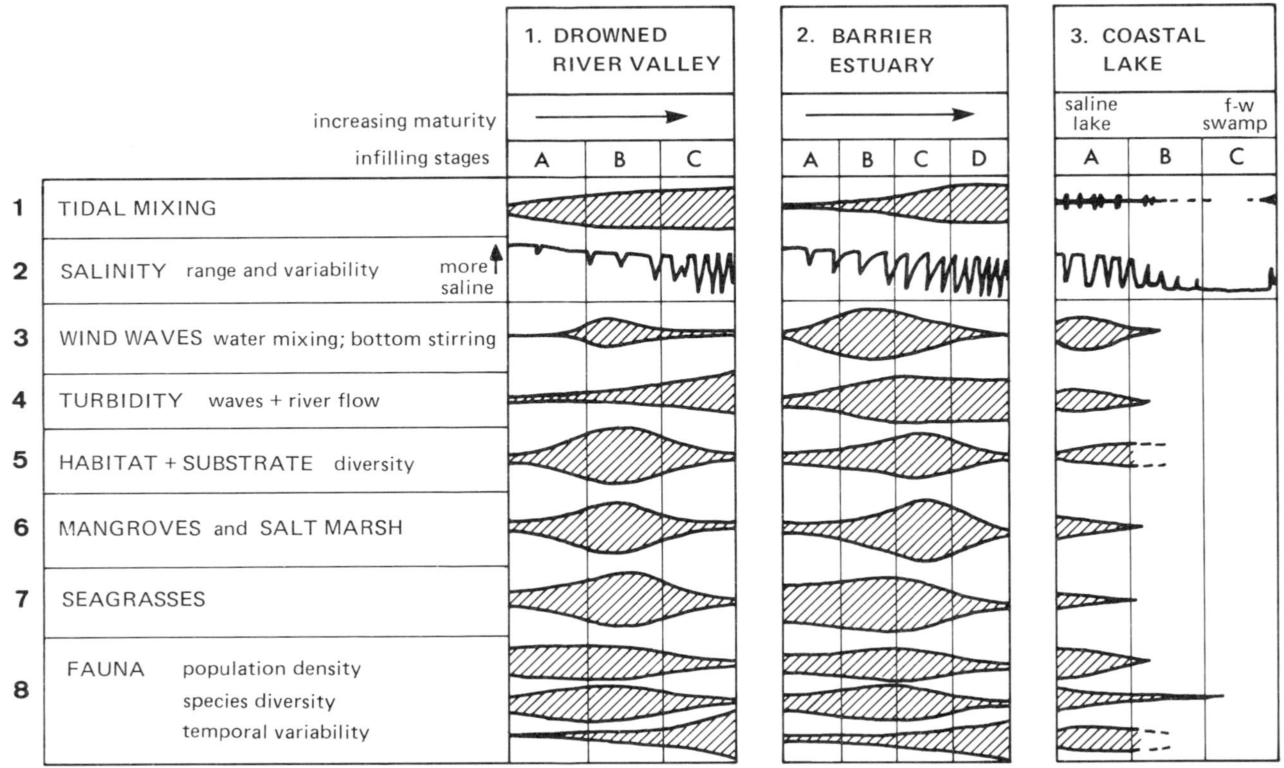

FIGURE 6. *Predicted changes in various properties of the water bodies and biota of the three estuary types as they evolve. Confidence levels of the trends decrease from 1 to 8 and no absolute correlations are implied between estuaries of different type.*

restricted habitat development, but perhaps to a lesser degree. In some barrier estuaries (e.g. Tuggerah Lake, Lake Macquarie) they colonise the surfaces of relict tidal deltas; in coastal lakes they reach maximum development at this stage because of relatively high salinities.

In their youthful stages all estuaries have relatively deep, uniform basin environments with organic-rich mud. The mud supports a rich infauna of molluscs and polychetes and as a result is extensively bioturbated. Despite reworking, reducing conditions exist a few centimetres below the surface (Roy, 1981). Except in saline coastal lakes, salinities are consistently high although bottom waters may occasionally become deoxygenated (MacIntyre, 1959). As a result faunal population densities could be expected to be high and relatively stable, but species diversity is somewhat reduced by the limited range of habitats and substrate types. In coastal lakes, estuarine fauna reach maximum development at this time and are limited to species tolerant of salinity variations (e.g. *Notospisula*). As the saline lakes infill and shoal, tolerance levels are increasingly exceeded, fewer species remain viable and these commonly experience blooms and mass mortalities.

The biota maxima reached at intermediate stages of infilling is mainly associated with an expansion of intertidal and shallow sub-tidal environments in the fluvial deltas. Faunal population densities and species diversity increases as shorelines become more complex and ecological habitats and sediment types diversify. In barrier estuaries increased tidal ranges contribute to peak mangrove development at late-intermediate stages of infilling. Bottom stirring by waves that cause turbid water conditions affect both sea grasses and estuarine fauna especially in early-intermediate stage estuaries such as Tuggerah Lake and Lake Illawarra. The loss of water clarity reduces the depth of which sea grasses grow and therefore their areal extent. It is likely, however, that estuarine animals benefit from a limited increase in bottom stirring as nutrients, otherwise bound-up in the muds, are released and dispersed through the estuary. In basins with weak circulations, excessive turbidity and addition of nutrients causes algal blooms that may cause oxygen levels to fall below the tolerance levels of most estuarine species. These extreme conditions are probably encountered more often in shallow, enclosed barrier estuaries than in drowned river valley estuaries with better tidal flushing characteristics. At this stage saline coastal lakes are only ephemerally saline.

Estuarine plants and animals experience a general decline in distribution, population numbers and species diversity in mature estuaries. This is primarily associated with the expansion of flood plains that constrict estuaries of both drowned river valley and barrier types to steep-sided channels. These limit intertidal (mangrove and salt marsh) zones despite relatively high tidal ranges. Habitats are less diverse; channel beds are composed mainly of clean sands although muds occur in deep scour holes at bends. High river discharges produce rapid salinity and turbidity fluctuations that cause faunal populations and species numbers to be lower overall and show more short-term variability.

Conclusion

The principal value of models such as the one presented here is to focus attention on possible inter-relationships between diverse phenomenon. It has been argued that estuarine geology, hydrology and biology form a hierarchical succession. Of course the proposed causal relationships remain to be rigorously tested. Despite the limitations of incomplete data, the classification has been applied to all N.S.W. estuaries (Roy, 1982). It accommodates virtually all estuaries except for 'interbarrier creeks'. These latter are small water bodies that occupy elongated, coast-parallel depressions between onlapping sand barriers. The depressions are essentially depositional features and, unlike most N.S.W. estuaries, are unrelated to the drowning of coastal river valleys. The data show that regional trends in estuary type are only weakly developed along the N.S.W. coast and appear to reflect broad differences in coastal relief (see Roy and Thom, 1981). Drowned river valley estuaries are the least common type and are found almost exclusively in central N.S.W. Barrier estuaries and saline coastal lakes occur coast-wide, but the former are larger and more mature in the northern half of the state.

The model's usefulness is to be measured by its ability to encompass and explain diverse estuarine characteristics. Specifically this applies to the hypothesis that estuary entrance conditions produce basically different types of estuaries whose subsequent hydrological and biological changes accure from unique styles of estuarine sedimentation. The predicted changes in estuarine biology are largely speculative. Because of the geological time-spans involved, even long-term biological and hydrological studies in single estuaries encompass only a minute part of estuary evolution and thus contribute little to test the model. A broader perspective may be gained from studies of estuaries of similar type at different stages of infilling. However, local differences in estuary sizes and configuration and even climate (e.g. latitudinal differences in mangrove species and diversity) will allow only generalised comparisons. Alternatively, palaeontological studies may offer insights into estuarine faunal changes. Drill holes through sediment sequences in mature estuaries would encounter older deposits at depth that correspond to more youthful stages of estuary evolution. The sediments contain fossil remains of presently living estuarine mollusca and foraminifera which have known habitats. Variations in their type and abundance could be expected to reflect long-term trends in estuary evolution. This appears to be the case in Gippsland (B. Thom, pers. comm.).

The concept of evolutionary change provides an opportunity to examine specific environmental conditions in natural estuaries of similar type but at different stages of infilling. For example, West (1983) documents characteristic sea grass associations that appear to correspond to the various estuary types and stage of maturity proposed in this paper. Improved knowledge of natural directions of change will form a basis for successfully managing those estuaries under developmental pressure. For example, to assess the effect of deepening a shallow estuary by dredging, it is appropriate to look at a less mature estuary of the same type. Another possible application arises from the suggested role of marine processes acting at the estuary mouth. In the past, man-made changes in this area (e.g. training walls, breakwaters etc.) have been designed to improve navigation or reduce flooding. Studies by Nielsen and Gordon (1981) indicate that more far-reaching, albeit inadvertent, changes to the tidal regime have occurred in Wallis Lake. The model suggests that changes in this, or in other directions can be controlled. Whether or not they should be implemented are value judgements, but decisions must be based on a sound understanding of estuarine systems.

Acknowledgements

Much of the geology discussed here is based on results of estuarine drilling and sampling carried out by the Geological Survey of New South Wales and the Department of Public Works over the last decade. The basic ideas on estuary types arose from collaboration with Bruce Thom and Don Wright and have been reported in Roy *et al.*, (1980). For the fruitful discussions that led to the expansion of these ideas to their present state, I am indebted to Lex Nielsen, Rob Williams, John MacIntyre, Alan Jones, Phil Gibbs, Ian Walsh, Mike Field and Bruce Thom. The paper is published with the permission of the Secretary, N.S.W. Department of Mineral Resources.

References

Australian Museum (1980). A survey of the estuarine fauna of Eurobadalla Shire. D. Milledge (Ed.). *Australian Museum publication*, 165 pp.

Boon, J.D. and Byrne, R.J. (1981). On basin hypsometry and the morphodynamic response of coastal inlets. *Marine Geol.* 40, 27-48.

Bruun, P.M. (1977). Design of tidal inlets on littoral drift coasts. *Proc. 5th Symposium of the Waterway, Port, Coastal and Ocean Division of A.S.C.E.*, Charleston, S.C., 754-773.

Bruun, P.M. and Gerritsen, F. (1958). Stability of coastal inlets. Am. Soc. Civil Engrs. *J. of the Waterways, Harbours and Coastal Div.*, 84, No. WW3.

Davies, J.L. (1974). The coastal sediment compartment. *Aust. Geog. Studies*, 12, 139-151.

Gordon, A.D. and Cooke, R. (1977). Dee Why Lagoon, Investigation of dredging proposal. *N.S.W. Department of Public Works, Coastal Engineering Branch - Report*, PWD 77029.

Hesp, P.A. (1984). Aspects of the geomorphology of South Western Australian estuaries. *In* Hodgkin, E.P (ed.), *Estuarine Environments of the Southern Hemisphere*. Dept. Conservation and Environment, Perth, W.A. Bulletin 161, 61-8

Jennings, J.N. and Bird, E.C.F. (1967). Regional geomorphological characteristics of some Australian estuaries. *In* Lauff, G.H. (Ed.), *Estuaries*, Am. Assoc. Adv. Sci. Publ., 83, 121-128.

MacIntyre, R.J. (1959). Some aspects of the ecology of Lake Macquarie, N.S.W. with regard to an alleged depletion of fish, Part Vll. The benthonic macrofauna. *Aust. J. Mar. Freshwat. Res.*, 10, 341-359.

Nielsen, A.F. and Roy, P.S. (1982). Age contamination of radiocarbon dates on shell hash from coastal sand deposits: SE Australian examples *Proc. 5th Australian Conference on Coastal and Ocean Engineering*, Perth, 177-182.

Nielsen, A.F., and Gordon, A.D. (1981). Tidal inlet behavioural analysis. *Proc. 17th International Coastal Engineering Conference*, Sydney, pp. 2461-2480.

O'Brien, M.P. (1931). Estuary tidal prism related to entrance areas. *Civil Engr.*, 1(8) 738-793.

O'Brien, M.P. (1969). Equilibrium flow areas of inlets on sandy coasts. *A.S.C.E. Jour. Waterways and Harbours Division*, 95, WW1.

Pritchard, D.W. (1967). What is an estuary: physical viewpoint *In* Lauff, G.H. (Ed.), *Estuaries*. Am. Assoc. Adv. Sc. Publ. 83 Washington D.C., 3-5.

P.W.D. (1983). Lake Illawarra entrance study. Public Works Department of New South Wales; 128 pp. (unpubl.).

Rochford, D.J. (1951). Studies in Australian estuarine hydrology, 1: introduction and comparative features. *Aust. J. Marine Freshwater Res.*, 2, 1-116.

Rochford, D.J. (1959). Classification of Australian Estuarine Systems. *Arch. Oceanog. Lim.*, 11, 171-177.

Roy, P.S. (1981). Fossil shell assemblages in Lake Macquarie. *Quarterly Notes, Geol. Survey, N.S.W.*, 42, 12-17.

Roy, P.S. (1982). Evolution of estuaries in N.S.W. Symposium papers, Management of Estuaries. Water Research Foundation of Australia, April, 1982, Sydney. 27 pp.

Roy, P.S. (1984). Holocene sedimentations in estuaries - case histories from N.S.W., Australia. *In* Hodgkin, E.P. (Ed.), *Estuarine Environments of the Southern Hemisphere.* Dept. Conservation and Environment, Perth, W.A., Bulletin 161, 23-59.

Roy, P.S. and Peat, C. (1976). Bathymetry and bottom sediments of Tuross estuary and Coila Lake. *Rec. Geol. Surv. N.S.W.*, 18, 103-134.

Roy, P.S. and Stephens, A.W. (1981). Geological controls on process-response, S.E. Australia. *Proc. 17th International Coastal Engineering Conference*, Sydney, 913-933.

Roy, P.S., Thom, B.G. and Wright, L.D. (1980). Holocene sequences on an embayed high-energy coast; an evolutionary model. *Sedimentary Geol.*, 26, 1-19.

Roy, P.S. and Thom, B.G. (1981). Late Quaternary marine deposition in New South Wales and southern Queensland - an evolutionary model. *Jour. Geol. Soc. Aust.*, 28, 471-489.

Thom, B.G. and Chappell, J. (1975). Holocene sea levels relative to Australia. *Search*, 6 (3), 90-93.

Thom, B.G. and Roy, P.S. (1983). Sea level change in New South Wales over the past 15,000 years. *In* Hopley, D. (ed.), *Australian Sea Levels in the Last 15,000 Years: A Review.* Department of Geography, James Cook University, Monograph Series, Occasional Paper No. 3, 64-84.

Thom, B.G., Polach, H.and Bowman, G. (1978). *Holocene age structure of coastal sand barriers in N.S.W., Australia.* University of New South Wales Faculty of Military Studies, Department of Geography, Duntroon, 86 pp.

Thom, B.G., Bowman, G.M., Gillespie, R., Temple, R. and Barbetti, M. (1981). *Radiocarbon dating of Holocene Beach-ridge sequences in Southeast Australia.* University of New South Wales Faculty of Military Studies, Department of Geography, Duntroon, Monograph 11, 36 pp.

West, R.J. (1983). The sea grasses of New South Eales estuaries and embayments. *Wetlands*, 3 (1), 34-44.

Wright, L.D. (1977). Morphodynamics of a wave-dominated river mouth. *Proc. 15th International Coastal Engineering Conference*, Honolulu, 1721-1737.

6

Facies Variation and Holocene Reef Growth in the Southern Great Barrier Reef

JOHN F. MARSHALL

P.J. DAVIES

Bureau of Mineral Resources
Canberra, ACT

Introduction

The development of Holocene reefs in the Great Barrier Reef region has until recently been inferred from generalised reef morphological schemes (e.g. Maxwell, 1968; Maiklem, 1968; Flood and Orme, 1977). However, over the last decade it has come to be realised that to comprehend the variation in reef development, both in time and space, reefs must be viewed as three dimensional entities, constrained in growth by the shape and depth of their substrate and by the air-sea interface towards which they grow (Stoddart, 1975; Hopley, 1982).

Presently, the emphasis in coral reef research has shifted away from morphological/sedimentological studies of the reef's surface to multiple shallow drill holes of the Holocene section of the reef. This has eventuated because of the need to develop three-dimensional models of reef growth. It has been made possible by the development of relatively inexpensive and portable drilling rigs that can be used for coring the reef, even in submerged environments (e.g. Macintyre, 1978). As a result, the thickness, accumulation rate and internal structure of many modern coral reefs throughout the world have been investigated.

COASTAL GEOMORPHOLOGY
IN AUSTRALIA
ISBN 0 12 687880 3

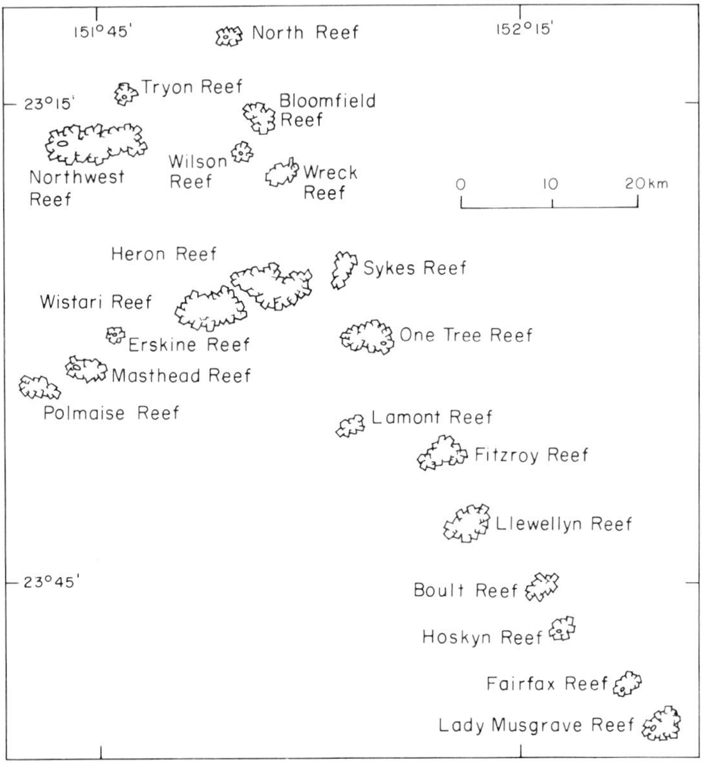

FIGURE 1. Locality diagram showing reef distribution in the Capricorn and Bunker Groups, southern Great Barrier Reef.

A fairly extensive programme of shallow drilling has been conducted in the Great Barrier Reef (e.g. Thom *et al*., 1978; Hopley *et al*., 1978; Rhodes, 1981; Grimes, 1982), in particular by the Bureau of Mineral Resources and the Department of Geography, James Cook University (Davies and Marshall, 1980; Marshall and Davies, 1982; Hopley *et al*., 1983; Davies and Hopley, 1983; Davies and Marshall, in press. Up until the end of 1983 a total of 69 holes had been drilled in the Great Barrier Reef. These holes have penetrated to depths ranging between 6 and 76 m. This paper discusses the results from sixteen holes drilled by the Bureau of Mineral Resources into windward, leeward, and lagoonal environments on four reefs in the southern Great Barrier Reef (Figs. 1, 2). At present, this is the only area where analysis of drill core has been completed from more

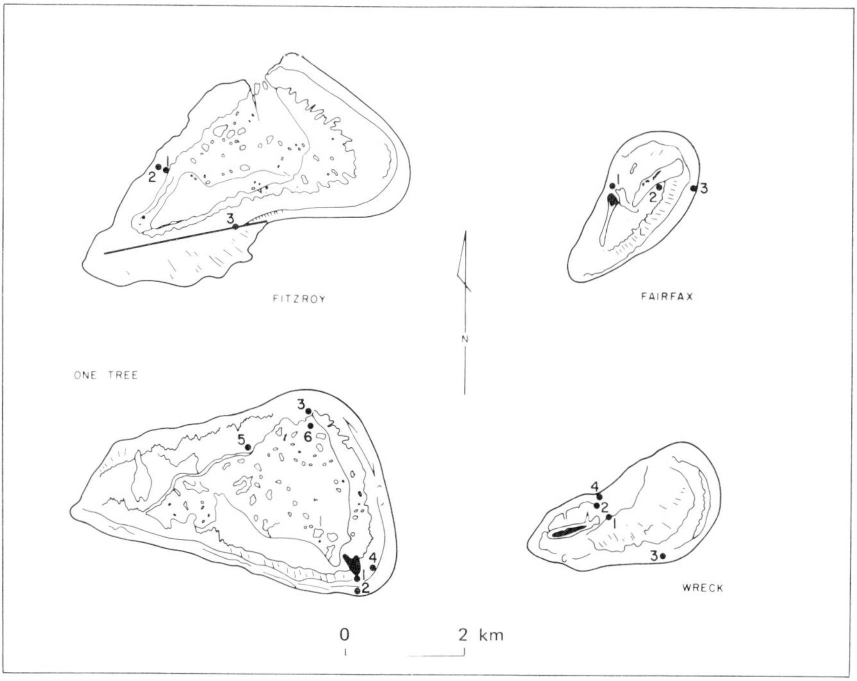

FIGURE 2. Physiographic outlines of Fitzroy, One Tree, Fairfax and Wreck Reefs showing position of drill hole sites.

than one reef. However, on the basis of existing and preliminary data from other reefs, it can be assumed that models of reef growth developed in the southern region are widely applicable to the rest of the Great Barrier Reef.

Drilling Results

Of the sixteen holes that were drilled in the southern Great Barrier Reef, six were on One Tree, three on Fairfax, four on Wreck and three on Fitzroy Reef (Fig. 2). Depths of the holes varied between 5.8 and 23.0 m with recovery varying between 12 and 95 percent (Fig. 3). Six of the holes penetrated the entire Holocene sequence and into the Pleistocene. These holes show that the Holocene reef thickness varies between 7.4 and 14.3 m, whereas more widespread coverage using seismic refraction techniques indicate thicknesses of between 7 and 23 m (Harvey *et al.*, 1979). The lithology of the Pleistocene sections is dominated by corals

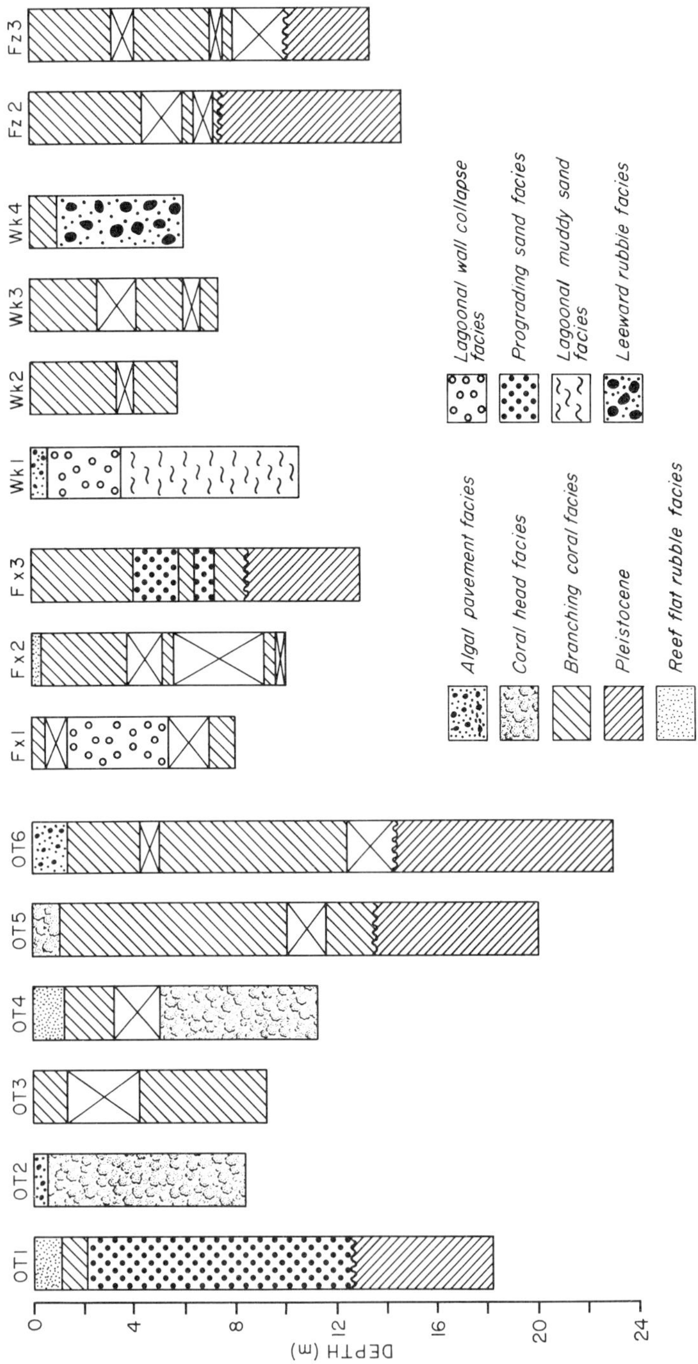

FIGURE 3. Drill hole logs showing the distribution of the various facies.

and *Halimeda* limestone, indicating that the Holocene reef has developed on an older reef surface that had previously grown during the Last Interglacial (Marshall and Davies, 1984.

Reef Facies and their Distribution

Up to eight bio-lithofacies (Fig. 3) have been identified within the drill holes from the four reefs (Davies and Marshall, in press). These facies can be broadly grouped into (i) a framework assemblage, and (ii) a detrital assemblage. The framework assemblage consists of those elements of the reef that have remained in their growth position, whereas the detrital assemblage consists of material that has been broken-down and transported to its depositional site.

Framework Assemblage

Following the scheme of Marshall and Davies (1982), the framework assemblage can be divided into three facies: a) algal pavement, b) coral head, and c) branching coral facies. The *algal pavement facies* is a relatively thin surface feature. It consists of coral and mollusc fragments that have been encrusted and bound by crustose coralline algae with associated encrusting organisms, such as vermetid gastropods, foraminifera and bryozoans. This facies is prominent on the windward margins where it forms the distinctive algal rim so often referred to in coral reef literature. However, drilling at One Tree Reef indicates that it is only 0.4 m thick, while at Fairfax and Wreck Reefs it is less than 0.1 m in thickness. Apart from the windward margin, relatively narrow linear or sinuous algal pavements are present on leeward margins, such as at Wreck and Fairfax Reefs. These features are considered to be analogous with the coralline algal cappings of the leeward lagoonal walls at One Tree and Fitzroy Reefs, which at Wreck and Fairfax are indicative of the extent of previous lagoons which have been partly or wholly infilled (Davies and Marshall, in press). The thickest development of this facies is a 1.4 m sequence at the top of the patch reef at One Tree Reef (hole 6; Fig. 3).

The *coral head facies*, interpreted from One Tree Reef as indicative of a high energy environment (Marshall and Davies, 1982) was not encountered in any of the drill holes from the other reefs. At One Tree Reef this facies occurs beneath the windward margin (holes 2 and 4; Fig. 3). It has been subdivided into a lower coralline algal encrusted subfacies and an upper non-encrusted subfacies (Marshall and Davies, 1982). The encrusted subfacies consists of coral heads 5-25 cm high separated by coralline algal crusts 2-10 cm thick, whereas the non-encrusted subfacies consists of heads 6-52 cm high, but with relatively thin (<1 cm) algal crusts.

The *branching coral facies* is the dominant framework facies, being encountered within every drill hole with the exception of hole 2 on the windward margin of One Tree Reef. Branching corals form the major framework facies at Wreck, Fairfax and Fitzroy Reefs where they occur beneath both windward and leeward margins. At One Tree Reef the leeward margin is dominated by branching corals, as is also the patch reef (hole 6; Fig. 3). On the windward margin of this reef the branching coral facies is present as a 4.5 m thick sequence on top of the coral head facies (hole 4; Fig. 3), whereas at the junction of the windward reef flat and the sand wedge (hole 1; Fig. 3) it is only 0.6 m thick. In this area the branching coral facies is related to the development of the windward reef flat.

Detrital Assemblage

The detrital assemblage has been subdivided into five lithofacies. These include 1) reef-flat rubble facies, 2) lagoonal-wall collapse facies, 3) leeward rubble facies, 4) prograding sand facies, and 5) lagoonal muddy sand facies. The first three facies are very coarse-grained and have similar lithologies, but occur within distinctly different environments. The intertidal *reef-flat rubble facies* is a prominent surface feature on windward margins. This is considered to be a high energy deposit, derived principally from corals growing on the windward reef slope that have been tossed up onto the reef-flat during storms. The other two coarse-grained facies are essentially subtidal. Drill holes sited on leeward algal pavement features at Wreck Reef (hole 1) and Fairfax Reef (hole 1) encountered a rubble facies of dominantly tabular corals that are interpreted as a quiet water *lagoonal-wall collapse breccia* that developed at the leeward edge of a now infilled lagoon (Davies and Marshall, in press). This interpretation is borne out by radiocarbon dating of this material which show age reversals and unrealistically high accumulation rates. The *leeward rubble facies* occurs in the lower part of hole 4 at Wreck Reef (Fig. 3). This rubble is interpreted to have formed in about 5 m of water at the edge of the reef and was then grown over by the reef. Examination of the present-day leeward reef slope at Wreck Reef shows coral gravel infilling shutes between the framework buttresses which in places have coalesced at their tops.

The *prograding sand facies*, similar to the 11m section encountered in hole 1 at One Tree Reef (Fig. 3), is a major sediment body that has developed behind the windward reef flats of those reefs with lagoons (e.g. One Tree and Fitzroy Reefs). Sediment dynamics studies indicate that this facies is rapidly prograding into the lagoon at the present time. A *muddy sand facies* encountered in the lower half of hole 1 at Wreck Reef (Fig. 3) is considered to be indicative of a lagoonal environment, since similar sediments are presently only found at the leeward ends of reef lagoons (e.g. Davies *et al.*, 1976). The presence of these muddy sediments in this hole indicates that a lagoon, possibly as deep as 9 m, was originally present at Wreck Reef, but that it was gradually infilled and eventually grown over by the reef flat.

Vertical Accretion Rates

The results of extensive and careful radiocarbon dating (136 dates) of corals from the drill holes in the southern Great Barrier Reef has allowed us to make estimates of the vertical accretion rates of these reefs during the Holocene. Overall, measured rates vary between 0.6 and 46.0 m per 1000 yrs (Table 1). Some of the measured rates exceed that for

TABLE 1. Ranges of vertical accumulation rates (in metres per thousand years) for various reef environments in the southern Great Barrier Reef.

	Windward margin	*Leeward margin*	*Lagoonal patch reef*
One Tree	*1.8 - 7.3*	*0.6 - 8.3*	*3.0 - 7.9*
Wreck	*10.8*	*3.4 - 8.0*†	
Fitzroy	*1.7 - 7.4*	*0.8 - 4.6*	
Fairfax	*2.2 - 46.0*	*0.7 - 4.3*†	

†indicates environments where unrealistic rates were measured (see text).

actual biological growth and appear to be related to accumulation of coarse coral detritus or synchronous growth of corals on vertical walls (e.g. lagoonal-wall facies). Using only realistic estimates, vertical accretion rates have varied between 0.7 and 10.7 m per 1000 yrs beneath windward margins, whereas beneath leeward margins the variation is 0.6 to 8.3 m per 1000 yrs. Average rates for both environments are approximately the same (4.8 m per 1000 yrs for windward margins and 4.0 m per 1000 yrs for leeward margins).

A common feature of the accretion curves for Wreck, Fitzroy and Fairfax Reefs is a decrease in the accretion rate as the reef approaches sea level (Fig. 4). This occurs on both the windward and leeward margins of these reefs. While the timing of this decrease varies slightly between the reefs, the depth at which it occurs is reasonably constant (2.2-3.2m). However, it does tend to be slightly deeper on windward margins. Such inflexions in the accumulation rate are considered to be a function of the vertically growing reef being subjected to a higher energy regime as it approaches sea level. This type of growth is not shown to be the case at One Tree Reef. While inflexions in the accretion curve do occur, they are generally deeper (7.6-10.2 m) and the growth rate may increase as well as decrease (Marshall and Davies, 1982; Fig. 4).

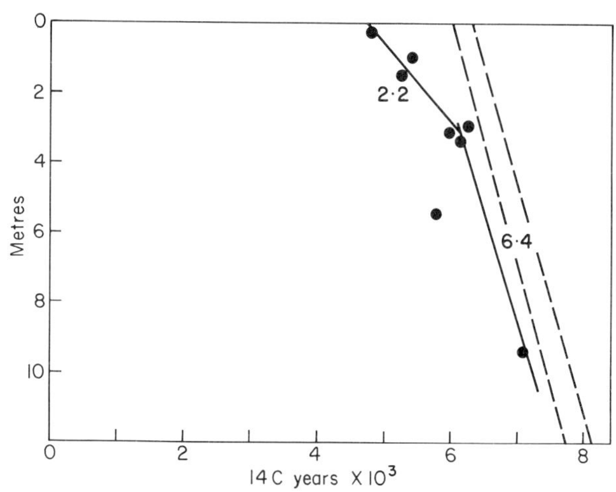

FIGURE 4. Positions of dated samples and vertical accretion rates (in metres per thousand years) in hole 2, Fairfax Reef. Note the decrease in accretion rate towards the top. Dashed lines represent the sea level envelope for eastern Australia (Thom and Chappell, 1975).

Another interesting facet of the dating program is the timing of initiation of reef growth between the four reefs. Holocene growth began at One Tree approximately 8000 yrs BP, at Fitzroy close to 7500 yrs BP, at Fairfax 7000 yrs BP, and 6000 yrs BP at Wreck. These differences appear to be related to the general level of the substrate on which the Holocene reef has grown. In particular, this is the case for the level of the reef's perimeters. Both drilling and seismic refraction results show that the Pleistocene substrate is marginally, but progressively deeper on those reefs that initiated their growth earlier. Therefore, because of the general difference in the depth of their substrates, One Tree Reef had established itself as a subtidal but well developed entity by 6000 yr BP (Marshall and Davies, 1982; Fig. 6A), while Wreck Reef had only begun to grow. However, because of its shallower substrate Wreck Reef had established itself at sea level by 4000 yrs BP, whereas at One Tree Reef the windward reef flat and parts of the leeward margin had not.

Aspects of Holocene Reef Growth

Drilling results from the southern Great Barrier Reef show that Holocene reefs of variable thickness (7-14 m) have grown on an older reef surface that developed during the last interglacial (Marshall and Davies, 1984). Initially growth appears to have been opportunistic. Relatively large cavities at the base of the Holocene sequences suggest that, during the early stage of colonisation, corals tended to grow as scattered patches. Once established, they began to grow vertically in response to rising sea level. This type of growth probably persisted until there was sufficient coalescence of the original patches to form a more widespread substrate for additional colonisation, and resulted in a more tighter framework that persisted throughout most of the transgressive growth stage.

A branching type of framework facies predominates on most reefs, beneath both windward and leeward environments. Only at One Tree Reef did a dominant assemblage of head corals develop and this was restricted to the windward margin. The absence of this facies below the windward margins of the other reefs is possibly related to lower energy conditions. Both Wreck and Fairfax are slightly protected by other reefs while the hole at Fitzroy is in a more protected environment. Alternatively, this facies was not detected from the limited amount of drilling on the windward margins of these reefs.

Within the branching coral facies there are isolated coral heads scattered throughout. This style of framework has persisted throughout the Holocene, and it is evident on windward reef flats and leeward reef slopes at the present time.

During construction of the framework assemblage, growth was dominantly vertical since the rate of sea level rise was faster than the average vertical accretion rate (6-10 m per 1000 yrs for sea level and 4.4 m per 1000 yrs for reef growth). Even though vertical accretion rates of up to 11 m per 1000 yrs were realised in some instances, by the time sea level stabilised (ca. 6000-6500 yr BP) most reefs were still some 4 to 10 m below sea level. However, much of the perimeters of the reefs had reached sea level by 4000 BP. This period between 6000 and 4000 years appears to have been a transitional one. As mentioned previously, many of the reefs exhibit a decrease in their accretion rates during this period as a result of their proximity to the surface wave and current regime. Since 4000 years BP the reefs have experienced additional vertical growth in some environments as well as the establishment of thin algal pavement cappings, particularly on windward margins. However, the major change to the reefs in this period has been the onset of lateral expansion resulting from the interaction of calcification with the surface hydrology. During this time the detrital assemblage has assumed the major role, as evidenced by the growth of the prograding sand facies, the infilling of lagoons by the muddy sand facies (both presently at One Tree and Fitzroy Reefs and in the past at Wreck Reef), and the lateral expansion of the leeward margins (such as at Wreck Reef) by the leeward rubble facies. The combination of these factors have, in time, produced a surface morphological expression that depicts the classical zonation of coral reefs. However, rather than highlighting their differences as a means of classification it is now possible to perceive their similarities within the continuum of reef development.

References

Davies, P.J., Radke, B.M. and Robison, C. (1976). The evolution of One Tree Reef, southern Great Barrier Reef, Queensland. *BMR Journal of Australian Geology and Geophysics*, 1, 231-240.

Davies, P.J. and Marshall, J.F. (1980). A model of epicontinental reef growth. *Nature*, 287, 37-38.

Davies, P.J. and Hopley, D. (1983). Growth facies and growth rates of Holocene reefs in the Great Barrier Reef. *B.M.R. Journal of Australian Geology and Geophysics*, 8, 237-252.

Davies, P.J. and Marshall, J.F. (in press). Age and Lithologic structure of Holocene reefs in the southern Great Barrier Reef. *Coral Reefs*.

Flood, P. and Orme, G.R. (1977). A sedimentation model for platform reefs of the Great Barrier Reef, Australia. *Proceedings of the Third International Symposium on Coral Reefs*, Miami, 2, 111-118.

Grimes, K.G. (1982). Drill-site report, AIMS Davies No. 1. *Geological Survey of Queensland Record*, 1982/81, 7 pp (unpublished).

Harvey, N., Davies, P.J. and Marshall, J.F. (1979). Seismic refraction - tool for studying coral reef growth. *BMR Journal of Australian Geology and Geophysics*, 4, 141-147.

Hopley, D. (1982). *The Geomorphology of the Great Barrier Reef: Quaternary Development of Coral Reefs*. John Wiley-*Interscience*, New York.

Hopley, D., McLean, R.F., Marshall, J. and Smith, A.S. (1978). Holocene-Pleistocene boundary in a fringing reef: Hayman Island, north Queensland. *Search*, 9, 323-325.

Hopley, D., Slocombe, A.M. Muir, F., Grant, C. (1983). Near-shore fringing reefs in North Queensland. *Coral Reefs*, 1, 151-160.

Macintyre, I.G. (1978). A hand-operated submersible drill for coring reef substrates. In: Stoddart, D.R. and Johannes, R.E. (editors), Coral Reefs: Research Methods. *UNESCO Monographs on Oceanographic Methodology*, No. 5, pp. 78-80.

Maiklem, W.R. (1968). The Capricorn Reef Complex, Great Barrier Reef, Australia. *Journal of Sedimentary Petrology*, 38, 785-798.

Marshall, J.F. and Davies, P.J. (1982). Internal structure and Holocene evolution of One Tree Reef, southern Great Barrier Reef. *Coral Reefs*, 1, 21-28.

Marshall, J.F. and Davies, P.J. (1984). Last interglacial reef growth beneath modern reefs in the southern Great Barrier Reef. *Nature*, 307, 5, 44-46.

Maxwell, W.G.H. (1968). Atlas on the Great Barrier Reef. Elsevier, Amsterdam, 258 pp.

Rhodes, E.G. (1981). Reef coring techniques, Great Barrier Reef. *Journal of Sedimentary Petrology*, 51, 650-652.

Stoddart, D.R. (1975). The structure and morphology of coral reefs. *Reef Biogenesis Symposium*, Australian Institute of Marine Science, Townsville, p. 5 (abstract)

Thom, B.G., Orme, G.R. and Polach, H. (1978). Drilling investigations at Bewick and Stapleton Islands. *Philosophical Transactions of the Royal Society of London*, 291, 37-54.

Thom, B.G. and Chappell, J. (1975). Holocene sea levels relative to Australia. *Search*, 3, 90-93.

7

The Holocene 'High Energy Window' on the Central Great Barrier Reef

DAVID HOPLEY

Department of Geography
James Cook University of North Queensland
Townsville, Qld

Introduction

The concept of a Holocene 'high energy window' was introduced by Neumann (1972). He suggested that on tropical coasts during the mid-Holocene, a period of higher wave energy may have existed during the period when the present level of the sea was being first approached by the Holocene transgression and prior to the protective development of hermatypic coral reefs. It has also been postulated (Stoddart *et al.*, 1978; Hopley, 1982) that the 'window' may have operated on a more local scale on individual reefs with waves breaking not on margins of an extensive reef flat as at the present time, but more extensively over a shallowly submerged reef top prior to the development of the reef flat. This paper examines the concept of a period of a higher wave energy on the central Great Barrier Reef in the mid-Holocene and considers some of the implications for coastal geomorphology.

Holocene Development of Coral Reefs

In general, modern reefs have formed from the re-colonisation of older Pleistocene foundations during the Holocene transgression (see Hopley, 1982, for discussion). In areas of tectonic stability the depth to the antecedent foundation is within the range of 5 to 20m, but they may be either shallower (even exposed) or deeper. Thus during the early part of the Holocene transgression over a continental shelf, prior

COASTAL GEOMORPHOLOGY
IN AUSTRALIA
ISBN 0 12 687880 3

to the level of the sea reaching the top of the Pleistocene reefs, a considerable degree of protection to the mainland coast would have been given by these older exposed reefs. They would have stood offshore as high limestone islands. Therefore, the time during which the high energy window could have been open would occur between the first submergence of antecedent reefs and the growth of modern reefs at a stable or still rising sea level surface. It presumes that the Holocene reefs were retarded in their growth or did not have the potential to maintain a vertical accretion rate equal to the rate of sea level rise.

The rate of sea level rise is an important factor in determining the possibility of a 'high energy window'. As shown by world and local scale hydro-isostatic studies (Walcott, 1972; Chappell, 1974; Clark *et al.*, 1978), the nature of the transgression has had distinctive regional differences, particularly during the last 7000 years. In the Caribbean, where Neumann developed the 'high energy window' concept, his own (1971) sea level curve from Bermuda shows sea level rising from ca -10m 7000 yrs BP at a declining rate right up to the present day. As suggested by Hopley (1982), this may have allowed many Caribbean reefs to catch up with the relatively slowly rising sea level, severely limiting the period of the 'open window'. In contrast, the sea level pattern for eastern Australia (Thom and Chappell, 1975; Hopley and Thom, 1983) shows a rapid rise from -20m at 9000 yrs BP to a more or less stable position at modern sea level between 6500 and 6000 yrs BP, an average rate of close to 10 mm/yr or almost twice that suggested for the latter part of the transgression in the Caribbean. Thus there is the greater potential for reefs to be left behind by the more rapidly rising sea level on a southern hemisphere continental shelf, like Australia and for the existence of a 'high energy window' over a longer period (Fig. 1).

The Present 'Closed Window' situated on the Great Barrier Reef

The Great Barrier Reef north of Cairns forms an almost total barrier to oceanic swells with a near continuous series of ribbon reefs situated on the shelf edge. To the south, the reefs become less continuous but still form an effective barrier to swell waves (Fig. 2). As an east coast swell environment (Davies, 1972, p. 44) the ocean off northern Queensland, is not normally a high energy one. Moreover, as Jennings (1965), and Hopley (1982) have shown, local wind strengths decline northwards along the Queensland coastline.

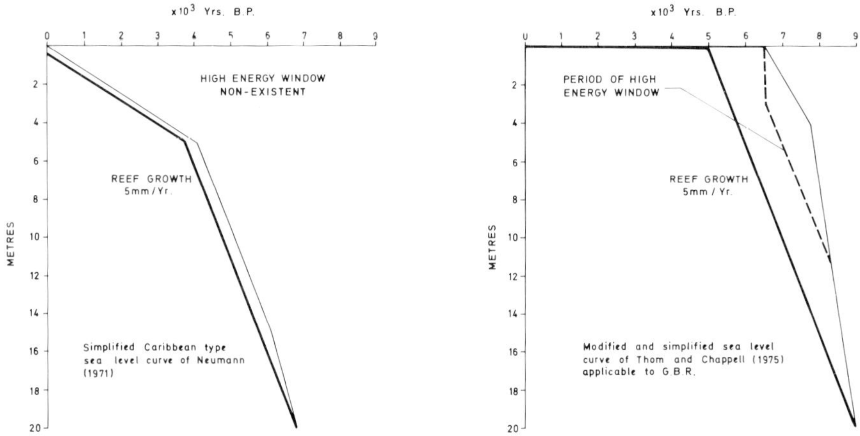

FIGURE 1. Sea level curves for eastern Australia and the Caribbean showing the time available for a 'high energy window' presuming a mean reefal upward accretion rate of 5 mm/yr from -20 m foundations. An arbitrary water depth of 5 m is taken for the window to be 'open' but it is realised that a proportion of wave energy would be lost at this depth.

Gale force winds occur 0.4 to 0.8% of the time off southern Queensland, but reduce to 0.1% of the time over the northern Great Barrier Reef.

The reduction in wave energy at the mainland, seen in Figure 3, is the combined effect of shelter from the Reef and declining wind speeds. The Reef is probably more important in reducing wave heights associated with tropical cyclones by restricting the fetch for the development of waves to the inner channel between the outer reefs and the mainland. Empirical methods suggest that in the central Coral Sea outside the Great Barrier Reef wave heights of up to 10m may be experienced as the result of the 100 year cyclone (Spillane and Dexter, 1976; Dexter and Watson, 1975). Estimates for inside the Great Barrier Reef are less than 8m. Actual records from the Gold Coast to the south of the Great Barrier Reef show maximum wave heights of over 8m recorded in an 18 month period (Delft Hydraulics Laboratory, 1970). In contrast, even at the southern end of the Reef at Yeppoon where reefal protection is minimal, cyclonic waves typically reach only 3-4m in height and exceed 5m for a duration of 12 hours with an average return rate of only 50-100 years (Beach Protection Authority, 1979). Cyclones are more common here and of greater strength than those experienced on the Gold Coast (Lourenz, 1977). At Cairns the protection given by the Reef is even greater and the highest significant wave height

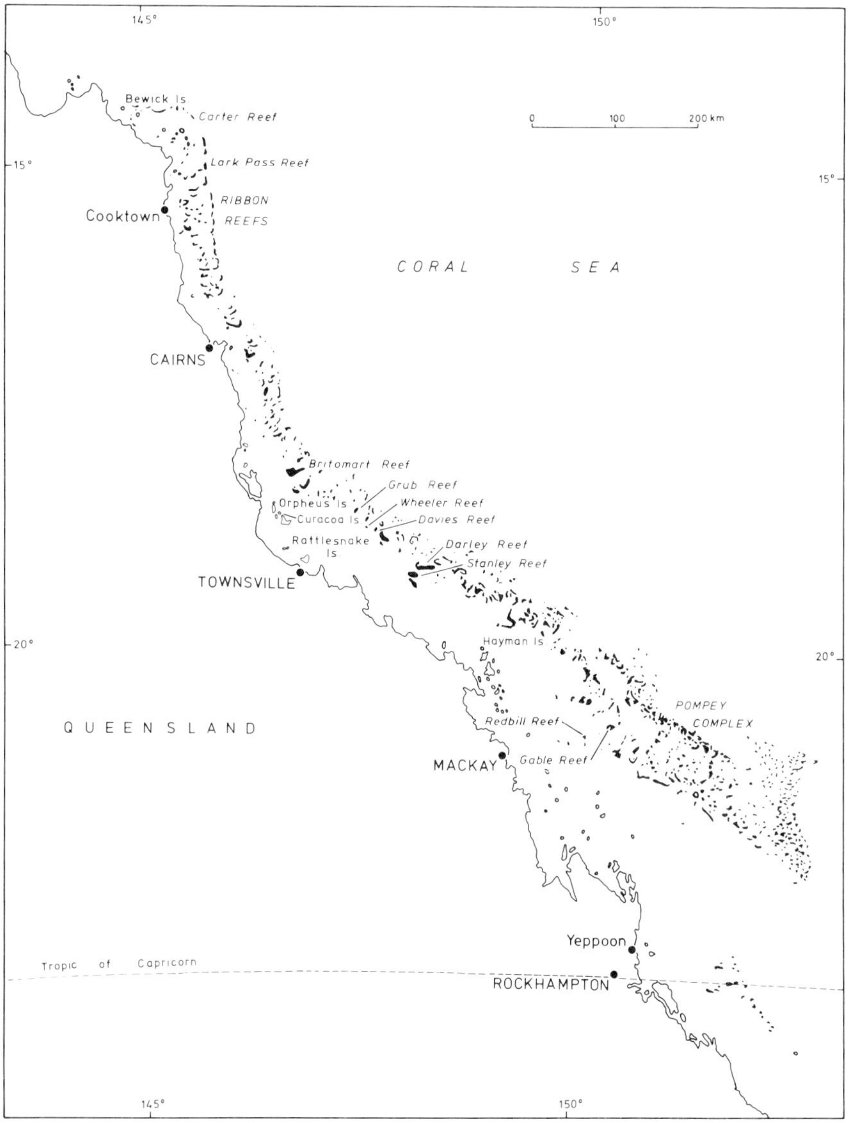

FIGURE 2. The Central Great Barrier Reef and adjacent mainland.

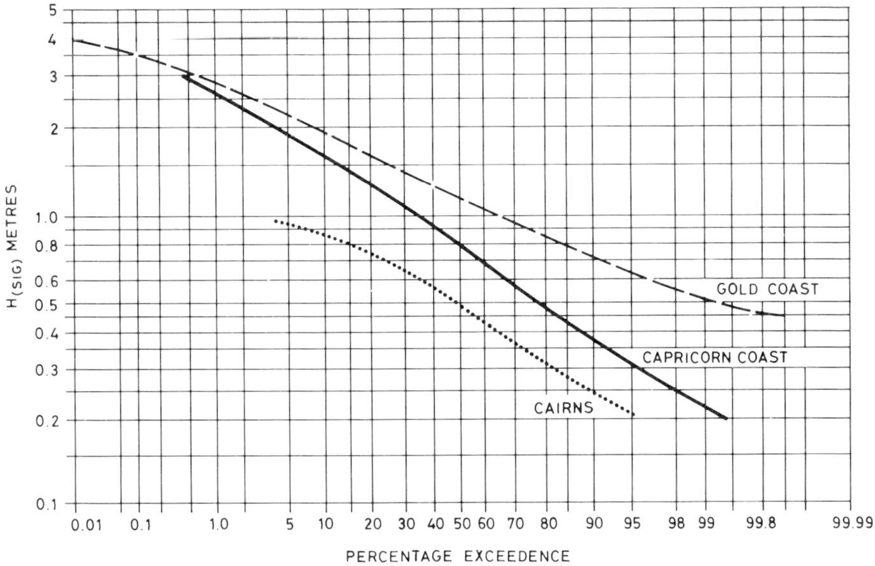

FIGURE 3. Percentage time exceedence of wave heights (H_{sig}) for all wave periods south of the Reef at the Gold Coast (Delft Hydraulics Laboratory, 1970), at the southern end of the Reef on the Capricorn coast (Beach Protection Authority, 1979), and on the northern Reef at Cairns (Beach Protection Authority, 1978).

recorded during Cyclone Keith in January, 1977, was only 1.91 m with a maximum wave height of 3.21m (Beach Protection Authority, 1978).

Thus whilst the dominant south-easterly trade winds may produce wind speeds of more than 41 km/hr for more than 10% of the time along the Queensland coastline, wave energy conditions are limited by the restriction of fetch and the exclusion of far travelled swells. Even during tropical cyclonic conditions, the Great Barrier Reef is clearly a barrier rather than a window at the present time.

The Mid-Holocene 'Window' of the Central Great Barrier Reef

Sufficient data are now available to evaluate the importance of the 'high energy window' in Great Barrier Reef waters. The depth of the Pleistocene antecedent foundations is known from drilling, seismic refraction and seismic reflection surveys for many localities. Reef development and rates of vertical reef accretion during the latter part of the Holocene transgression are also known from dated reef cores; further details of numerous reef top ages are given in Hopley (1982, see especially table 9.2). With the exception of the outermost shelf opposite Townsville where some downward flexuring has been postulated, there is nothing in this portfolio of data which suggests that the relative sea level curve for eastern Australia of Thom and Chappell (1975), or as partially revised in Hopley and Thom (1983), is not applicable to the area of the Queensland shelf under discussion.

Northern Great Barrier Reef

Older reefal foundations have been found beneath all Holocene reefs so far investigated. In Torres Strait they may actually outcrop (A. Barham, pers comm), and thus have given protection to the mainland and high continental islands throughout the Holocene transgression. North of 17°S, about the latitude of Cairns, the foundations also appear to be shallow, the unconformity having been discovered by drilling only 3m below reef flat level in Bewick Island (Thom *et al.*, 1978). Possibly even more important are the seismic refraction results (Harvey 1977, 1980) which suggest that the Pleistocene lies only 9-10m beneath the outer ribbon reefs. These old reefs have provided a complete protective barrier at the shelf edge until they were first submerged ca 8000 ^{14}C yrs BP. During the period of Holocene growth, these were probably never more than 5m beneath the rising sea surface. Not surprisingly radiocarbon dates of almost 6000 yrs BP have been obtained from just below the surface of Carter and Lark Pass ribbon reefs (Hopley 1977, 1982) clearly indicating that the ribbons were providing a barrier not long after modern sea level was first achieved. Therefore, an effective 'high energy window' on the northern Reef is unlikely to have been in existence at any stage during the Holocene.

South Central Great Barrier Reef

Similar conclusions may be made from data for the South Central Great Barrier Reef, south of about 19°S. Harvey's (1980) seismic refraction results for a number of reefs in this area show the Pleistocene at depths of only 6 to 12m beneath reef flat level. Drilling in this area is limited, but has shown the Pleistocene to be at 15m beneath the Hayman Island fringing reef (Hopley *et al.*, 1978) and 13m beneath Redbill Reef (see below). Reef cap radiocarbon dates (Hopley

1982, table 9.2) show some reefs in this area being within at least 1m of modern sea level by 6000 ^{14}C yrs BP. Dates (conventional ^{14}C yrs BP) include 6210 ± 140 yrs BP from 1.3m on Darley Reef (GaK-7021), 6420 ± 160 yrs BP from 0.62m on Stanley Reef (GaK-8931), and 6550 ± 150 yrs BP from 3.1m depth (but only just below tidal datum) on Redbill Reef (GaK-7288). A date of 7160 ± 170 yrs BP (GaK-8937) from a depth of only 2.2m beneath Stanley Reef would suggest that reef growth in this area was able to keep up with or only just lag behind the sea level rise. Again, an effective 'high energy window between 19°S and 21°S is highly unlikely within the very short span of time between the first submergence of the Pleistocene limestone islands which now form the foundations of the modern reefs, and the time at which these modern reefs first commenced the development of reef flat surfaces at the level of modern sea level. This is illustrated in Figure 4 where the time available for a high energy window on Redbill and Gable Reefs is seen to be negligible.

North Central Great Barrier Reef

There is greater possibility of a 'high energy window,' though of short duration, between 17° and 19°S. In contrast

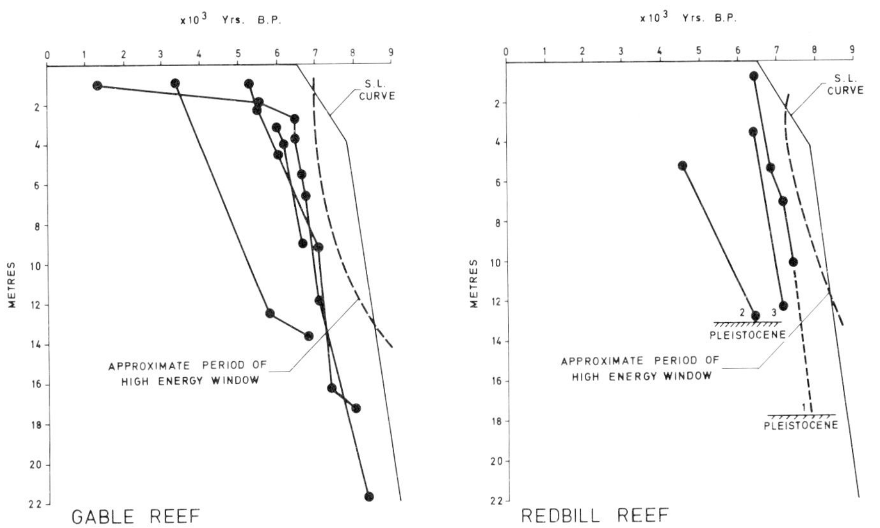

FIGURE 4. Reef growth at Gable and Redbill Reefs off Mackay. Radiocarbon dating was undertaken by Sydney University Radiocarbon Laboratory. Dates have been amended for reservoir effect (-450 yrs) and for past variation in atmospheric ^{14}C according to Clark (1975) and Barbetti (1980).

to the areas to both north and south, Harvey's (1980) seismic refraction data show that nowhere in this area (between Cairns and Townsville) is the seismic discontinuity less than 10m below the surface. It may be greater than 20m beneath some reefs. Although further work is required and drilling results to date are limited (>14m on Grub Reef, >19m on Wheeler Reef, 20m on Britomart Reef and >20m on Davies Reef), there is at least the suggestion in this area that the antecedent foundations of modern reefs may have been drowned earlier (as early as 9500 ^{14}C yrs BP). This means that reef accretion in this area lagged behind the rise of sea level. Generally reef flat development commenced later than 5000 ^{14}C yrs BP and locally as late as 3500 ^{14}C yrs BP.

Younger reef flat ages in this part of the Great Barrier Reef are reported by Hopley (1982) and Hopley and Harvey (1982). They were interpreted as indicating not only deeper foundations from which the modern reefs were growing, but also a degree of cross shelf warping as the result of both hydro-isostasy (see also Chappell *et al.*, 1982) and tectonic downwarping on the margins of the active Halifax Basin. This is likely to have most affected the shelf edge reefs where the relative sea level curve may indicate modern sea level being achieved as much as 3000 years later than on adjacent parts of the shelf. Although this factor makes the greater depth of water over these reefs in mid-Holocene times more apparent than real, the continued rise of local sea level probably resulted in more patchy reef development and less horizontal reef flat extension in mid and late Holocene times. The more open nature of the reef would thus provide less of a barrier to the open ocean swells.

Wheeler and Grub Reefs are thought to be marginal to the zone of downwarping and their growth trends as shown in Figure 5 may be representative of Holocene reef development in the area between Cairns and Townsville. If so, then there may have been a period between ca 8500 yrs and 6000 yrs BP when water depths over the reefs was greater than 5m and the 'high energy window' was at least partially open.

Regional Limitation of the 'High Energy Window'

Drilling results from the Great Barrier Reef suggest that the maximum vertical accretion rate for Holocene reef growth over an extended time period has been between 6 and 7 mm/yr, a figure which compares favourably with Kinsey's (1982) carbonate production figure based on alkalinity depletion for shallowly submerged reef surfaces. As the rate of sea level rise during the time that most antecedent platforms were being

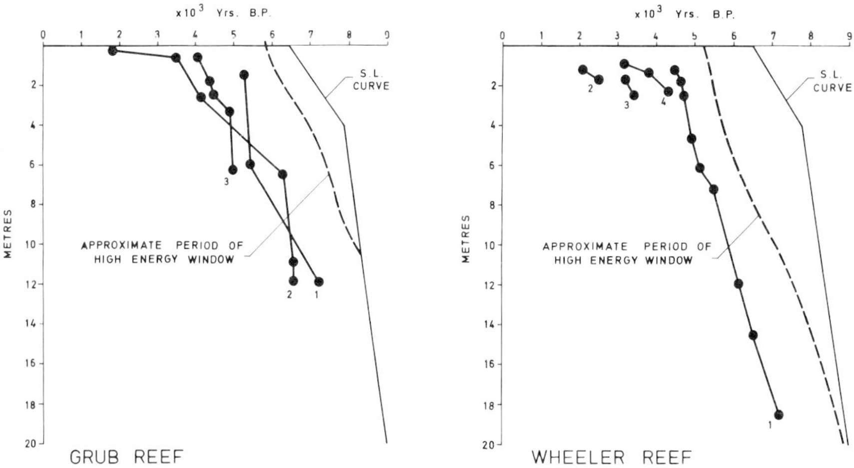

FIGURE 5. Reef growth at Wheeler and Grub Reefs off Townsville. Radiocarbon dating was undertaken by Sydney and Gakushuin Radiocarbon laboratories. Dates have been amended for reservoir effect (-450 yrs) and for past variation in atmospheric ^{14}C according to Clark (1975) and Barbetti (1980).

submerged was at least 10 mm/yr, then reefs would have trailed the rise in sea level by ca 3m every thousand years up to the time of sea level stabilisation. This may have been greater because there are difficulties of colonisation of the antecedent surfaces by modern reefs, possibly due to early high turbidity, which have produced initially slow rates of growth (Davies and Marshall, 1979). As a result overall accretion rates were probably closer to 5 mm/yr. However, translation of these accretion rates to the pattern of sea level rise and to the known depths of the Pleistocene and reef top ages clearly reinforces the concept of little or no high energy period for the Great Barrier Reef to the north of Cairns and to the south of Townsville. Up to about 8000 yrs BP the antecedent platforms provided effective protection to the coastal waters behind them. After 6000 yrs BP the modern Holocene reefs were developing reef flats at a rate fast enough to give similar protection. In the intervening 2000 years it is highly unlikely that the reefs of these areas were ever greater than 5m below the rising sea surface, and thus they were probably sufficiently close to the surface to provide a significant degree of protection.

In contrast, in the area between Cairns and Townsville, the north central region noted above, the development of the Great Barrier Reef from deeper foundations may have resulted in the following sequence:

i) up to 9000 ^{14}C yrs BP, prior to the drowning of the antecedent platforms, adequate protection was given to the expanding waters between the outer reefs and the transgressing coastline;

ii) between ca 8500 ^{14}C yrs BP and 6000 ^{14}C yrs BP, the rise in sea level outstripped the capacity of the reefs for vertical growth; the 'high energy window' was open with water depths greater than 5m over the reefs and up to 8m depth shortly after 8000 yrs BP when sea level may have stood as little as 5m below its present position;

iii) since 6000 ^{14}C yrs BP, the 'window' has been progressively closed as reefs have grown towards modern sea level and developed reef flats.

Implications for Coastal Geomorphology in the Cairns to Townsville Region

It has been suggested that the 'high energy window' was restricted in space to the area between Townsville and Cairns, and in time to the last part of the Holocene transgression and the very early part of the Holocene stillstand period. Most of the characteristics of the present coastal zone have developed since the closure of the 'window'. These include beach-ridge sequences and mangroves on the mainland coast (Hopley and Murtha, 1975), and rocky foreshores, complex sand spits and fringing reef flats on continental islands (Hopley, 1971). However, it is possible that some vestiges of the higher energy period can still be seen.

Seismic reflection surveys of this shelf area (Searle *et al.*, 1980, 1982) show features which may be related to the opening and closing of the 'window'. The shelf generally has only a thin cover of Holocene sediments, but below depths of 20 to 25m there are numerous relict features of presumably depositional origin. They include positive relief features over buried fluvial channels which have been interpreted as estuarine and deltaic lobes formed during the transgression (Johnson *et al.*, 1982); low amplitude, relict, sandy bedforms interpreted as transgressional strandline deposits at water depths of 25-37 m have also been observed (Searle *et al.*, 1980, 1982). Such features could have formed during the early 'closed window' period, and would have been in sufficient water depth during the succeeding 'open window' period to have avoided extensive reworking.

In contrast, the inner shelf shallower than -20m is almost devoid of Holocene sediment and the Pleistocene surface may outcrop in places. During the Holocene transgression, the sea rose over this area during the 'open window' phase, and it is highly likely that the terrigenous deposits being brought down to the coast at this time were reworked and pushed shorewards. They now form the inner parts of the extensive beach ridge sequences of the mainland, a high proportion of which appear to have been deposited very rapidly after the initial stabilisation of sea level. Subsequent to the still-stand and to the reclosure of the 'window', an extensive mud wedge up to 18m thick and extending 8 to 20km offshore has prograded across the shelf from the mainland. Although no dates are yet available for these sediment units, the nature and distribution of sediments on the continental shelf may be very closely related to the energy levels determined by the opening and closing of the reefal barrier.

The 'open window' period was also the time in which the fringing reefs of the continental islands were developing (Hopley *et al.*, 1978, 1982). Inshore water depths around the islands are generally not much greater than 20m. Hopley *et al.*, (1982) have noted that many fringing reefs appear to have grown over a foundation of largely terrigenous sediments and to have colonised the rocky shores of the islands very quickly after they were drowned by the rising sea level. Higher energy conditions may help to explain both features. The terrigenous materials, like those of the earliest sand barriers of the mainland, were derived from the shelf together with material reworked from the regolith which probably covered these hillslopes prior to submergence. Higher energy levels, either during 'normal' conditions or during cyclonic weather would have helped in the rapid removal of fines from the regolith reducing the length of time of high turbidity which had slowed down the rate of colonisation on the outer reefs. Thus as soon as sea level stabilised about 6500 years ago (apart from relatively minor fluctuations), reef flat extension was immediately possible. From the radiocarbon dates available inner flats had developed as early as 6000 yrs BP (see Hopley *et al.*, 1982; Chappell *et al.*, 1982). This is seen at Orpheus and Rattlesnake Islands (Fig. 6), but at Hayman, where a Pleistocene antecedent reef is present at -20m depth, modern reef flat development appears to have commenced later and a short high energy phase on the island shores is indicated.

Explanation for the nature of rocky shorelines on both high continental islands and on mainland headlands may also lie in having a short period of higher wave energy during the phase of the 'high energy window'. Such shores currently

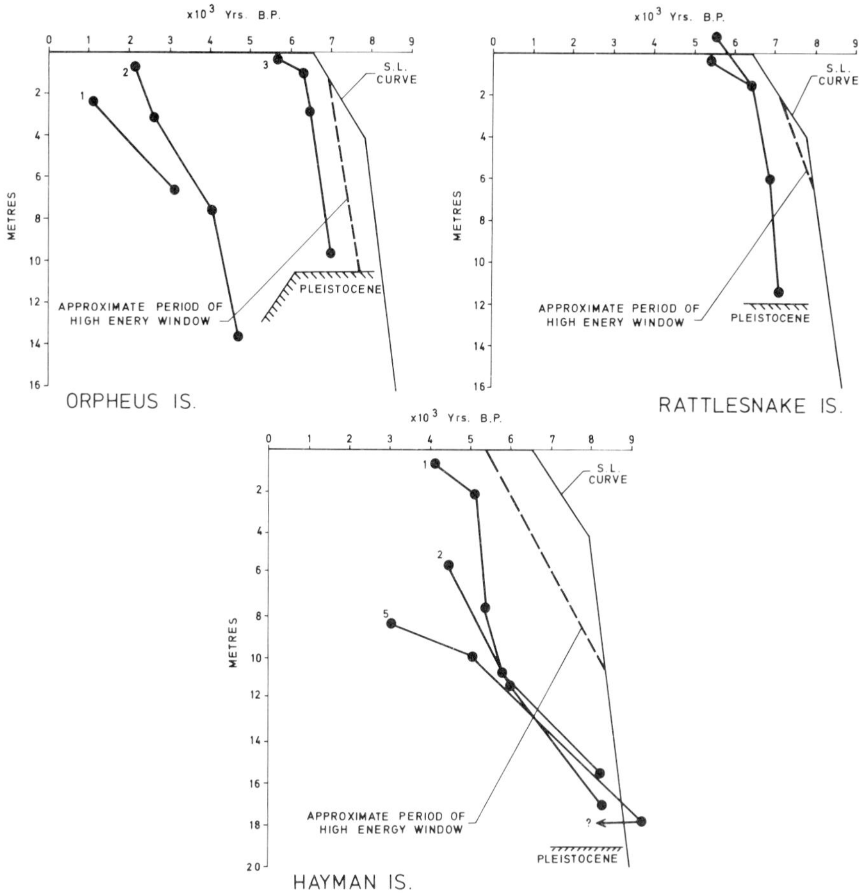

FIGURE 6. Reef growth on the fringing reefs of Orpheus, Rattlesnake and Hayman Islands (see Hopley et al., 1978, 1982 for details).

consist of convex bare rocky slopes to several metres above high tide level. At the base of these slopes in many locations are large, poorly sorted, boulder beaches in relatively sheltered locations. Well sorted, boulder beaches (mean size up to 1.0m) occur in more open situations (see descriptions by Hopley 1968, 1971, 1975, 1982). Based on radiocarbon dates from material incorporated within the beaches or relationships with other dated material, the majority of these boulder beaches appear to have been emplaced prior to 5000 yrs BP. Previously a combination of higher Holocene sea levels and cyclonic storm activity have been quoted as explanation for the features. However, reassessment would suggest that two further factors need consideration.

First, even during cyclonic weather of the present time these beaches are stable. During cyclone Althea in 1971 for example, even in the zone of highest wind speeds (exceeding 100 kts), little or no movement took place in the boulder beaches (Hopley, 1974). Yet many beaches are well sorted and even the coarsest show a degree of sorting (see Hopley, 1982, table 6.1). Certainly, cyclonic conditions are required for movement of these materials, but the absence of a complete reefal barrier prior to 5000 yrs BP would have been conducive to greater mobility. Although detailed surveys have not been undertaken, it may be pertinent that massive terrigenous boulder beaches have not been found north of the southern end of the ribbon reefs.

The second factor requiring reassessment is the volume of material in some of these beaches. Curacoa Island in the Palm Group (Hopley, 1968, 1971) is indicative of the problem. Boulder beaches make up a foreland area of over 30 ha, terrigenous material accounting for approximately 70% of the total. Conservatively, the thickness of the deposit is at least 10m giving a massive terrigenous volume of 2,100,000 m^3. Even allowing for the formation of the feature during the whole 2500 yrs of the 'high energy window's' existence, this appears to represent an enormous amount of geomorphic work. Re-evaluation of the evidence, based on observation of boulder beaches elsewhere in the world would suggest that much of the material had accumulated in earlier Pleistocene high stands of the sea. The Holocene transgression merely re-worked the deposit, though even for this to occur high energy conditions would appear to have been necessary.

Conclusion

The concept of a Holocene 'high energy window' appears valid on the central Great Barrier Reef and is helpful in explaining a number of otherwise anomalous coastal features. It is, however, limited in time to the period between the first complete submergence of the older reefal platforms and the time at which modern reef flat development commenced. The depth of these foundations in relation to the nature and rate of sea level rise during the latter part of the transgression is critical to the period during which a 'high energy window' is open. This restraint is such that is has probably limited the 'open window' on the central and northern Great Barrier Reef to a relatively short period between 8500 and 6000 yrs BP, and to a relatively small area between Townsville and Cairns. In searching for comparable areas in the world where the 'high energy window' may have been an effective influence, an Australian type relative sea level curve (southern hemisphere Zone V of Clark *et al.*, 1978) rather than a Caribbean type curve with its declining rate of sea level rise, would appear more suitable. Alternatively, reefal areas where the Pleistocene foundations are deep (>20m) are more likely to provide a period of high energy than where such foundations are at only shallow depth.

Acknowledgements

Research reported in this paper has been supported by the Australian Marine Sciences and Technology Committee whose help is gratefully acknowledged. Field and laboratory assistance has been provided by Clive Grant, Frazer Muir and Trevor Shearn of the Department of Geography, James Cook University of North Queensland.

References

Barbetti, M. (1980). Geomagnetic strength over the last 50,000 years and changes in atmospheric ^{14}C concentration. *Radiocarbon*, 22, 192-199.

Beach Protection Authority (1978). Wave data recording programme, Cairns region. *Report*, W 01.1, 15 pp.

Beach Protection Authority (1979). *Capricorn Coast Beaches*, Brisbane, 238 pp.

Chappell, J. (1974). Late Quaternary glacio- and hydro-isostasy on a layered earth. *Quat. Res.*, 4, 429-440.

Chappell, J., Rhodes, E.G., Thom, B.G. and Wallensky, E. (1982) Hydro-isostasy and the sea level isobase of 5500 BP in north Queensland, Australia, *Marine Geol.*, 49, 81-90.

Clark, R.M. (1975). A calibration curve for radiocarbon dates. *Antiquity*, 49, 251-266.

Clarke, J.A., Farrell, W.E. and Peltier, W.R. (1978). Global changes in post glacial sea level: a numerical calculation. *Quat. Res.*, 9, 265-287.

Davies, J.L. (1972). *Geographical Variation in Coastal Development*. Oliver and Boyd, Edinburgh, 204 pp.

Davies, P.J. and Marshall, J.F. (1979). Aspects of Holocene reef growth - substrate age and accretion rate. *Search*, 10, 276-279.

Delft Hydraulics Laboratory (1970). Gold Coast, Queensland, Australia: Coastal erosion and related problems. *Report*, 257, 2 vols, 99 pp.

Dexter, P.E. and Watson, K.B. (1976). A climatology of extreme wave heights in the Australia tropics. *Austr. Meteorol. Mag.*, 24, 2, 59-72.
Grimes, K.G. (1982). Drill-site report, AIMS Davies No. 1, *Geol. Surv. Qld Record*, 1982/18.
Harvey, N. (1977). The identification of subsurface solution disconformities on the Great Barrier Reef, Australia between 14°S and 17°S, using shallow seismic refraction techniques. *Proc. Third Int. Coral Reef Symp.*, 2, 45-51.
Harvey, N. (1980). Seismic investigations of a pre-Holocene substrate beneath modern reefs on the Great Barrier Reef Province. Unpubl. PhD thesis, James Cook University of North Queensland, 329 pp.
Hopley, D. (1968). Morphology of Curacoa Island spit, north Queensland. *Austr. J. Sci.*, 31, 122-123.
Hopley, D. (1971). The origin and significance of north Queensland island spits. *Zeit. fur Geom. NF*, 15, 371-389.
Hopley, D. (1975). Contrasting evidence for Holocene sea levels with special reference to the Bowen-Whitsunday area of Queensland. *In* Douglas I., Hobbs, J.E. and Pigram, J.J. (eds.), *Geographical Essays in Honour of Gilbert J Butland*, Dept. of Geography, Univ of New England, Armidale, 51-84.
Hopley, D. (1977). The age of the outer ribbon reef surface, Great Barrier Reef, Australia: implications for hydro-isostatic models. *Proc. Third. Int. Coral Reef. Symp.*, 2, 23-28.
Hopley, D. (1982). *The Geomorphology of the Great Barrier Reef: Quaternary Development of Coral Reefs*, John Wiley-Interscience, New York, 453 pp.
Hopley, D. and Harvey, N. (1982). Radiocarbon ages and morphology of reef tops in the Great Barrier Reef between 14° 39'S and 20°45'S: Indicators of shelf neotectonics? *Proc. 4th Inter. Coral Reef Symp.*, Manila, 1980, 1, 523-530.
Hopley, D. and Murtha, G.G. (1975). The Quaternary deposits of the Townsville coastal plain. *Dept. of Geog. James Cook Univ. Monograph Ser.*, 8, 30 pp.
Hopley, D. and Thom, B.G. (1983). Australian sea levels in the last 15,000 years: a review. *In* Hopley D (ed.), Australian sea levels of the last 15,000 years, *Dept. of Geog. James Cook Univ., Monograph Ser., Occ. Paper* 3, 3-26.
Hopley, D., McLean, R.F., Marshall, J.F. and Smith, A.S. (1978). Holocene-Pleistocene boundary in a fringing reef: Hayman Island, north Queensland. *Search*, 9, 323-325.
Hopley, D., Slocombe, A.M., Muir, F. and Grant, C. (1983). Nearshore fringing reefs in north Queensland. *Coral Reefs*, 1, 151-160.

Jennings, J.M. (1965). Further discussion of factors affecting coastal dune formation in the tropics. *Austr. J. Sci.*, 28, 166-167.

Johnson, D.P., Searle, D.E. and Hopley, D. (1982). Positive relief over buried post-glacial channels, Great Barrier Reef Province, Australia. *Marine Geol.*, 46, 149-159.

Kinsey, D.W. (1982). The Pacific/Atlantic reef growth controversy. *Proc. 4th Inter. Coral Reef Symp.*, Manila, 1980, 1, 493-498.

Lourenz, R.S. (1977). *Tropical Cyclones in the Australian Region July 1909 to June 1975.* Bureau of Meteorology, Canberra, 111 pp.

Neumann, A.C. (1971). Quaternary sea level data from Bermuda. *Quaternaria*, 14, 41-43.

Neumann, A.C. (1972). Quaternary sea level history of Bermuda and the Bahamas. *Amer. Quat. Assoc. Second Nat. Conf. Abstr.*, 41-44.

Searle, D.E., Harvey, N. and Hopley, D. (1980). Preliminary results of seismic profiling in the Great Barrier Reef Province between 16°10'S and 19°20'S. *Geol. Surv. Qld. Record*, 1980/23.

Searle, D.E., Harvey, N., Hopley, D. and Johnson, D.P. (1982). Significance of results of shallow seismic research in the Great Barrier Reef Province between 16°10'S and 20°05'S. *Proc. Fourth Inter. Coral Reef Symp.*, Manila, 1980, 1, 531-

Spillane, K.T. and Dexter, P.E. (1975). Design waves and wind in the Australian tropics. *Austr. Meteorol. Mag.*, 24, 2, 37-58.

Stoddart, D.R., McLean, R.F., Scoffin, T.P., Thom, B.G. and Hopley, D. (1978). Evolution of reefs and islands, northern Great Barrier Reef: synthesis and interpretation. *Phil. Trans. Roy. Soc. London B*, 284, 149-159.

Thom, B.G. and Chappell, J. (1975). Holocene sea levels relative to Australia. *Search*, 6, 90-93.

Thom, B.G., Orme, G.R. and Polach, H.A. (1978). Drilling investigation of Bewick and Stapleton Islands. *Phil.Trans. Roy.Soc. London A*, 291, 37-54.

Walcott, R.I. (1972). Past sea levels, eustasy and deformation of the earth. *Quat. Res.*, 2, 1-14.

8

Holocene Sea-Level History of a Tropical Estuary: Missionary Bay, North Queensland

JOHN GRINDROD

Department of Biogeography and Geomorphology
Australian National University
Canberra, ACT

E. G. RHODES

Gulf Oil Company
Houston, USA

Introduction

The project reported here was initially developed by E.G.R. in conjunction with the Australian Institute of Marine Science as part of their broader research program into the dynamics of tropical estuaries. It is aimed at the reconstruction of a Holocene sea-level record for Missionary Bay in northeastern Queensland, with emphasis on the marine transgressive period, roughly prior to 6000 radiocarbon years Before Present (^{14}C yrs BP). The study site has been the focus of intensive ecological research specifically regarding mangroves and related ecosystems (Bunt, 1982; Bunt *et al.*, 1979; Duke and Bunt, 1979; Tomlinson *et al.*, 1978). It provides a suite of depositional environments which bear predictable relationships to tidal datum.

Earlier reconstructions of the course of the Postglacial Marine Transgression (P.M.T.) in eastern Australia are based on either meagre observations or summaries of observations from widely spaced sites (see Hopley, 1983). Thom and Roy (1983) construct a transgressional sea-level history for southeastern Australia based on a synthesis of data from a spread of sites from southern Queensland to Tasmania. The data include revised

COASTAL GEOMORPHOLOGY
IN AUSTRALIA
ISBN 0 12 687880 3

radiocarbon dates formerly reviewed by Thom and Chappell (1975). Radiocarbon determinations on a range of sea level-related features are employed. An 'approximate zone of sea-level' for the period 12,000 to 6000 radiocarbon years B.P. is fitted to the data; the upper and lower limits of this area are defined by freshwater facies and shallow water marine/estuarine facies, respectively. Belperio (1979) provides a Holocene sea-level curve for the Townsville area in northeastern Queensland based entirely on radiocarbon dating of inferred mangrove deposits, including 5 observations from 3 separate sites for samples collected from transgressive deposits. Although both studies provide valuable information regarding transgressive sea levels in eastern Australia, their application is limited for two main reasons. Firstly, in the former study the nature of the relationship between palaeo-sea level and many of the samples dated is imprecisely known (Thom and Chappell, 1975, p. 91). Secondly, as neither study relies on data from a single location, precision in sea-level reconstruction is potentially limited by between-site variations in the relative heights of land and sea due to tectonic and/or isostatic adjustments in the earths crust, and perhaps by differential rates of sediment compaction.

Clearly the need exists for detailed, site-specific studies of the trace of the P.M.T. in eastern Australia. Mangroves and related vegetational environments provide potentially useful depositional facies for this purpose as they bear a definable relationship to mean tide level. They are also likely to contain fossil material which is distinctive and suitable for radiocarbon dating. Facies samples collected by drilling and shown by palynological techniques to be of intertidal origin provide the basis for the reconstruction of Holocene transgressive sea level at Missionary Bay.

Regional Setting

Missionary Bay ($18^{0}17'$S and $146^{0}14$E'), located at the northern end of Hinchinbrook Island (Fig. 1), is a broad, shallow bay characterized by extensive development of tropical mangroves. It is protected from the Coral Sea by the Ramsay Bay sand barrier, which connects the main Hinchinbrook range with the smaller bedrock outlier of Cape Sandwich. The recent evolution of this barrier is detailed by Pye and Rhodes (in prep). Although rainfall for the area is apparently between 2000 and 3000 mm per year (Bunt, 1982, p. 225), Missionary Bay receives only limited direct freshwater inflow from upland catchments.

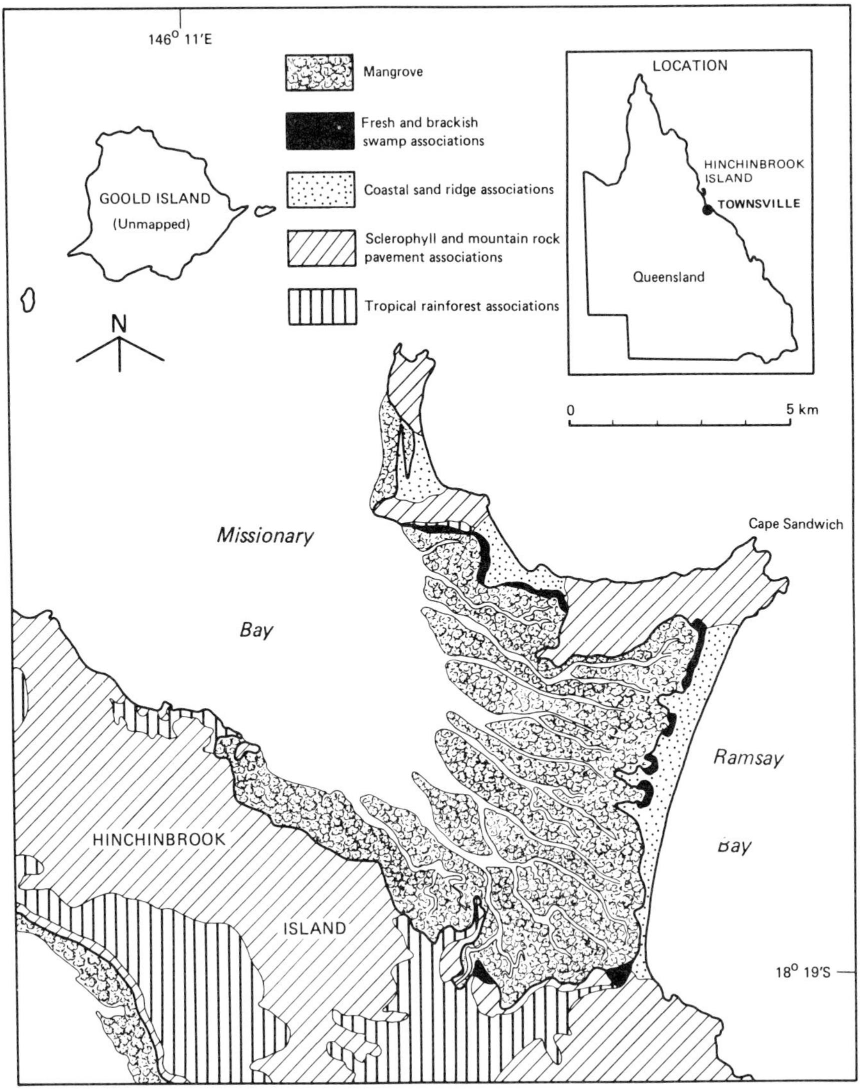

FIGURE 1. Location of study area on Hinchinbrook Island and vegetation units in the vicinity of Missionary Bay (after Tracey and Webb, 1978).

A general account of the vegetation of Hinchinbrook Island is given by Tracey and Webb (1978), while structural and floristic details of mangroves of the area are described by Bunt (1982) and Bunt *et al.*, (1982). The brief descriptions presented here of the vegetation of the Missionary Bay area are primarily based on these sources, and are intended only as a basis for the interpretation of modern and fossil pollen assemblages. For more detailed vegetation descriptions the above references should be consulted.

Figure 1 is a simplification of part of the vegetation map presented by Tracey and Webb (1978). Sclerophyll and mountain rock pavement communities dominate the terrestrial vegetation on Hinchinbrook Island where soils are relatively poorly developed. Sclerophyll communities include a diverse array of open forests and woodlands commonly dominated by *Eucalyptus acmenoidies, E. intermedia, E. tereticornis, E. tessellaris, Syncarpia glomulifera* and *Tristania suaveolens*. Small tree and shrub layers include numerous species of *Acacia, Grevillea, Melaleuca, Persoonia* and *Cycas*. Grasses such as *Themeda australis* and *Imperata cylindrica* are common ground cover taxa (Tracey and Webb, 1978, p. 33). Mountain rock pavement communities take the form of scrubs, shrublands and heaths, and occur on the high, rugged ridges of the Hinchinbrook main range and Cape Sandwich. These communities vary from monospecific scrubs of *Casuarina littoralis* to heaths of *Xanthorrhoea, Banksia, Gahnia, Dicranopteris, Gleichenia,* and stunted *Syncarpia, Tristania* and *Eucalyptus*(Tracey and Webb, 1978, p. 34).

Extensive communities of tropical rainforest (including complex mesophyll vine forest, mesophyll vine forest and closed forests with sclerophyll emergents and co-dominants), occur south of Missionary Bay on the lower slopes of the Hinchinbrook Range and along major sheltered stream gullies. Isolated pockets of this type also occur around the northern edge of the bay. The communities are of closed canopy structure and are characterized by a high diversity of non-sclerophyll tree taxa and numerous vines and epiphytes (Tracey and Webb, 1978, p. 31-32).

Coastal sand ridge associations occupy the sandy soils of barrier complexes east and north of Missionary Bay. These communities are dominated by sclerophyll elements, particularly *Eucalyptus, Acacia, Melaleuca* and *Casuarina*, which typically form open woodlands and shrublands. Less extensive communities also include vineforests and thickets and tall to stunted paperbark forests and woodlands (Tracey and Webb, 1978, p. 33). Ground covers are generally sparse, though sedges (Cyperaceae) and grasses (Gramineae) may be locally abundant in inter-dune swales.

Fresh and brackish swamp communities occur in narrow bands on poorly drained sites transitional between dryland vegetation and mangroves in Missionary Bay. These communities vary floristically according to local conditions, such as water depth and salinity, but typically include a tree canopy of *Melaleuca quinquenervia* and a ground layer including numerous sedges, grasses and ferns (Tracey and Webb, 1978, p. 32). The palm-like tree *Pandanus* may also be locally abundant.

Mangroves within Missionary Bay cover an area of approximately 50 km^2, and are both structurally and floristically diverse. Although no fewer than 27 of the 40 or so mangrove species recorded for northeastern Australia are known to occur in Missionary Bay (Bunt 1982, p. 223), mangrove communities are overwhelmingly dominated by three genera of the Rhizophoraceae, namely *Rhizophora, Bruguiera* and *Ceriops,* with *Avicennia* the next most common taxon. Other common mangrove trees include *Lumnitzera, Xylocarpus* and *Excoecaria.* Detailed studies of the litter fall within a variety of mangrove forest types and of two brackish swamp taxa (*Melaleuca sp.* and *Pandanus sp.*) at Missionary Bay are described by Bunt (1982). These studies indicate the high productivity of these vegetation communities, particularly during the humid summer period. In brackish swamps the great majority of deposited organic material is likely to remain within local sediments, due to the sheltered and poorly drained nature of these environments. However, mangroves experience regular tidal innundation which to some extent achieves the redistribution and/or removal of the locally produced organic component, a point which will be further developed in the sediment description of this facies type (see below).

The vertical ranges of brackish swamp and mangrove communities relative to tide datum in Missionary Bay are shown in Figure 2. The diagram is based mainly on survey and tide recorder observations made by the Australian Survey Office. The upper and lower limits of *Rhizophora* and non-*Rhizophora* mangrove types respectively are estimates based on an average of spot heights produced by comparison of air photos and detailed contour mapping.

Methods

Field

All surface sample, hand core and drill hole locations are shown in Figure 3. Surface sediment samples for pollen analysis were collected from a range of brackish swamp, mangrove

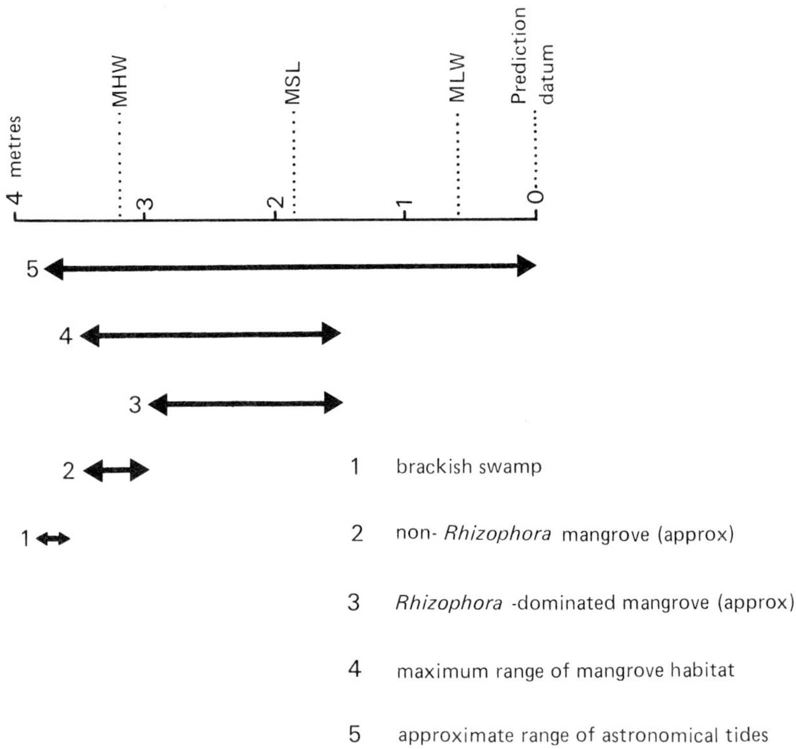

FIGURE 2. Range of elevations for different tidal levels and plant habitats, Missionary Bay, Hinchinbrook Island.

swamp and low-tide mud bank environments in and around Missionary Bay. Summaries of the vegetation at each surface sample location are given in Table 1. With one exception (sample S1), all surface samples consisted of the top half centimetre of sediments at the respective sites. Sample S1 is derived from a moss polster collected at the edge of Bowen Creek beneath mesophyll vine forest. Analysis of these samples provides a basis for the interpretation of the fossil pollen assemblages. Sediment samples for dating the regressive sequence were collected with a D-section hand corer. Deeper stratigraphic data and sediment samples (for sediment, pollen and radiocarbon analyses) were collected with a Jacro 350 barge-mounted hydraulic drilling rig, using both solid auger and continuous coring techniques. In every instance drilling continued until the Holocene - Pleistocene boundary was encountered. This was marked by a well defined transition into an oxidized terrigen-

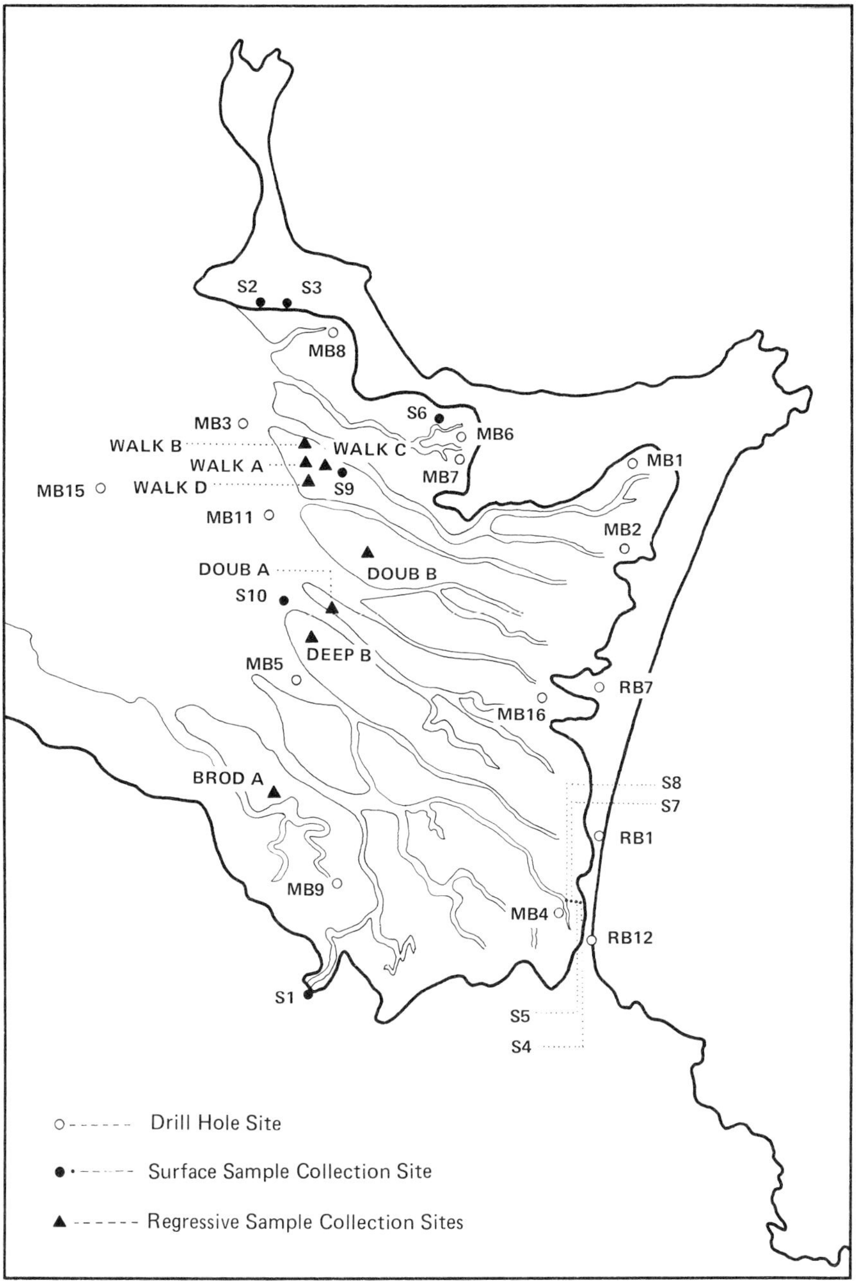

FIGURE 3. Location of samples and drill holes referred to in the text.

TABLE 1. Sample type and vegetation descriptions for surface pollen samples

Sample No.	Sample description	Vegetation at surface sample site
S1	moss polster	closed mesophyll vine forest
S2	organic mud	brackish swamp forest immediately landward of *Rhizophora* mangrove forest
S3	organic mud	brackish swamp forest immediately landward of *Rhizophora* mangrove forest
S4	intertidal organic mud	mixed mangrove/dune scrub interface
S5	intertidal organic mud	mixed mangrove and *Rhizophora* mangrove forest transition
S6	intertidal organic mud	Rhizophoraceae mangrove forest
S7	intertidal organic mud	*Rhizophora* mangrove forest
S8	intertidal organic mud	*Rhizophora* mangrove forest
S9	intertidal organic mud	low, closed *Ceriops* forest
S10	low tide mud	sample collected 100m. seaward of *Rhizophora* mangrove forest

ous soil horizon. Elevations of boreholes and of modern facies environments in Missionary Bay were established relative to a tide prediction datum (-1.55 metres A.H.D.) by the Australian Survey Office.

Laboratory

Analyses to determine organic content, particle size and mineralogy (by X.R.D. analysis) of sediments were performed at the Australian Institute of Marine Science following standard techniques. Radiocarbon determinations were performed on organic samples at the Beta Analytical Laboratory in Florida.

All samples analysed for pollen were prepared by standard chemical procedures (see Erdtman, 1969; Faegri and Iversen, 1966) to remove extraneous organic and inorganic material, and to render the morphology of the pollen grains clearly visible under the microscope. Prepared samples were mounted in silicon oil on microscope slides. Pollen counts were undertaken on a light microscope at X600 magnification. With one exception counts for each sample continued until a pollen sum of at least 150 angiosperm and gymnosperm pollen grains had been achieved. Due to a sparseness of pollen, surface sample S10 was counted to 100 grains in the pollen sum only. Identification of pollen was facilitated by comparison with reference material held in the Monash University Geography collection. The majority of relevant mangrove pollen types had been sampled from flowering material of determined plant species held in the herbarium of the Australian Institute of Marine Science. Identification to genus level was possible for most mangrove plants and woody taxa. The major exceptions are *Ceriops/Bruguiera* and *Melaleuca/Eucalyptus*. The pollen of herbaceous taxa and many dry land plants could only be identified to family level.

Facies Types

Figure 4 summarises stratigraphic relationships between major facies units recorded in Missionary Bay. Sedimentological descriptions of estuarine and subtidal facies which relate to both the transgressive and regressive history of the bay (units 1 to 5 in Figure 4) are given below.

1. *Brackish swamp facies.* This facies is a sandy, highly organic (40-60%) deposit on which marginal forests are developed. As described earlier, such forests are transitional between the upland communities and mangroves. They display a restricted and well-defined topographic relationship to tidal inundation in the range 3.60 to 3.85 metres above prediction datum. The sand fraction of this facies is coarse, and rich in both quartz and feldspar. Dominated by material derived from hillslope deposits, the sediments are poorly sorted and angular. This facies pinches out against upland talus deposits, fluvial gravels and in the case of the eastern margin of the swamp, against aeolian sands of the Ramsay Bay barrier. The brackish swamp facies generally overlie upland deposits except where they have regressed over intertidal organic muds. In both cases, they grade laterally into intertidal organic muds in the seaward direction. Thickness rarely exceeds 0.75-1.0 metre, and the horizontal extent of the facies varies from a few metres to several hundred metres where it extends up stream valleys on the southern margin of the bay.

2. *Intertidal organic muds.* This facies is the substrate on which mangroves are growing in Missionary Bay. Other authors (Thom, 1975, and Coleman *et al.*, 1970) use the term mangrove peat and mangrove mud for similar deposits. However, in north-eastern Australia, such deposits lack sufficient organic material to be called peats and the application of a term such as organic mud is more appropriate. Intertidal organic muds are deposited within the upper portion of the tidal spectrum

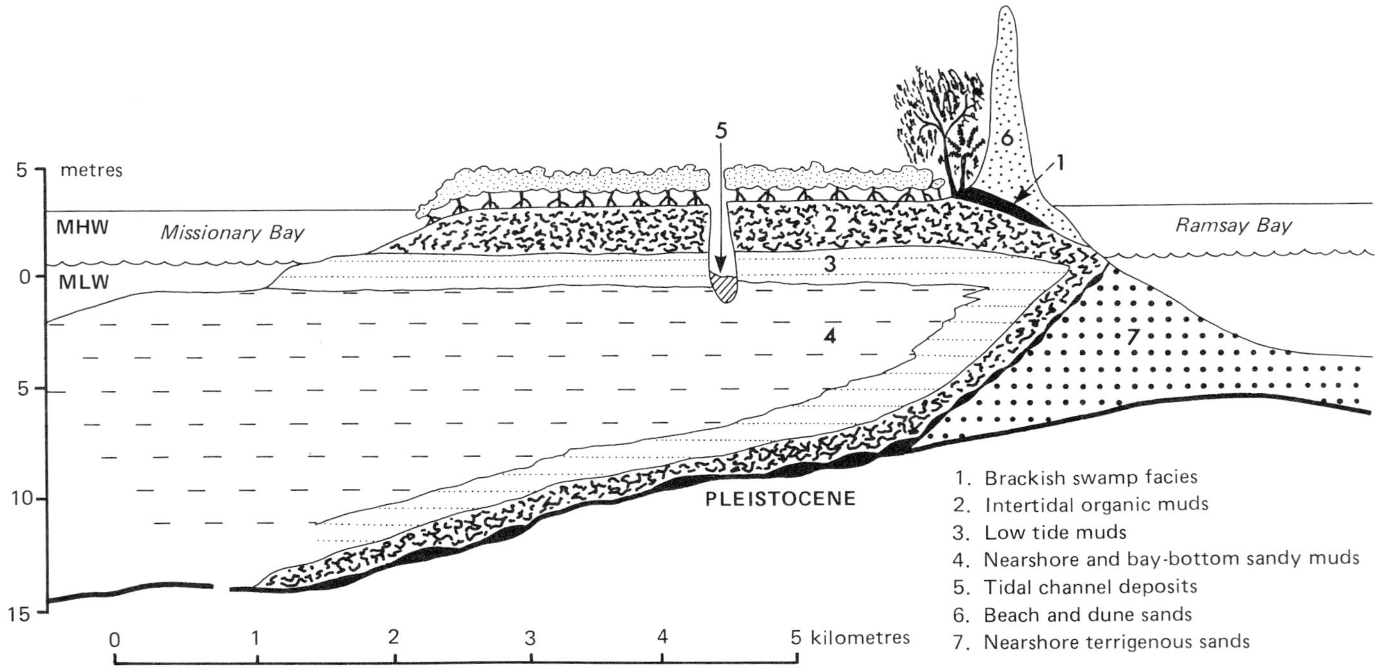

FIGURE 4. Generalized stratigraphic section from Ramsay Bay to Missionary Bay showing occurrence of various facies types.

generally within 1.5 to 3.0 metres above prediction datum. They overlie low-tide muds and are laterally continuous for up to 10 kilometres. Containing up to 10% organic carbon, most of which is autochthonous, these muds are a mixture of organic detritus and fine silts and clays. Shell detritus is nearly absent in the organic muds. There is a diverse molluscan fauna, but observations of living specimens show that extensive etching and degradation proceeds rapidly. Skeletal material is dissolved soon after death of the organism. Although minor amounts of fine quartz, calcite and feldspar are present in the XRD analysis, the clay fraction is largely composed of kaolin, illite and montmorillonite with minor amounts of chlorite.

Autochthonous carbon is provided mainly from the mangrove forest growing on the organic muds. Detritus is derived from leaf litter, branches, twigs, roots and fallen trees. Although Wolanski *et al.*, (1980) have shown an active exchange of swamp waters with the adjacent bay, especially during spring tides, it appears that much organic material remains *in situ*. Results reported by Boto and Bunt (1981) suggest that measured export rates from the tidal forest only account for 45%-70% of estimated primary production. Although input of allochthonous carbon into the intertidal organic muds is probably slight, it may in some cases be sufficient to reduce the reliability of radiocarbon dating of this facies (see below).

3. *Low-tide muds*. This unit occurs from slightly below prediction datum to 1.6 metres above prediction datum. It grades laterally into the nearshore mud facies with the facies change sometimes indistinct in terms of either texture or mineralogy. The low-tide mud facies is brownish-black in colour, and is composed of clay and silt with a variable calcareous sand content. Median particle size values average 7.3 phi, although deposits forming shoals or middle grounds have median values in the fine sand range due to their high content of shell fragments.

Calcium carbonate and organic carbon values show an inverse relationship in the low-tide muds. Calcium carbonate levels are generally low, averaging 5.3% although ranging as high as 48% on middle grounds. Organic carbon values average 3%. Carbon values probably undergo significant annual variation due to algal growth and seagrass communities. Kaolin and illite are dominant clay minerals with only minor montmorillonite. X-ray diffractometry shows that both quartz and feldspar are present in the clay fraction.

Low-tide mudflats are heavily colonized by crustacean which maintain a high degree of bioturbation in the upper 1 to 1.5

metres of this facies. The result is a homogeneous structure, broken only by concentrations of finely comminuted molluscan fragments near crab burrows.

4. *Nearshore and bay-bottom sandy muds.* This is a subtidal facies extending over several square kilometres of sea floor in Missionary Bay. It is deposited in water depths from prediction datum to at least -12 metres. The facies may be up to 15 metres thick and comprises most of the Holocene sedimentation in the bay. Greyish olive to olive grey in colour, nearshore and bay-bottom sandy muds are relatively low in organic carbon (maximum average 1.4%), and contain moderate to high amounts of $CaCO_3$ plus a significant fine terrigenous sand component. The latter varies from 5 to 40% and consists of a poorly sorted to moderately sorted quartz sand with a variable mean grain size, ranging from 1.0 to 3.5 phi. Some fine quartz particles extend into the clay size fraction as shown by its nearly ubiquitous occurrence on XRD traces. Kaolin and illite are dominant, with minor or accessory montmorillinite and trace occurrence of chlorite.

5. *Tidal channel deposits.* Three types of channel deposits are associated with tidal channels within Missionary Bay. These are:

- Type A - terrigenous sand
- Type B - organic lag
- Type C - shell lag

Type A deposits are present in those tidal creeks which are extensions of perennial creeks draining the upland surface. The channels of these creeks, although incised into a finer-grained facies such as low-tide muds or intertidal organic muds, have a carpet of well-sorted sand 1.0 to 3.0 metres thick in their thalwegs. Ebb-oriented mega-ripples, 0.2 to 0.4 metres in amplitude and 0.5 to 1.5 metres in length, are common on the intertidal portion of the channel. In deeper, sub-tidal sections, sand waves up to 1.0 metre in amplitude and 5 to 8 metres in length are visible on fathometer traces. Although these channel sands are coarse, they are finely skewed or nearly symmetrical with a mesokurtic distribution. Channel sands contain little or no calcium carbonate and are dominantly angular to sub-angular quartz and feldspar deposits with minor rock fragments. Mean grain size lies in the range 0.10 to 0.60 phi with median values averaging 0.3 phi. Grainsize decreases toward the centres of the estuary due to dilution of these coarse sands by fine estuarine sands and silts mixed with fine carbonate.

Type B channel deposits are unique to tidal channels with little or no input from the upland surface. These deposits consist of organic muds 0.5 to 1.25 metres thick which blanket the channel bottom. They may be continuous over several hundred metres in length and extend across channels which vary from 10 to 100 metres in width. Dark brown to black in colour, these deposits are highly organic averaging 4.5% carbon. They contain up to 10% finely comminuted shell (average 6 to 7%) with the remainder of the sediment being fine quartz silt and various clays.

Type C deposits are greyish olive shelly muds which occur both within the swamp and beneath the open bay. Ranging up to 60% $CaCO_3$, and averaging 42%, these deposits are assemblages of broken shell, mostly estuarine molluscan. Several hundred metres long and over 100 metres in width, these deposits are approximately 2.5 metres in thickness and remain subtidal even on spring-tide lows, except at the mouths of some tidal creeks. Despite the high calcium carbonate content, 30-40% of most samples consist of clay. This clay fraction is dominantly kaolin and illite with minor occurrences of chlorite. The presence of large quantities of broken shell raises the median grain size to approximately 4.5 phi suggesting that the facies consists of fine silt. However, this facies is bi-modal with a very coarse to coarse carbonate sand fraction and a fine terrigenous silt and clay fraction.

Point-bar muds. This facies type is not indicated in Figure 4, but it is also recorded in Missionary Bay estuarine environments. Point-bar muds are common in tidal environments with moderate to high rates of sedimentation. There is apparent modification to tidal channels within the Missionary Bay mangrove swamp through a sequence of meander cut-bank erosion and point-bar deposition. The intertidal portion of point bars in Missionary Bay is comprised of homogeneous brown to olive black mud with less than 10% calcium carbonate and over 5% organic carbon. These muds are fine-grained with a median value averaging 7.6 phi. Kaolin and illite are dominant clay minerals with accessory occurrence of montmorillonite. Observations at several sites showed that the muds become increasingly sandy down the ramp of the point bar.

Pollen Analysis

Pollen Content of Estuarine Facies

The results of all pollen analyses are shown in Figure 5a where values for individual taxa are shown as percentages of the pollen sum. These data are summarized in Figure 5b. The surface samples, ordered along a gradient from brackish swamp

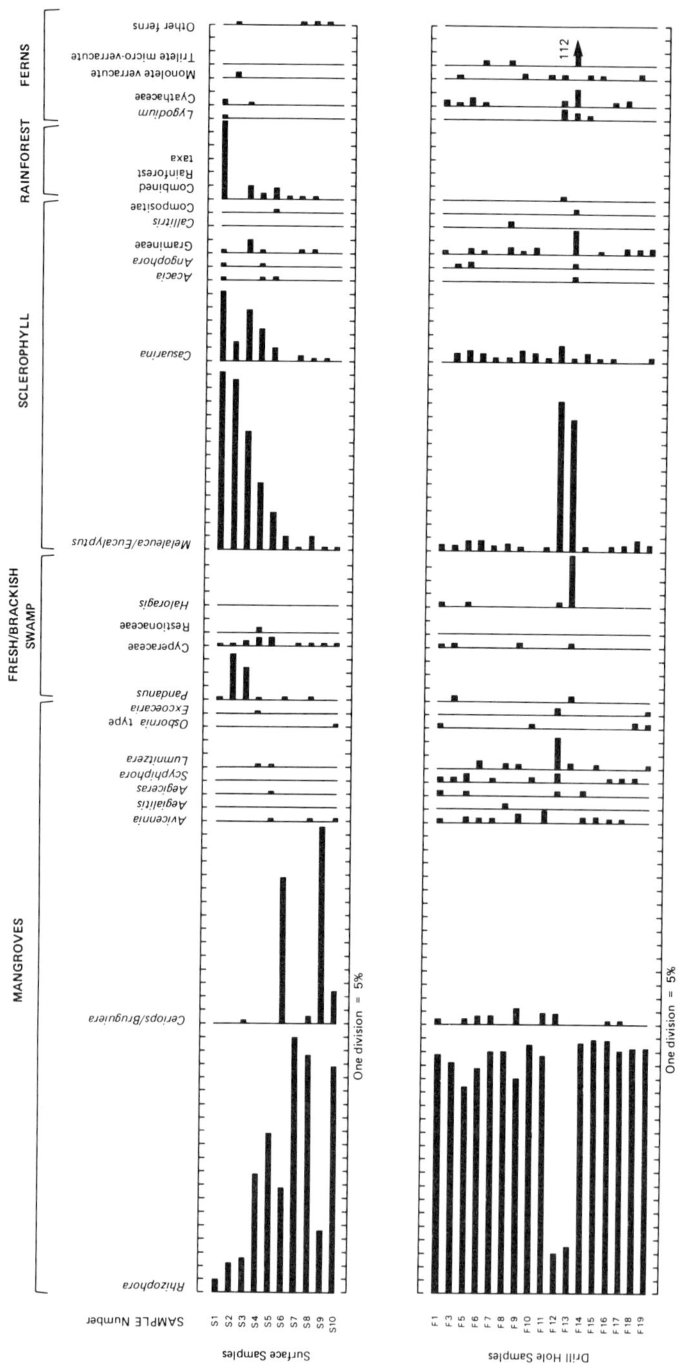

FIGURE 5a. Pollen diagram for surface (S) and drill hole (F) samples.

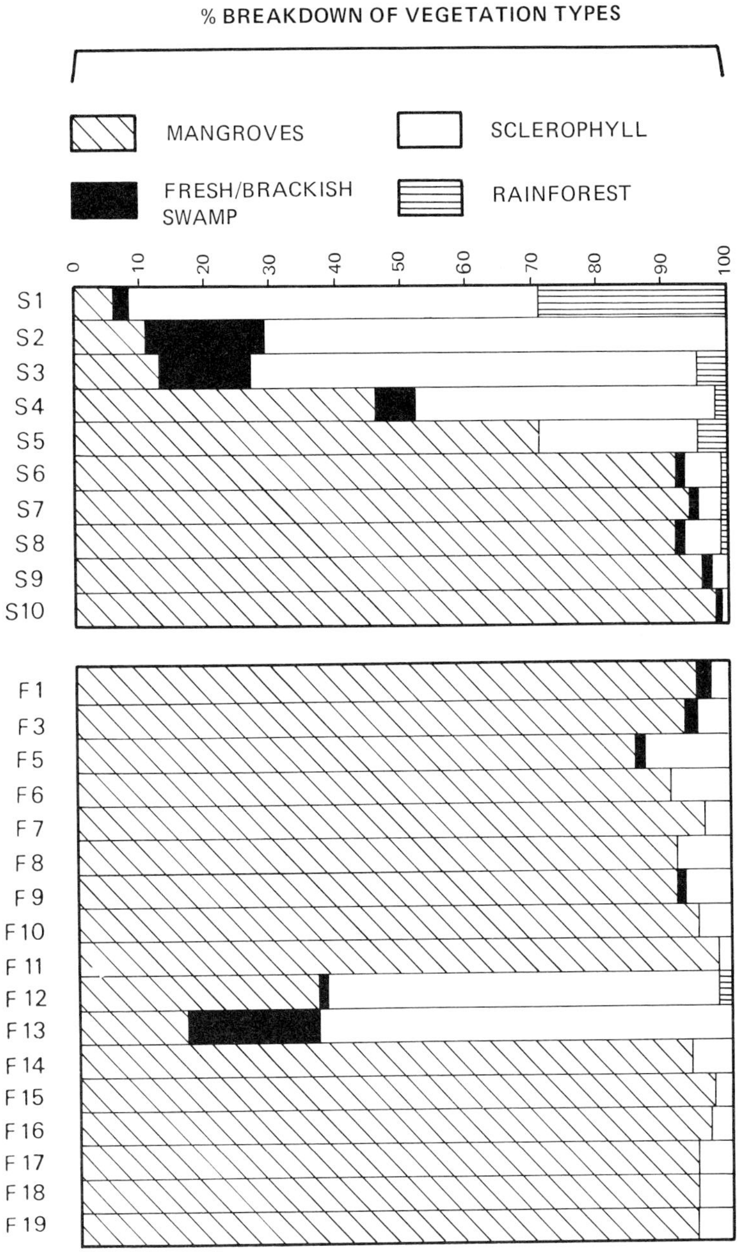

FIGURE 5b. Summary diagram of vegetation types based on pollen data shown in Figure 5a.

through mangrove swamp to low tide muds, show: (i) a progressive increase in Rhizophoraceae (Rhizophora and Ceriops/Bruguiera), and (ii) a decrease in fresh/brackish swamp and sclerophyll taxa. All samples from closed canopy Rhizophoraceae - dominated mangrove forests (samples S6 to S9) contain at least 90 percent Rhizophoraceae pollen. In general, these samples display relatively low mangrove pollen diversity when compared to the number of plant species recorded in the parent vegetation throughout the bay. This is not surprising when the overriding dominance in terms of percentage occurrence of Rhizophoraceous taxa at the site is considered, and with regard to other mangrove palynological studies (e.g. Caratini *et al.*, 1973; Grindrod, 1983). The low-tide mud bank sample (S10) is dominated by *Rhizophora* pollen; it contains very little else. Rainforest taxa are poorly represented except in sample S1, which as stated earlier was collected beneath mesophyll vine forest.

Fossil pollen assemblages, labelled F, are ordered in relation to depth below present sea level. All except F12 and F13 are dominated by mangrove pollen particularly *Rhizophora* which ranges from 79 to 96 percent of the pollen sum. A larger range of other mangrove taxa is recorded than in the surface samples though some, including *Ceriops/Bruguiera*, are common to both modern and fossil assemblages.

Samples F12 and F13 are dominated by *Melaleuca/Eucalyptus* pollen but differ in their minor taxa assemblages. Sample F13 has very low mangrove values but shows strong representation of Gramineae and *Haloragis* and extremely high pteridophyte percentages. Sample F12 contains a diverse mangrove assemblage within which *Lumnitzera*, *Scyphiphora* and *Excoecaria* in addition to *Rhizophora* are prominent.

Comparison of the surface and fossil pollen data shows that the fossil samples contain a greater variety of mangrove taxa and higher proportions of pteridophytes. The surface samples generally contain higher percentages of *Pandanus* and *Casuarina* and a greater rainforest component. *Ceriops/Bruguiera* which dominates in two surface samples never achieves high representation in the fossil samples.

Interpretation of Fossil Pollen Assemblages

Despite some obvious differences in the range of variation recorded in the two data sets, there are sufficient similarities for the surface samples to be used as a basis for interpretation of the conditions under which the fossil samples were deposited.

Surface pollen data indicate that samples taken immediately landward of the Rhizophoraceae - dominated tidal forest (samples S4 and S5) show relatively low representation of Rhizophoraceae pollen (44 and 59 percent respectively), while samples from within the forest consistently give values above 90%. Fifteen fossil samples have a *Rhizophora* pollen content in the range 79 to 96% indicating that these samples represent sites within or peripheral to the *Rhizophora*-dominated zone. The major terrestrial pollen types represented in these samples - *Melaleuca/Eucalyptus*, *Casuarina* and Gramineae - are known to be efficiently dispersed by wind (Grindrod, 1983), and may therefore be expected to show at least minor representation in intertidal and offshore situations. The relatively high representation of pteridophyte spores (particularly Cyatheaceae) is less easily explained, but probably represents both wind-born and stream-born components from wetter terrestrial sites achieving relatively even dispersal, albeit in low concentrations, before finally settling in bay sediments.

Sample F12 is taken to represent a site landward of the *Rhizophora*-dominated mangrove forest and in near proximity to the terrestrial forest/mangrove ecotone. This is suggested by the diverse mangrove pollen flora, which includes upper intertidal mangrove taxa such as *Scyphiphora, Lumnitzera* and *Excoecaria*. Relatively low representation of Rhizophoraceae pollen types compared to *Casuarina* and *Melaleuca/Eucalyptus* also suggest a landward orientation for this sample.

The pollen assemblage for sample F13 suggests a fresh or brackish swamp habitat in near proximity to the mangrove forest. This interpretation is based on the relatively high representation of *Haloragis* whose pollen, like that of other low growing swamp taxa, is likely to display highly localized distribution (Kershaw 1979, p. 260). The very strong representation of pteridophyte taxa also supports this interpretation.

Holocene Stratigraphy and Sea-Level History

Transgressive Sequence

The stratigraphic summary in Figure 4 of Missionary Bay is based on data from the drill holes located in Figure 3, and on sedimentological and palynological descriptions of sediment samples. It details the transgressive and regressive history at the site. The transgressive sequence includes brackish swamp facies overlain by intertidal organic muds, and then by low-tide muds which were deposited within the bay during the P.M.T. Bay-bottom sandy muds have accounted for much of the sedimentary infill subsequent to sea level attaining approximately its present level, although it is possible for this facies to have been deposited in deeper water during the transgression.

The time-depth plot presented in Figure 6 is based on radiocarbon determinations for 19 organic facies samples from transgressive sediments in Missionary Bay. Radiocarbon results for all samples are given in Table 2. The pollen content and inferred origin of 17 of these samples is described earlier. No pollen analysis was performed on samples F2 and F4, and these are inferred to be of mangrove origin from their macrofossil components. These samples carry a larger vertical error term than those determined to be of specific mangrove (i.e. *Rhizophora* or non-*Rhizophora*) or brackish swamp origin (see Fig. 2).

TABLE 2. ^{14}C results and environments of deposition for transgressive samples from Missionary Bay. All dates are on decomposed plant organics.

Sample No.	Drill Hole No.	Sample Depth*	^{14}C Lab. No Beta. Anal.	^{14}C Age	Environment
F1	MB9	0.50	B1476	7405 ± 110	*Rhizophora* mangrove
F2	RB12	1.75	B1196	6275 ± 90	mangrove (no pollen)
F3	MB6	2.50	B1472	7860 ± 120	*Rhizophora* mangrove
F4	MB6	2.70	B1473	6845 ± 85	mangrove (no pollen)
F5	MB1	3.50	B1184	7765 ± 110	*Rhizophora* mangrove
F6	MB4	3.90	B1468	7950 ± 155	*Rhizophora* mangrove
F7	MB8	4.75	B1475	8045 ± 110	*Rhizophora* mangrove
F8	MB2	6.45	B1185	7855 ± 110	*Rhizophora* mangrove
F9	MB16	7.30	B1479	7760 ± 115	*Rhizophora* mangrove
F10	MB7	7.35	B1474	7345 ± 525	*Rhizophora* mangrove
F11	MB5	8.10	B1469	8530 ± 120	*Rhizophora* mangrove
F12	RB1	8.10	B1189	8555 ± 135	mixed mangrove
F13	RB7	8.60	B1194	8590 ± 150	brackish swamp
F14	MB15	9.85	B1478	8415 ± 120	*Rhizophora* mangrove
F15	MB3	9.90	B1464	8155 ± 200	*Rhizophora* mangrove
F16	MB3	12.0	B1465	8045 ± 85	*Rhizophora* mangrove
F17	MB3	12.65	B1466	8935 ± 145	*Rhizophora* mangrove
F18	MB3	12.95	B1467	7815 ± 165	*Rhizophora* mangrove
F19	MB11	15.80	B1457	8300 ± 125	*Rhizophora* mangrove

* mean depth of sample below prediction datum (m)

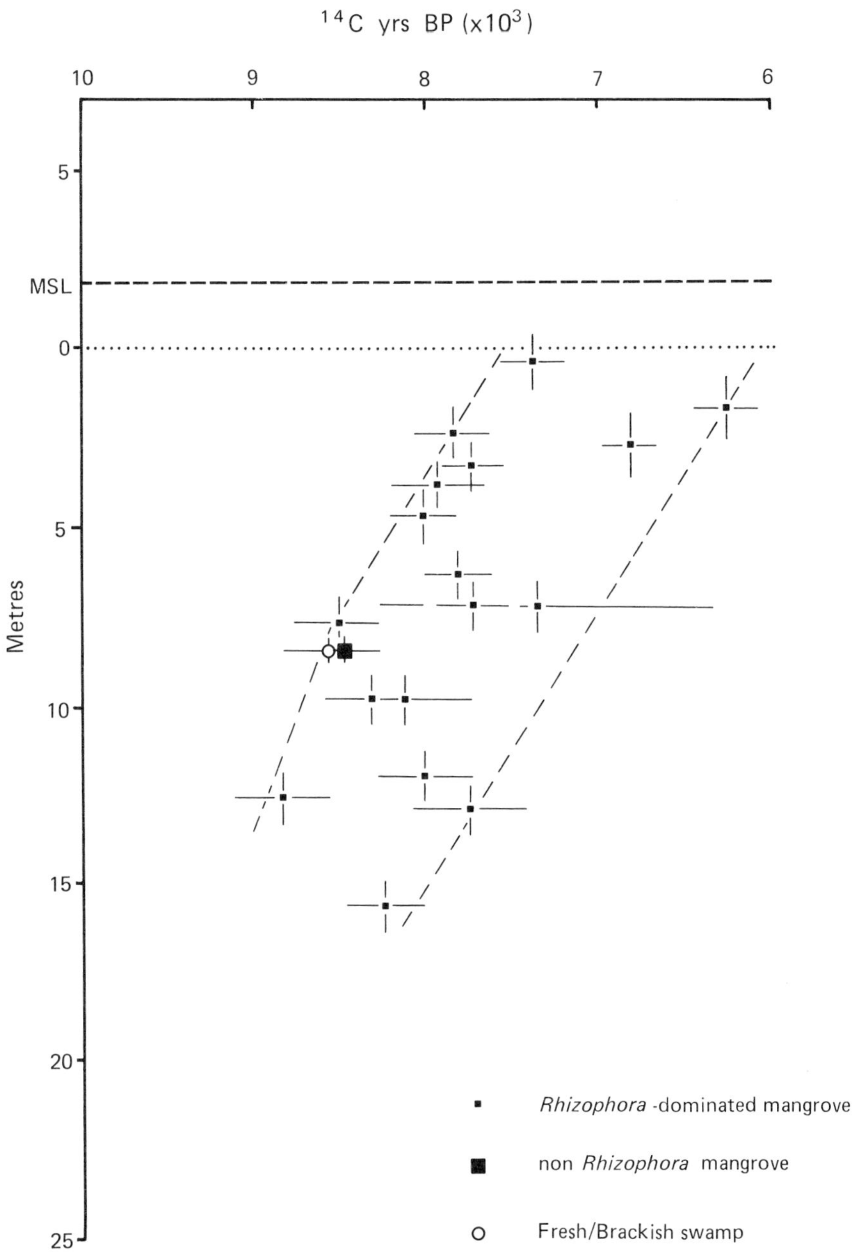

FIGURE 6. Time/depth plot of radiocarbon dates showing zone occupied by relative sea level during the Postglacial Marine Transgression.

The record of relative sea level shown in Figure 6 suggests a rapid rise from approximately 15 metres lower than present at about 8500 ^{14}C yrs BP to a point slightly lower than present between 6200 and 7500 ^{14}C yrs BP. The broad scatter within the observation set restricts precision in the reconstruction of relative sea level.

Regressive Sequence

The upper, regressive sequence in Figure 4 shows the reverse order of the transgressive sequence. Continued infilling of the bay has provided suitably shallow conditions for the seaward encroachment of intertidal organic facies over low-tide environments. Modern brackish water peats are deposited above the level of mangrove habitation around the landward margins of the bay.

Table 3 provides descriptions and radiocarbon ages for samples, labelled R, collected from regressive sediments in Missionary Bay (see Fig. 3 for position of collection sites). These data are also presented in the time-depth plot in Figure 7. A general increase in radiocarbon age with increasing depth of sample is indicated. However, the wide scatter of points within the data set suggests inconsistencies in dating the regressive sequence. For instance, organic material selected at or near the surface of the modern swamp (samples R7, R13 and R17) return ages between 2000 and 3000 ^{14}C yrs, while organic and shell samples from the same sampling point (R4 and R5) return disparate ages. The observations indicate that problems exist in the accurate dating of stratigraphic horizons in mangrove and related sedimentary systems.

Discussion

As mangrove and brackish swamp environments maintain a consistent and easily definable relationship to tidal datum, their *in situ* fossil legacies provide useful indicators of former sea level. Pollen analysis of organic sediments collected from modern environments and by drilling at Missionary Bay has enabled the confident reconstruction of vegetation types in depositional environments relative to the transgressive shoreline. In particular, the technique has enabled the distinction between non-*Rhizophora* and *Rhizophora*-dominated mangrove facies, as well as the confident recognition of fresh or brackish swamp facies. In transgressive sequences brackish swamp facies may not be distinguishable from mangrove facies by

TABLE 3. Sample types, ^{14}C results and environments of deposition for regressive samples from Missionary Bay. All shell samples have been corrected for environmental effect.

Sample No.	Sample Site	Sample Depth*	^{14}C Lab. No (Beta. Anal.)	^{14}C Age	Sample Type	Environment
R1	MB3	−0.8	B1461	4710 ± 105	shell hash > 1 cm	estuarine clay
R2	MB3	−0.8	B1462	4605 ± 90	shell hash < 1 cm	estuarine clay
R3	WALK A	−0.4	B1900	4600 ± 95	shell	low-tide muds **
R4	WALK B	+0.5	B2025	2935 ± 115	organic material	base of intertidal organic muds
R5	WALK B	+0.5	B2026	5070 ± 190	broken shell	base of intertidal organic muds
R6	WALK B	−0.3	B2027	5240 ± 190	shell hash	low-tide muds **
R7	WALK C	+3.0	B2028	2860 ± 90	organic material	top of intertidal organic muds
R8	WALK C	+2.1	B2029	3500 ± 125	organic material	intertidal organic muds
R9	WALK C	+0.5	B2030	3830 ± 100	broken shell	low-tide muds **
R10	WALK D	+1.75	B1898	690 ± 70	organic detritus	intertidal organic muds
R11	WALK D	+0.1	B1988	4370 ± 110	broken shell	low-tide muds
R12	DOUB A	+1.3	B2014	4010 ± 180	broken shell	intertidal organic muds
R13	DOUB A	+3.0	B2017	2990 ± 210	organic material	intertidal organic muds
R14	DOUB A	+2.8	B2018	2410 ± 110	organic material	intertidal organic muds
R15	DOUB B	+0.5	B2019	4390 ± 115	shell hash	low-tide muds **
R16	DEEP B	+0.5	B2021	3950 ± 125	shell hash	low-tide muds **
R17	BROD A	+3.0	B2022	2100 ± 170	organic material	intertidal organic muds
R18	BROD A	+1.7	B2023	4160 ± 120	organic material	intertidal organic muds
R19	BROD A	+0.6	B2024	4080 ± 90	shell hash	low-tide muds **

* vertical relationship to prediction datum (m) ** beneath intertidal organic muds

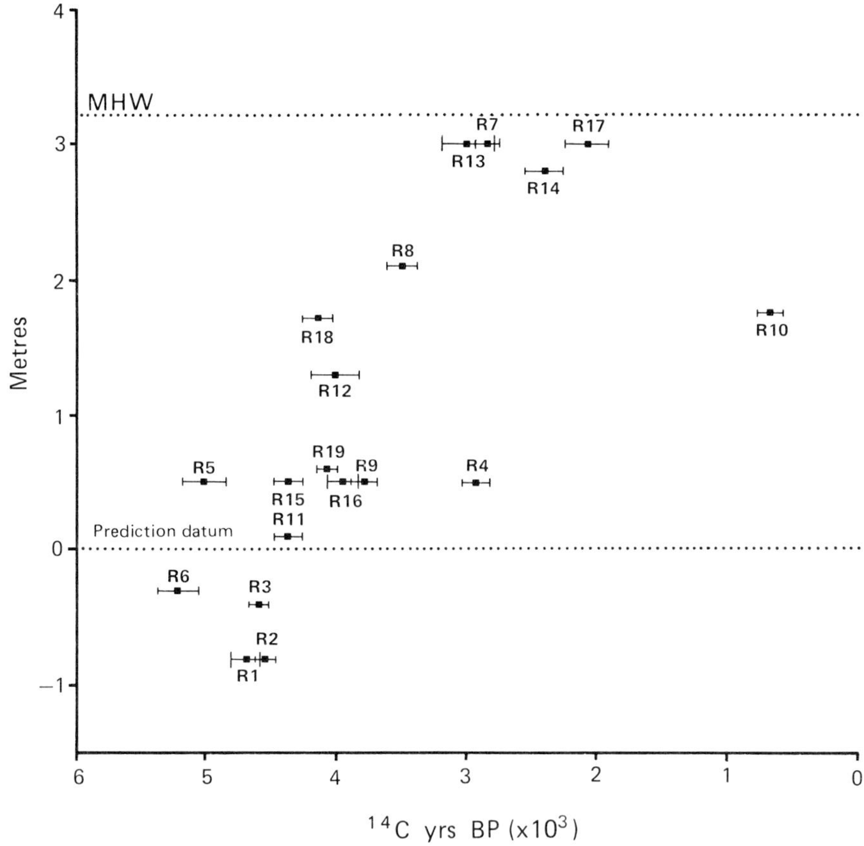

FIGURE 7. Plot of radiocarbon dates on samples which have accumulated since sea level reached approximately its present position (see text for explanation).

macrofossil analysis alone, as they are likely to contain mangrove root material from subsequent stages in plant succession.

The time-depth plot for regressive samples indicates potential problems for the accurate dating of given periods of sedimentation in mangrove and related environments. These problems, which are discussed by Woodroffe (1981), may arise from one or a combination of the following effects:

1. Sediment mixing and intrusive carbon. Agents of bioturbation in mangrove sediments include burrowing organisms such as crabs, worms and clams. Disused burrows subsequently fill with tidally distributed sediments (Seminuik *et al.*, 1978, p. 51). Intrusive carbon may also arise through

penetration into older sediment layers by plant root material. Disparate ages for shell and plant organics from equivalent sedimentary horizons within Missionary Bay are probably the result of this process.

Allochthonous carbon. Although most carbon within intertidal organic facies is known to be autochthonous, some allochthonous material especially fine particulate organics, is probably introduced from elsewhere in mangrove environments and from terrestrial and estuarine sources. This may account for anomalously old ages for organic material presently being deposited in the mangrove swamp.

When attempting to reconstruct former sea levels from tertidal organic facies the processes outlined above provide tential error sources. Further uncertainty may be introduced rough:

Reworking of sediments. This may occur by wave and tide action, particularly on transgressive shores. However, at Missionary Bay the presence in drill hole sequences of mangrove or brackish swamp facies directly overlying the Pleistocene surface suggests that redeposition on a large scale has not occurred. Detailed analyses of transgressional sediment sequences are required before a reliable picture of conditions prevailing in the sedimentary prism along transgressional shores can be developed.

Vertical range of sea-level indicators. Features most useful as palaeosea-level indicators should occupy a narrow vertical range. In the present study the degree of precision in the use of organic deposits in this regard is enhanced by recognition of different facies types in intertidal and associated organic deposits. Of those organic sediments which were recovered, the brackish swamp facies has the narrowest vertical range relative to tidal datum, but was detected only once in drill hole samples. However, *Rhizophora*-dominated mangrove facies which displays a relatively broad relationship to tidal datum, is ubiquitous. Reliance on the later facies type therefore decreases the certainty with which former sea levels can be pinpointed. It is envisaged that through pollen and other microfossil analysis of continuous core sequences from intertidal environments more precise data relating to the palaeo-sea levels will emerge.

Variation in sea level and tidal range. The spread in sea level observations may also relate to actual variations in sea level or tidal range. At present there are no data relating to palaeotidal range at Missionary Bay, but work along these lines may prove invaluable to site specific reconstructions of relative sea level. Recent work suggests that post transgressional sea levels of the mid to late Holocene relative to the mainland coast and inner islands of north Queensland have fallen smoothly from a position approximately one metre higher than present about 6000 ^{14}C yrs BP (Chappell, 1983; Chappell *et al.*, 1983). However, the possibility that short-term oscillations in sea level occurred during the Holocene transgressive phase remains, and this problem will only be resolved through continued studies of this period.

The ^{14}C chronology for transgressive environments is based on mangrove organic samples, while the regressive chronology derives from mangrove organic and shell hash samples. Problems inherent in relating the ages of samples such as these to the time of deposition of a sedimentary unit are outlined above and have led to a wide scatter in the observation sets. A ^{14}C chronology constructed for regressive sediments at Princess Charlotte Bay in North Queensland (Chappell and Grrindrod, this volume) displays a much higher degree of internal consistency. This data set is based primarily on the radiocarbon analysis of marine shell samples. Dated samples for mangrove and lower-intertidal muds at Princess Charlotte Bay consisted of articulated shells which were interpreted as remaining *in situ* since the time of death of the organism. Such careful selection of dateable material was possible as samples were collected from stratigraphic trenches dug by backhoe. In the drill hole samples from Missionary Bay positive detection of *in situ* shell was generally not possible. Comparison of the ^{14}C results from the two sites indicates the value of selecting *in situ* shell rather than shell which is possibly reworked or organic material for the dating of mangrove sedimentary units and, in particular, the reconstruction of relative sea level.

Comparison with other sea-level studies

Figure 8 shows the relationship between the suggested envelope of relative sea level from Missionary Bay and results of the composite reconstructions of transgressional sea levels in eastern Australia by Belperio (1979) and Thom and Roy (1983) The observations by Belperio relate to radiocarbon dated mangrove sediments collected at three sites at Townsville and Hinchinbrook Island. It can be seen in Figure 8 that each observation falls within the suggested sea-level envelope constructed from the Missionary Bay data, indicating that a relatively reliable though imprecise record of Holocene transgressive sea levels is emerging for this region. The envelope of relative sea level proposed by Thom and Roy for southeastern Australia is based on a variety of primary data types and covers the period approximately 12,000 to 6300 ^{14}C yrs BP.

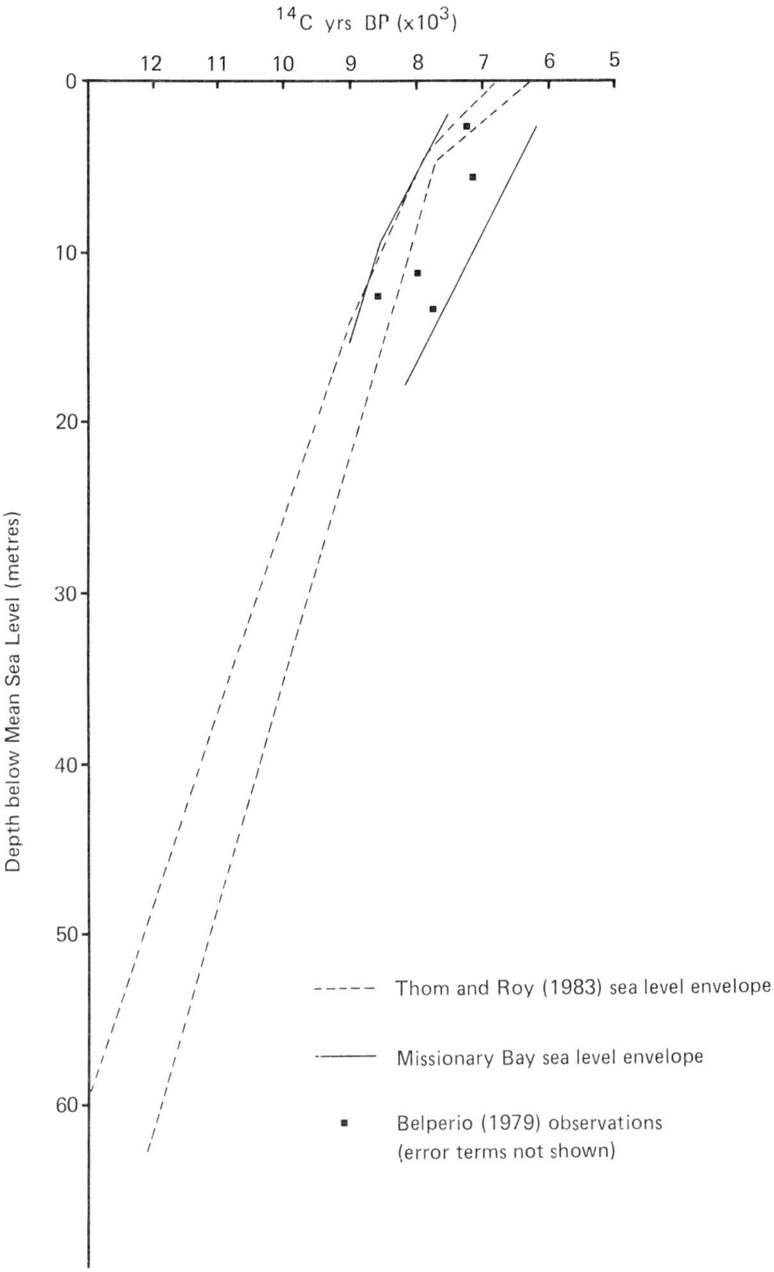

FIGURE 8. Comparison of relative sea levels during the Postglacial Marine Transgression for different sites in eastern Australia.

The upper portion of this record falls within the sea-level envelope from Missionary Bay data, indicating that at least general agreement in Holocene transgressional sea-level reconstruction in eastern Australia may be reached on a continental scale.

The record of relative sea level from Missionary Bay is one of the few site-specific studies of the Holocene transgressive period in Australia. Although mangrove and related environments provide convincing sedimentary evidence of palaeo-sea levels, precision in sea-level reconstruction is marred by numerous error sources. It is envisaged that greater accuracy will be achieved by detailed microfossil analysis of continuous core sequences from intertidal environments of the transgressive period (see Grindrod, 1983). If such histories can be provided for transgressive sequences, precision in the reconstruction of sea level from intertidal organic sediments will be greatly enhanced.

Acknowledgements

The project reported here was made possible by funding and logistic support from the Australian Institute of Marine Science. A number of individuals deserve thanks for their generous assistance during the course of this work. Ken Pye and Peter Kershaw assisted in field work and maintained keen interest throughout the study. Bruce Thom, John Chappell and Peter Kershaw entered into valued discussion and provided critical comment on the text. Norman Duke furnished information regarding mangrove ecology and permitted sampling for reference pollen from specimens held in the Australian Institute of Marine Science herbarium. To these people the authors extend their sincere thanks.

References

Belperio, A.P. (1979). Negative evidence for Mid-Holocene high sea level along the coastal plain of the Great Barrier Reef Province. *Marine Geology*, 32, M1-M9.

Boto, K.G. and Bunt, J.S. (1981). Tidal export of particulate organic matter from a Northern Australian mangrove system. *Estuarine, Coastal and Shelf Science* 13, 247-255.

Bunt, J.S. (1982). Studies of mangrove litter fall in tropical Australia. In B.F. Clough (ed.) *Mangrove Ecosystems in*

Australia: Structure, function and management, A.N.U. Press, 223-238.

Bunt, J.S., Boto, K.G. and Boto, G. (1979). A survey method for estimating potential levels of mangrove forest primary production. *Marine Biology*, 52, 123-128.

Bunt, J.S., Williams, W.T. and Duke, N.C. (1982). Mangrove distributions in north-east Australia. *Journal of Biogeography*, 9, 111-120.

Caratini, C., Blasco, F. and Thanikaimoni, G. (1973). Relation between the pollen spectra and the vegetation of a south India mangrove. *Pollen et Spores*, Vol. XV, No. 2, 281-292.

Chappell, J. (1983). Evidence for smoothly falling sea-level relative to north Queensland, Australia, during the past 6,000 years. *Nature*, 302, 406-408.

Chappell, J., Chivas, A., Wallensky, E., Polach, H. and Aharon, P. (1983). Holocene palaeo-environmental changes, central to north Great Barrier Reef inner zone. *BMR Journal of Australian Geology & Geophysics*, 8, 223-235.

Coleman, J.M., Gagliano, S.M. and Smith, W.G. (1970). Sedimentation in a Malaysian high tide tropical delta. In J.P. Morgan (ed.) *Deltaic Sedimentation: Modern and Ancient*. Society of Economic Paleontologists and Mineralogists Special Pub., No. 15, pp. 185-197.

Duke, N. and Bunt, J.S. (1979). The genus *Rhizophora* (Rhizophoraceae) in north-eastern Australia. *Aust. J. Bot.*, 27, 657-678.

Erdtman, G. (1969). *Handbook of Palynology. An Introduction to the Study of Pollen Grains and Spores*. Munksgaard.

Faegri, K. and Iversen, J. (1966). *Textbook of Pollen Analysis*. Second revised edition, Munksgaard, Copenhagen.

Grindrod, J. (1983). *Holocene History of Mangroves on a Prograding Tropical Coast*. M. Sc. Thesis, Australian National University, Canberra.

Hopley, D. (1983). Evidence of 15,000 years of sea level change in tropical Queensland. In D. Hopley (ed.) *Australian Sea Levels in the Last 15,000 Years. A Review*. Monograph Series, Geog. Dept., James Cook University, Occasional Paper 3, 93-104.

Kershaw, A.P. (1979). Local pollen deposition in aquatic sediments on the Atherton Tableland, North-Eastern Australia. *Australian Journal of Ecology* 4, 253-263.

Pye, K. and Rhodes, E.G. (in prep.). Holocene development of an episodic transgressive dune barrier, Ramsay Bay, North Queensland, Australia.

Semeniuk, V., Kenneally, K.F. and Wilson, P.G. (1978). *Mangroves of Western Australia* Western Australian Naturalists Club, Handbook No. 12, Perth.

Thom, B.G. (1975). Mangrove ecology from a geomorphic viewpoint. In G.E. Walsh (ed.) *Int. Symp. on Biology and Management of Mangroves.* Uni. Florida Press, 469-481.

Thom, B.G. and Chappell, J. (1975). Holocene sea levels relative to Australia. *Search*, 6, No. 3, 90-93.

Thom, B.G. and Roy, P.S. (1983). Sea level change in New South Wales over the past 15,000 years. In D. Hopley (ed.) *Australian Sea Levels in the last 15,000 Years. A Review.* Monograph Series, Geog. Dept., James Cook University, Occasional Paper 3, 64-84.

Tomlinson, P.B., Bunt, J.S., Primack, R.B. and Duke, N.C. (1978). *Lumnitzera rosea* (Combretaceae) - its status and floral morphology. *Journal of the Arnold Arboretum*, 59, No. 4, 342-351.

Tracey, J.G. and Webb, L.J. (1978). The vegetation of Hinchinbrook Island. *Queensland Naturalist*, 22, 1-4.

Wolanski, E., Jones, M. and Bunt, J.S. (1980). Hydrodynamics of a tidal creek-mangrove swamp system. *Australian Journal of Marine and Freshwater Research* 31, 431-450.

Woodroffe, C.D. (1981). Mangrove swamp stratigraphy and Holocene transgression, Grand Cayman Island, West Indies. *Marine Geology*, 41, 271-294.

9

The Holocene Marine Transgression as a Forcing Function in Episodic Dune Activity on the Eastern Australian Coast

K. PYE

Department of Earth Sciences
University of Cambridge, UK

G. M. BOWMAN

Division of Soils
CSIRO
Glen Osmond, SA

Introduction

Several recent Australian studies of coastal morphostratigraphy have suggested that dune formation during the Quaternary has been episodic, with periods of dune instability alternating with stabilization and weathering (Thom, 1978; Thom *et al.*, 1978, 1981b; Thompson, 1981). In other parts of the world, episodic dune development has been extensively documented (see Cooper, 1958, and references), although the ages of the unstable phases and their causes remain poorly understood and a matter of considerable debate. Some authors have correlated major episodes of dune instability during the Pleistocene with periods of glacial low sea level, with wide areas of sandy sediment exposed on continental shelves and generally windier conditions (Sayles, 1931; Crocker, 1946; Wright, 1963; Coetzee, 1975/6; Hobday, 1977; Roy and Thom, 1981). Others have concluded that periods of dune formation corresponded with interglacial high sea levels, while low sea level phases were associated with dune stabilization and

COASTAL GEOMORPHOLOGY
IN AUSTRALIA
ISBN 0 12 687880 3

weathering (Bretz, 1960; Battistini, 1964; Land *et al.*, 1967; Harmon *et al.*, 1981).

Similarly contrasting interpretations have been made concerning episodes of dune instability during the Holocene. In New Zealand, for instance, Schofield (1975) concluded that dune building episodes could be correlated with periods of minor sea-level regression and shoreline progradation, while minor transgressions caused breaks in aeolian accumulation. However, an opposite view, originally suggested by Cooper (1958) in the context of the Oregon coast, maintains that transgressive dune instability phases are associated with sea-level rise and shoreline erosion, whereas periods of relative emergence are characterized by the vegetative fixation of littoral sand close to the beach. According to this model, the period of rapid sea-level rise during the Postglacial Marine Transgression (P.M.T.) of the early Holocene should have been accompanied by major transgressive dune activity. Previously, however, no radiocarbon evidence has been available to substantiate this hypothesis. In this contribution we present radiocarbon dates from several dunefield locations in eastern Australia which lend support to the Cooper model.

Geologic Setting of Coastal Dunefields in Eastern Australia

The coastline of eastern Australia consists of a series of broad embayments and intervening bedrock headlands. Wider depositional plains occur where the continental shelf is broadest and flattest, and where Cenozoic fluvial sediment input has been greatest (Roy and Thom, 1981). Much of the continental shelf has probably experienced slow subsidence throughout Cenozoic time (Phipps, 1966; Marshall, 1979; Cook and Mayo, 1977). The present coastal configuration came into existence during the marine transgression which followed the last glaciation, when the lower reaches of river valleys and many broad embayments were drowned by the rising sea. Since that time, large quantities of mixed fluvial and quartzose marine sediments have accumulated, partially infilling embayments and estuaries (Bird and Hopley, 1969; Roy and Crawford, 1977). In contrast to the coasts of western and southern Australia, biogenic carbonate sediment provides only a minor contribution to most beaches and other Holocene deposits in Queensland and New South Wales (Bird, 1978).

Along the entire eastern coastline, dune formations are best developed in those embayments or headland/island situations where exposure to the prevailing high energy south or southeast winds and waves is maximized (Pye, 1983a). Many of the bedrock headlands and islands have acted as sediment 'traps', limiting the northerly transport of sediment from one 'compartment' to the next (Roy and Thom, 1981). Open ocean, high energy embayments in which the transgressive dunes occur are dominated by ocean swell and wind waves (Wright, 1976), although the former are considerably reduced in North Queensland due to the protective effect of the Great Barrier Reef. The high energy embayments generally have steeper offshore gradients than the sheltered bays (Roy *et al.*, 1980) and have been classified as 'dissipative beaches' (Wright *et al.*, 1979), in which large quantities of sand are stored in wide surf zones.

The morphology of transgressive dunefields varies considerably. In many New South Wales embayments, for example Newcastle Bight, both parabolic and long-walled transgressive dunes have migrated across beach-ridge plains, backbarrier swamps, and have climbed low bedrock hills (Thom *et al.*,1981b). In southern Queensland, several generations of transgressive dunes are superimposed and comprise much of the Cooloola-Noosa area, Fraser Island, Moreton Island, and North Stradbroke Island, extending to depths of more than 70 m below sea level (Laycock, 1978). In North Queensland, the largest dunefields occur near Cape Flattery, Temple Bay and Newcastle Bay and consist almost exclusively of parabolic and elongate parabolic dunes. These have migrated up to 20 km inland over low hills and alluvial deposits (Pye, 1980, 1982a, 1983b). This varying dunefield morphology reflects the interaction of such factors as coastal configuration and orientation, sand availability, and differences in wind regime and vegetation.

Radiocarbon Age Determinations

Seven radiocarbon dates, three from North Queensland and four from New South Wales, suggest that dunes were active during the early Holocene. The locations from which the dated samples were obtained are shown in Figure 1. Data relevant to the radiocarbon age determinations used in this study are presented in Table 1. Analyses were carried out by the Cambridge University Radiocarbon Laboratory, Beta Analytic Inc., and the Sydney University Radiocarbon Laboratory.

With the exception of SUA-252, which was an analysis of fibrous peat, the other dates all relate to charcoal samples. All samples were cleaned by repeated HCl-NaOH treatment to

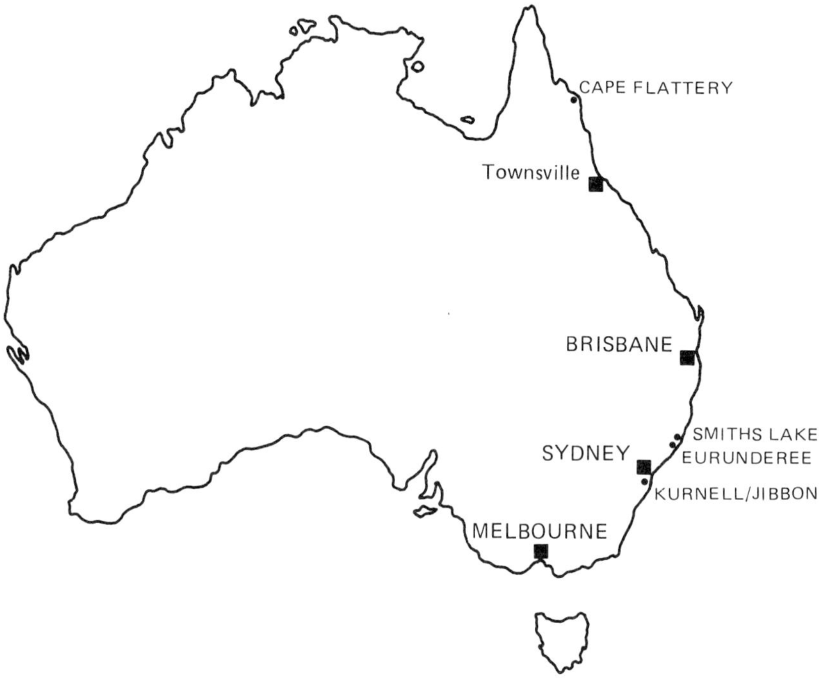

FIGURE 1. Location map showing radiocarbon dated early Holocene transgressive dune sites.

remove humic acid contaminants before assays were carried out. The indicated measured or estimated $\delta^{13}C$ values were used to correct the relative ^{14}C activities of the samples for environmental carbon isotopic fractionation (Craig, 1957; Polach, 1976). Results were calculated using a Libby half-life of 5568 years and 95% of the activity of NBS Oxalic Acid as the modern standard. The dates are reported in Table 1 as *conventional radiocarbon ages*, expressed as years before 1950AD (Stuiver and Polach, 1977), with one standard deviation error terms.

Evidence for Dune Instability during the PMT

Documentation of the age structure of coastal dunes in eastern Australia is still incomplete, but the radiocarbon data currently available suggest that dune ages fall into three broad groups: (1) pre-Holocene; (2) early Holocene;

TABLE 1. Radiocarbon dates related to transgressive dune activity in eastern Australia during the Postglacial Marine Transgression (P.M.T.)

Lab. No.	Location	Material	$\delta^{13}C‰$	$D^{14}C‰$	Age ^{14}C yrs BP
Q-2081	Cape Flattery N.Q.	Ch.	-26.39	-605.9	7480±75
Q-2082	Cape Flattery N.Q.	Ch.	-25.50	-639.7	8200±85
Beta-2494	Cape Flattery N.Q.	Ch.	nda	nda	7560±90
SUA-277	Jibbon, NSW	Ch.	Est.-25±2	-704.4±5.8	9790±160
SUA-278	Jibbon, NSW	Ch.	Est.-25±2	-490.6±14.4	5420±230
SUA-207	Kurnell, NSW	Ch.	Est.-24±2	-654.0±5.0	8530±115
SUA-41	Smiths Lake NSW	Ch.	Est.-24±2	-667.7±4.8	8850±115
SUA-252	Eurunderee NSW	P.	Est.-24±2	-596.1±11.1	7280±220

Ch = charcoal
P = peat
nda = no data available

(3) mid-to late Holocene. The majority of dates on pre-Holocene dunes have indicated 'background' radiocarbon ages (Coaldrake, 1962; Grimes 1979; Pye, 1981). At the present time it is not possible to specify the number or age of pre-Holocene, east Australian coastal dune units. Some authors have suggested that as many as six pre-Holocene episodes of dune formation might be represented in the coastal morphology of southern Queensland (Ward, 1977; Ward *et al.*, 1977; Thompson, 1981, 1983). Thom *et al.*, (1981b) have recognized pre-Last Interglacial, Last Interglacial and late-Last Glacial episodes of aeolian activity in the Myall Lakes-Port Stephens region of New South Wales.

North Queensland Holocene Examples

The three dates from the Cape Flattery dunefield, North Queensland, refer to a single stabilized dune located about 4 km west of Cape Flattery (Fig. 2). This dune is deeply

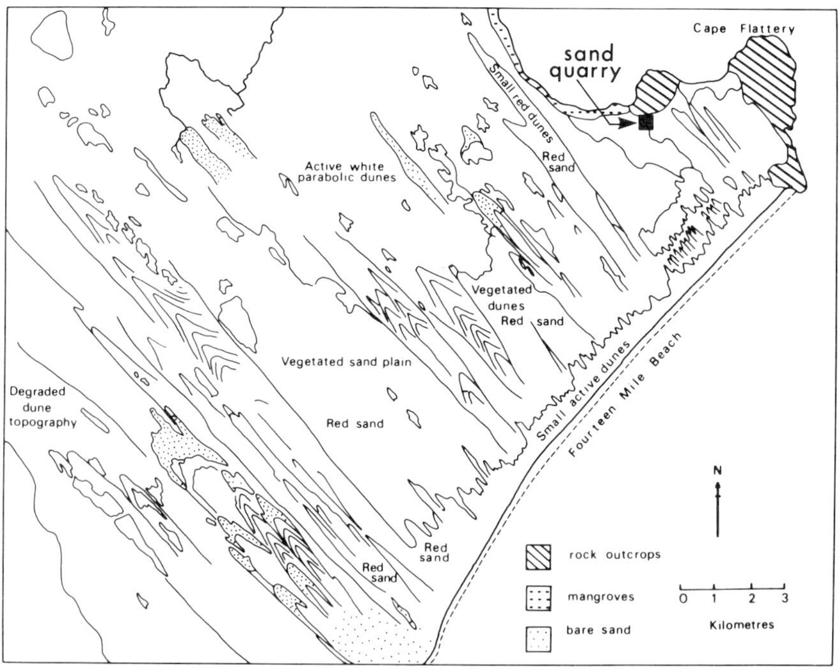

FIGURE 2. Map of Cape Flattery dune field showing location of the sand quarry from which the dated charcoal samples were obtained.

weathered and reddened in the subsurface horizons (Pye, 1981, 1983c). Two of the dated charcoal samples (Q-2081 and Q-2082) were obtained from depths of 2 m and 4 m, respectively, below the dune surface in a fresh quarry exposure. The third charcoal sample (Beta-2494) was obtained from the same dune at a depth of 8 m, about 20 m to the southeast of samples Q-2081 and Q-2082. The dates are mutually consistent and suggest that as the charcoal was incorporated within the sand while the dune was active, it must have been migrating between 7500 and 8000 ^{14}C years ago. Thus stabilization and weathering have taken place since 7500 ^{14}C years BP.

It is evident from drilling and air photograph interpretation (Pye, 1980) that dunes exhibiting a similar degree of morphological degradation and soil development are extensively developed in the Cape Flattery and Cape Bedford areas. These dunes are quite distinct from the weakly podzolized late Holocene dunes and from a much more weathered and degraded dune sand body which underlies the reddened dunes in many places (Pye, 1981; 1983c). The oldest dunes are indurated by humate and have been radiocarbon dated as older than 48000 ^{14}C years (Pye, 1982b).

Sea-level history in the Cape Flattery area is not well documented, but radiocarbon dating of corals from neighbouring reefs and islands suggests that the postglacial sea level had reached its present position by about 6000 years BP (McLean *et al.*, 1978; Chappell *et al.*, 1982). Dated microatolls on Three Isles and Nymph Island suggest that a maximum sea level, in excess of 1 m above present, may have been recorded around 3700 ^{14}C years BP. However, the possibility that the corals grew in the moated situation, and therefore do not reflect the true position of sea level, cannot be ruled out (McLean *et al.*, 1978, p. 174).

Morphostratigraphic evidence from the mainland suggests that the maximum inland extent of the P.M.T. was achieved about 6000 years ago. At this time shorelines existed up to 1.5 km inland of the present Fourteen Mile Beach and Cape Bedford Beach. There followed a period of coastal progradation when barrier spits formed parallel to the shore, and brackish water peats accumulated in barrier lagoons. Four radiocarbon dates indicate that the peats were forming between 5500 and 6000 ^{14}C years ago (Pye and Switsur, 1981). At that time there can have been little or no direct sand supply from the ocean beaches to the transgressive dunes further inland, tending to favour their stabilization. Coastal erosion and landward movement of the barriers during the later Holocene, possibly in response to a minor sea level rise, led to the exposure of the lagoonal peats on the foreshore and their partial burial by modern transgressive dune sand.

N.S.W. Holocene Examples

In New South Wales, two dates were obtained on charcoal pieces from a stabilized dune ridge which mantles a broad cliffed headland at Jibbon, on the southern outskirts of Sydney. Although the quartz sand which constitutes this cliff-top dune (Jennings, 1967) may have originated at either Marley or Jibbon Beach (Fig. 3), the former is the more likely source beach because of the predominance of southerly winds in Holocene aeolian sand transport on this coast (Roy and Crawford, 1979; Thom *et al.*, 1978). The charcoal pieces were sampled from the upper contact of each of two superimposed palaeosols exposed in a sand quarry near Jibbon Trig at an elevation of 70 metres above MSL (Fig. 3). These well-developed palaeosols are overlain by a contemporary *podzol* soil profile which is closely related to the present stabilized dune surface. Radiocarbon date SUA-277, relating to the lower palaeosol (7 m below the present dune surface and 3 m above bedrock), indicates that the sand ridge was actively forming just after 10,000 ^{14}C years BP, during the P.M.T. The date SUA-278, from the upper palaeosol (at 4 metres below present groundsurface) shows that a second episode of aeolian activity occurred about 5500 ^{14}C years BP, a time of maximum seaward progradation for many New South Wales sand barrier systems (Thom *et al.*, 1978).

At Kurnell Peninsula, located about 10 kilometres north of Jibbon on the opposite side of Bate Bay (Fig. 3), an extensive, largely active dunefield blocks the former mouth of the Georges River and extends over the bedrock of Kurnell Headland (Griffin,1963; Pickard, 1972; Roy and Crawford, 1979, Fig. 13). A well-developed *podzol* palaeosol, similar to the lower palaeosol at Jibbon, underlies much of the Kurnell Headland dunefield and is exposed in a sand quarry near the western edge of the area. A date of 8530 ± 115 ^{14}C years BP (SUA-207) was obtained by Bowman on a charcoal sample from the upper contact of this palaeosol, about 15 metres below the present ground-surface. No other palaeosols were visible in the quarry, but the contemporary soil profile was well-developed. On morphological criteria (Bowman, 1979) the latter was estimated to be about 6000 years old. The sampling site for SUA-207 has been erroneously depicted by Roy and Crawford (1979, Fig. 6, Section D-D^1) as being stratigraphically above an organic mud, previously dated at 4100 ± 125 yrs BP (0-2169) by Thom *et al.*, (1969). However, SUA-207 is regarded by Roy and Crawford (1979, p. 187) as indicating a maximum age for transgressive dune development at Kurnell during the P.M.T., an interpretation in agreement with that presented here.

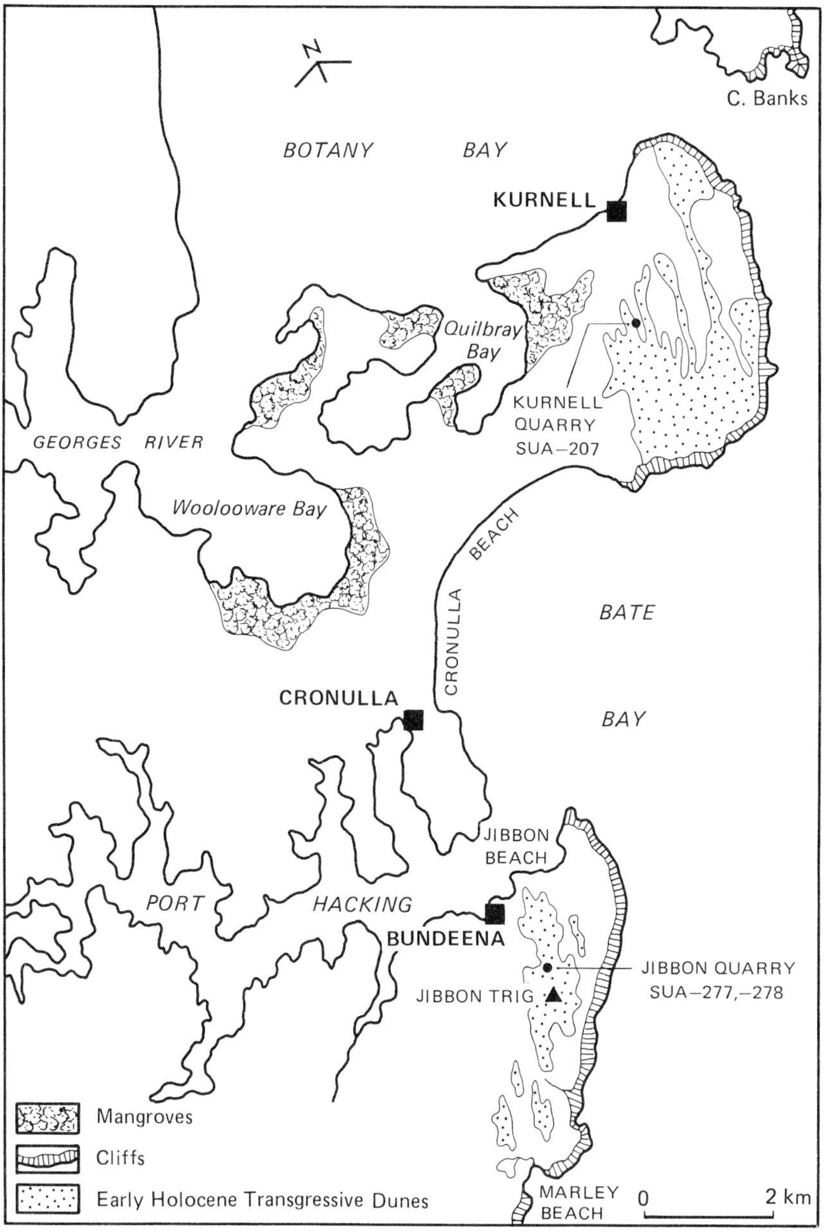

FIGURE 3. Map showing distribution of early Holocene transgressive dunes in the Jibbon and Kurnell headland areas and the location of sampled sand quarries.

A very similar age to SUA-207 has been obtained on a charcoal sample from transgressive dunes at Smiths Lake, in the Myall Lakes area of the central N.S.W. coast (Fig. 1). At Smiths Lake an exposure revealed a well-developed contemporary *podzol* soil profile in quartz dune sands, overlying an older dune surface (Fig. 4). The buried dune contained a humate-impregnated, sandrock horizon (Langford-Smith and Thom, 1969; Swanson and Palacas, 1965). A date on charcoal from near the upper limit of the buried unit (12 m above MSL) yielded an age of 8850 ± 115 ^{14}C years BP (SUA-41). Both the radiocarbon date and the degree of soil development in the overlying sands are consistent with dune initiation, deposition and stabilization during the P.M.T. (Thom *et al.*, 1981a). In fact, the close agreement between the radiocarbon ages from Smiths Lake, and Kurnell provides some indication of widespread transgressive dune activity along the southeast Australian coast between about 9000 and 8500 ^{14}C years BP.

A somewhat younger date (7280 ± 220 ^{14}C years BP; SUA-252) indicates that transgressive dune deposition in Eurunderee Embayment, located about 17 kms southwest of Smiths Lake, occurred towards the end of the P.M.T. The Eurunderee Holocene Outer Barrier is composed of elongated, active sand blowouts as well as stabilized, nested and juxtaposed parabolic dunes. These all overlie an indurated, humate-impregnated Inner Barrier sand unit (Fig. 5). Although the dated sample was obtained from near present mean sea level, the unit is interpreted by Thom *et al.* (1978, 1981a & b) as being of transgressive aeolian origin because of the lack of shell in several hundred heavy mineral exploration drillholes

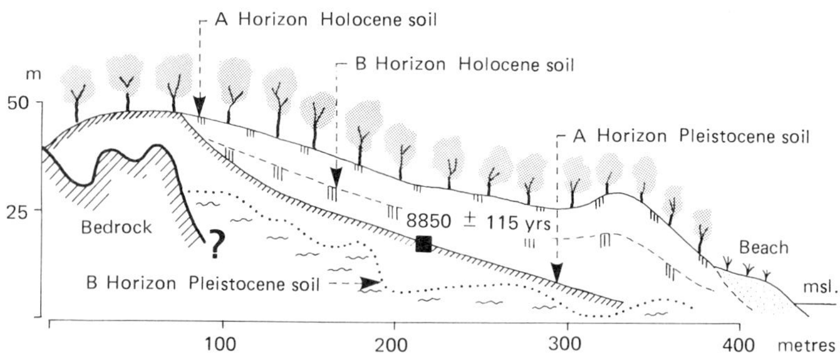

FIGURE 4. Cross-section at Smiths Lake, N.S.W. showing an early Holocene dune sheet overlying a Pleistocene dune and buried soil. Radiocarbon age obtained on charcoal at contact (SUA-41) after Thom et al., (1981a).

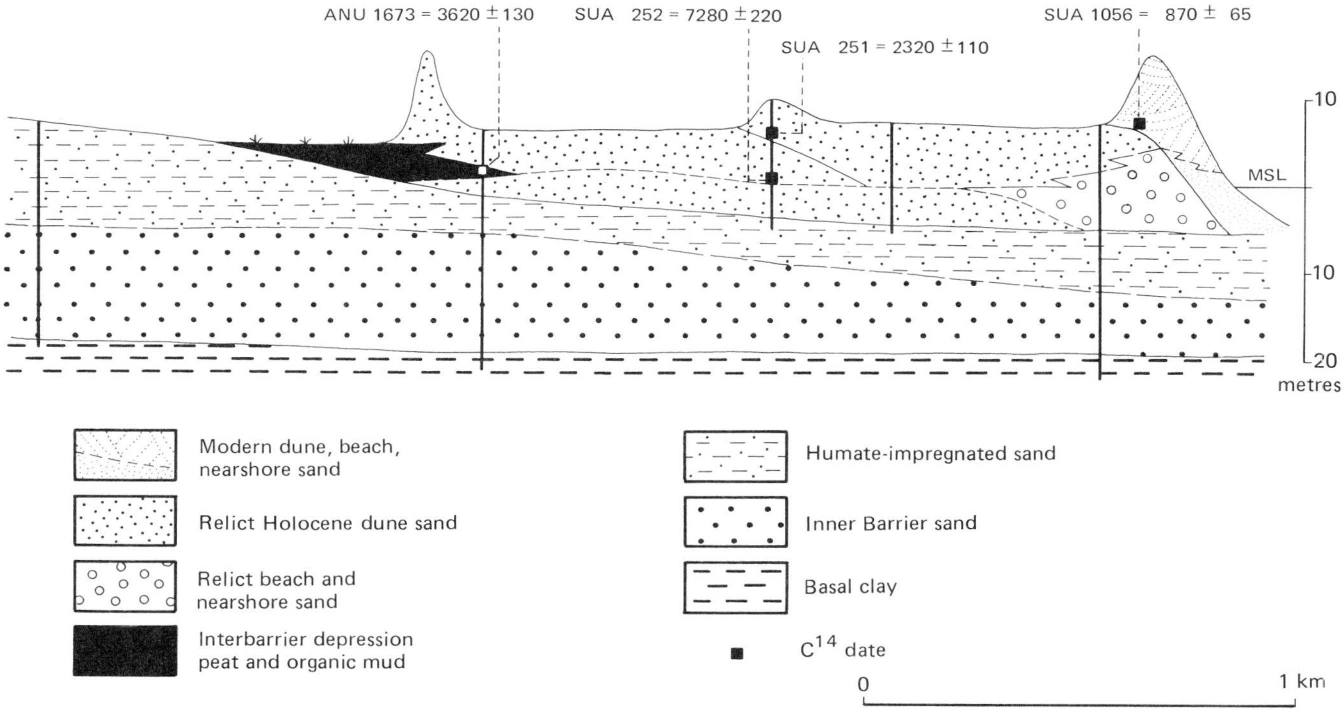

FIGURE 5. Eurunderee Outer Barrier cross-section after Thom et al., (1981a) showing sedimentary units and location of dated samples.

which intersect the unit, the high degree of sorting of the sediments, and the aeolian surface morphology of the Eurunderee Outer Barrier.

Relationship between Early Holocene Dune Instability and Sea-Level

The preceding discussion has shown that there is evidence from five widely separated sites in eastern Australia of an episode of transgressive dune instability between 10000 and 6000 ^{14}C years BP. This corresponds with the later part of the P.M.T. when sea level rose from about -30m at 10000 BP, to reach its present position (±1m) by 6000 BP (Thom and Chappell, 1975; Belperio, 1979; McLean *et al.*, 1978). This paper therefore provides further support for Thom's (1978, p.201) conclusion that: 'the rate of sea level rise is seen as the "triggering mechanism" or forcing function in forming transgressive dunes between 6000 and 10000 years ago'. However, the problem of the nature of the mechanism linking sea-level change with dune instability remains.

There can be little doubt that periods of sea-level rise cause shoreline erosion (Bruun, 1962; Schwartz, 1965) and that sea level fall tends to encourage shoreline progradation. However, a major question concerns whether transgressive dune instability is most favoured by shoreline erosion, shoreline progradation, or whether it can take place under both conditions. According to Schofield (1975), falling sea level requires that the offshore profile of equilibrium be adjusted by removal of sediment either offshore into deeper water, or shoreward where it becomes available for beach accretion and aeolian deflation, thereby feeding active transgressive dunes. However, although this model may be applicable to northern New Zealand, and possibly to other areas where littoral sand-binding vegetation is ineffective, there is little evidence that it applies to eastern Australia. In the latter area, most embayments which have experienced recent progradation display series of multiple beach ridges or parallel foredune ridges (Bird, 1960; Thom, 1974; Thom *et al.*, 1978, 1981a). In contrast, shorelines which are currently eroding display blowout development, destruction of foredunes, and aeolian reworking of beach-ridge sequences by transgressive sand sheets (Sless, 1958; Pickard, 1972; Pye, 1980). In such eroding embayments, destruction of foredunes is often most apparent at the exposed northerly end, where wind scour of the sides of blowouts leads to their lateral enlargement and the formation of remanié dunes (Thom, 1974). In the most extreme examples, shoreface erosion has led to the formation of active transgressive dunes along the entire length of embayments, for example on the eastern side of North Stradbroke Island (McGrath and Barr, 1971), along the shoreline of the Burdekin Delta (Pye, 1980), and in several embayments in the Newcastle Bight-Myall Lakes region (Thom *et al.*, 1981b). Thus, as noted by Thom *et al.*, (1978, p.41), 'present day conditions of wave and wind erosion provide the modern analog for removal of all or part of the vegetation which binds barrier sand'.

During the P.M.T. not all of the sand eroded from the shoreface would have been deposited on the floor of the adjacent nearshore zone as implied by the Bruun model (Schwartz, 1965; 1967). High aeolian energy conditions at exposed locations would have ensured that a substantial proportion of the shoreface sand was transferred landwards. Thus, to some extent, transgressive dune activity may have been self-reinforcing in helping to maintain disequilibrium of the nearshore profile and accelerating the rate of shoreline retreat (Roy and Thom, 1981). Only after sea level ceased to rise is sediment input from fluvial or offshore sources likely to have allowed the achievement of a new equilibrium nearshore profile, thereby leading to a cessation of both shoreline erosion and transgressive dune instability.

Conclusion

We have presented evidence from five widely separated locations in New South Wales and Queensland which indicates that between about 10000 and 6000 years ago, when sea level was rising at an average rate of 10-15 mm per year (Thom and Chappell, 1975), transgressive dunes were active in exposed coastal embayments in eastern Australia. In some embayments the dunes became partially submerged by the rising sea, but in others the dunes climbed bedrock hills and transgressed older deposits. Many of these dunes subsequently became stabilized by vegetation and experienced *podzol* pedogenesis, particularly after about 7000 ^{14}C years BP, as the rate of sea-level rise slowed and shoreline erosion and aeolian activity diminished. This evidence suggests that episodes of transgressive dune instability can be initiated by shoreline erosion which, in turn, can be caused by a rising sea level or other factors.

We do not propose that this model has worldwide applicability, or that sea-level rise is the only mechanism capable of initiating dune instability in eastern Australia.

The fact that some embayments have experienced intermittent large-scale dune activity in the past 6000 years, when sea level is considered not to have varied by more than ±1 m in eastern Australia indicates that other factors may be important. However, the model of the Postglacial Marine Transgression acting as a 'forcing function', giving rise to widespread, synchronous transgressive dune activity, is most likely to apply to those parts of the world which are similar to eastern Australia in possessing an embayed coastal configuration and relatively high energy wind and wave regimes, such as the Oregon coast. Further radiocarbon-supported morph-stratigraphic studies are required before the general applicability of the model can be adequately assessed.

Acknowledgements

This research was financially supported by the Royal Society Leverhulme Trustees, the U.K. Natural Environment Research Council, the University of Sydney and the N.S.W. National Parks and Wildlife Service. Professor B.G. Thom kindly supplied the Myall Lakes dates based on work partly sponsored by A.R.G.C. (Thom *et al.*, 1981a).

References

Battistini, R. (1964). L'extreme sud de Madagascar, etude geomorphologique. Thesis. Univ. Paris, Cujas, Paris 636p.

Belperio, A.P. (1979). Negative evidence for mid-Holocene sea level change along part of the coastal plain of the Great Barrier Reef. *Marine Geology* 32, M1-M9.

Bird, E.C.F. (1960). The formation of sand beach ridges. *Aust. J. Sci.* 22, 349-350.

Bird, E.C.F. (1978). The nature and source of beach materials on the Australian coast. *In* J.L. Davies and M.A.J. Williams (eds.) *Landform Evolution in Australasia*. A.N.U. Press, Canberra, 144-157.

Bird, E.C.F. and Hopley, D. (1969). Geomorphological features on a humid tropical sector of the Australian coast. *Aust. Geog. Stud.* 7, 89-108.

Bowman, G.M. (1979). *Development of Podzol Soils in East Australian Coastal Sand Barriers*. PhD Thesis, Dept. of Geography, Univ. Sydney.

Bretz, J.H. (1960). Bermuda, a partially drowned, late mature, Pleistocene karst. *Bull. Geol. Soc. Am.* 71, 1729-1754.
Bruun, P. (1962). Sea level rise as a cause of shore erosion. *J. Proc. Am. Soc. Civ. Engrs.,* Waterways and Harbours Div., 88, 1117-1130.
Chappell, J., Rhodes, E.G., Thom, B.G. and Wallensky, E. (1982). Hydro-isostasy and the sea-level isobase of 5500 BP in North Queensland, Australia. *Mar. Geol.* 49, 81-90.
Coaldrake, J.E. (1962). The coastal dunes of southern Queensland. *Proc. Roy. Soc. Qld.,* 72, 101-116.
Coetzee, (1975/76). Coastal aeolianites at Black Rock, Northern Zululand. *Trans. Geol. Soc. S. Afr.,* 78, 313-322.
Cook, P.J. and Mayo, W. (1977). Sedimentology and Holocene history of a tropical estuary (Broad Sound, Queensland). *B.M.R. Geol. Geophys. Bull.* 170, 206p.
Cooper, W.S. (1958). Coastal Sand Dunes of Oregon and Washington. *Mem. Geol. Soc. Am.,* 104, 131pp.
Craig, H. (1957). Isotopic standards for carbon and oxygen and correction factors for mass-spectrometric analysis of carbon dioxide. *Geochim. et Cosmochim,* 12, 133-149.
Crocker, R.L. (1946). Post-Miocene climatic and geologic history and its significance in relation to the genesis of the major soil types of South Australia. *CSIRO Bull.* 193.
Griffin, R.J. (1963). The Botany Basin. *Geol. Surv. N.S.W. Bull.* 18, 110 pp.
Grimes, K.G. (1979). Carbon-14 dates and the evolution of Fraser Island. *Qld. Govt. Min. J.* 80, 79-82.
Harmon, R.S., Land, L.S., Mitterer, R.M., Garrett, P., Schwartz, H.P. and Larson, G.J. (1981). Bermuda sea level during the last interglacial. *Nature,* 289, 481-483.
Hobday, D.K. (1977). Late Quaternary sedimentary history of Inhaca Island. *Trans. Geol. Soc. S. Afr.* 80, 183-191.
Jennings, J.N. (1967). Cliff-top dunes. *Aust. Geog. Stud.* 5, 40-49.
Land, L.S., Mackenzie, F.T. and Gould, S.J. (1967). Pleistocene history of Bermuda. *Bull. Geol. Soc. Am.* 78, 993-1006.
Langford-Smith, T. and Thom, B.G. (1969). New South Wales coastal morphology. *J. Geol. Soc. Aust.* 16, 572-580.
Laycock, J.W. (1978). North Stradbroke Island. *In* G.R. Orme and R.W. Day, Eds., Handbook of recent geological studies of Moreton Bay, Brisbane River, and North Stradbroke Island. *Dept. Geol. Univ. Qld. Papers* 8 (2), 89-96.
Marshall, J.F. (1979). The development of the continental shelf of northern New South Wales. *B.M.R. J. Aust. Geol. Geophys.* 4, 281-288.

McLean, R.G., Stoddart, D.R., Hopley, D. and Polach, H.A. (1978). Sea level change in the Holocene on the northern Great Barrier Reef. *Phil. Trans. Roy. Soc. Lond. A*, 291, 167-186.

McGrath, B.L. and Barr, D. (1971). *Qld. Dept. Harbours and Marine Report*, August 1971.

Phipps, C.V.G. (1966). Evidence of Pleistocene warping of the New South Wales continental shelf. *Geol. Surv. Canada Paper* 66, 280-293.

Pickard, J. (1972). Rate of movement of transgressive sand dunes at Cronulla, New South Wales. *J. Geol. Soc. Aust.* 19, 213-216.

Polach, H.A. (1976). Radiocarbon dating as a research tool in archaeology - hopes and limitations. *In* N. Bernard, (ed.) *Proc. Symp. on Scientific Methods of Research in Study of Anc. Chinese Bronzes and SE Asian Metal and other Arch'l. Artifacts.* National Gallery of Victoria, Melbourne, 255-298.

Pye, K. (1980). Geomorphic evolution of coastal sand dunes in a humid tropical environment: North Queensland. PhD. Thesis, Univ. Cambridge.

Pye, K. (1981). Rate of dune reddening in a himid tropical climate. *Nature*, 290, 582-584.

Pye, K. (1982a). Morphological development of coastal dunes in a humid tropical environment, Cape Bedford and Cape Flattery, North Queensland. *Geografiska Annaler* 64A, 212-227.

Pye, K. (1982b). Charcteristics and significance of some humate-cemented sands (humicretes) at Cape Flattery, Queensland, Australia. *Geol. Mag.* 119, 229-242.

Pye, K. (1983a). Formation and history of Queensland coastal dunes. *Zeit. Geomorph Suppl. Bd.* 45, 175-204.

Pye, K. (1983b). The coastal dune formations of northern Cape York Peninsula, Queensland. *Proc. Roy. Soc. Qld.* 94, 33-39.

Pye, K. (1983c). Post-depositional reddening of late Quaternary coastal dune sands, Northeastern Australia. *In* R.C.L. Wilson, (ed.) *Residual Deposits.* Geol. Soc. Lond. Special Publication 11. Blackwell, Oxford (in press).

Pye, K. and Switsur, V.R. (1981). Radiocarbon dates from the Cape Bedford - Cape Flattery dunefield, North Queensland. *Search* 12, 225-226.

Roy, P.S. and Crawford, E.A. (1977). Significance of sediment distribution in major coastal rivers, northern New South Wales. *Third Aust. Conf. Coastal Engng.*, 177-184.

Roy, P.S. and Crawford, E.A. (1979). Holocene evolution of the southern Botany Bay-Kurnell region, central N.S.W. coast. *Rec. Geol. Surv. N.S.W.* 20, 159-250.

Roy, P.S. and Thom, B.G. (1981). Late Quaternary marine deposition in New South Wales and southern Queensland. *J. Geol. Soc. Aust.* 28, 471-490.

Roy, P.S., Thom, B.G. and Wright, L.D. (1980). Holocene sequences on an embayed high energy coast: an evolutionary model. *Sedimentary Geology*, 26, 1-19.

Sayles, R.W. (1931). Bermuda during the Ice Age. *Proc. Am. Acad. Arts Sci.* 66, 381-468.

Schwartz, M.L. (1965). Laboratory study of sea level rise as as cause of shore erosion. *J. Geol.* 73, 528-534.

Schwartz, M.L. (1967). The Bruun theory of sea level rise as a cause of shore erosion. *J. Geol.* 75, 76-92.

Schofield, J.C. (1975). Sea level fluctuations cause periodic post-glacial progradation, South Kaipara barrier, North Island, New Zealand. *N.Z. J. Geol. Geophys.* 18, 295-316.

Sless, J.B. (1958). Coastal sand drift. *Soil Conservation Journal, N.S.W.*, 14, 50-68.

Stuiver, M. and Polach, H.A. (1977). Reporting of ^{14}C data: A discussion. *Radiocarbon* 19, 358-363.

Swanson, V.E. and Palacas, J.G. (1965). Humate in coastal sands of northwest Florida. *Bull. U.S. Geol. Surv.* 1214-B.

Thom, B.G. (1974). Coastal erosion in eastern Australia. *Search* 5, 198-209.

Thom, B.G. (1978). Coastal sand deposition in southeast Australia during the Holocene. *In* J.L. Davies and M.A.J. Williams (eds.) *Landform Evolution in Australasia*. A.N.U. Press, Canberra, 197-214.

Thom, B.G. and Chappell, J. (1975). Holocene sea levels relative to Australia. *Search* 6, 90-93.

Thom, B.G., Hails, J.R. and Martin, A.R.H. (1969). Radiocarbon evidence against higher postglacial sea levels in eastern Australia. *Marine Geology* 7, 161-168.

Thom, B.G., Polach, H.A. and Bowman, G.M. (1978). *Holocene age structure of coastal sand barriers in New South Wales, Australia*. Univ. N.S.W. Faculty of Military Studies, Dept. of Geography, 86pp.

Thom, B.G., Bowman, G.M., Gillespie, R., Temple, R. and Barbetti, M. (1981a). *Radiocarbon dating of Holocene beach-ridge sequences in Southeast Australia*. Monograph 11, Geography Department, Univ. N.S.W. at RMC, Duntroon.

Thom, B.G., Bowman, G.M. and Roy, P.S. (1981b). Late Quaternary evolution of coastal sand barriers, Port Stephens-Myall Lakes area, central New South Wales, Australia. *Quaternary Research* 15, 345-364.

Thompson, C.H. (1981). Podzol chronosequences on coastal dunes of eastern Australia. *Nature* 291, 5810, 59-61.

Thompson, C.H. (1983). Development and weathering of large parabolic dune systems along the subtropical coast of eastern Australia. *Zeit. Geomorph. Suppl. Bd.* 45, 205-225.

Ward, W.T. (1977). Sand movement on Fraser Island. *In* P.K. Lauer ed., *Occasional Papers in Anthropology* No. 8 Anthropology Museum, Univ. Qld., 113-126.

Ward, W.T., Stephens, A.W. and McIntyre, N. (1977). Brisbane's north coast and Fraser Island from the air. *In* R.W. Day, Ed., *1977 Field Conference, Lady Elliot Island - Fraser Island - Gayndah - Biggenden.* Geol. Soc. Aust., Qld. Div. Brisbane, 14-30.

Wright, H.E. (1963). Late Pleistocene geology of coastal Lebanon. *Quaternaria* 6, 525-539.

Wright, L.D. (1976). Nearshore wave-power dissipation and the coastal energy regime of the Sydney-Jervis Bay region, New South Wales: a comparison. *Aust. J. Marine Freshwater Res.* 27, 633-640.

Wright, L.D., Chappell, J., Thom, B.G., Bradshaw, M.P. and Cowell, P. (1979). Morphodynamics of reflective and dissipative beach and inshore systems: Southeastern Australia. *Marine Geology* 32, 105-140.

10

Chenier Plain Formation in Northern Australia

JOHN CHAPPELL

JOHN GRINDROD

Department of Biogeography and Geomorphology
Australian National University
Canberra, ACT

Introduction

Chenier plains essentially are coastal wetlands where long low ridges or 'cheniers' are separated by broad wet flats. The flats may be vegetated marshes as in Louisiana where cheniers were first described by Russell and Howe (1935), or they may be broad mudflats, hypersaline and largely supratidal, as described in Australia from Broad Sound by Cook and Mayo (1977), or the Gulf of Carpentaria by Rhodes (1980; 1982). The ridges vary in size, from small features less than 10 m wide and 100 m long, to much more extensive features used for settlement, agriculture or pasture. They are rather uniform in height, able to be overtopped by storm waves where they are close to the coast. Typically, they are built of shell debris of intertidal and shallow subtidal molluscs, although chenier ridges of mostly terrigenous sands or gravels have been described. Chenier plains are most common in hot-wet regions where rivers carry large quantities of fine sediment into relatively low-energy coastal provinces (Davies, 1972). Many variants have been recorded (see Curray, 1969; Otvos and Price, 1979). Typically, northern Australian examples occur on relatively sheltered coasts, as shown by the map of their distribution in Figure 1. The region has summer monsoonal rain (the 'wet season'), during which the mudflats between

COASTAL GEOMORPHOLOGY
IN AUSTRALIA
ISBN 0 12 687880 3

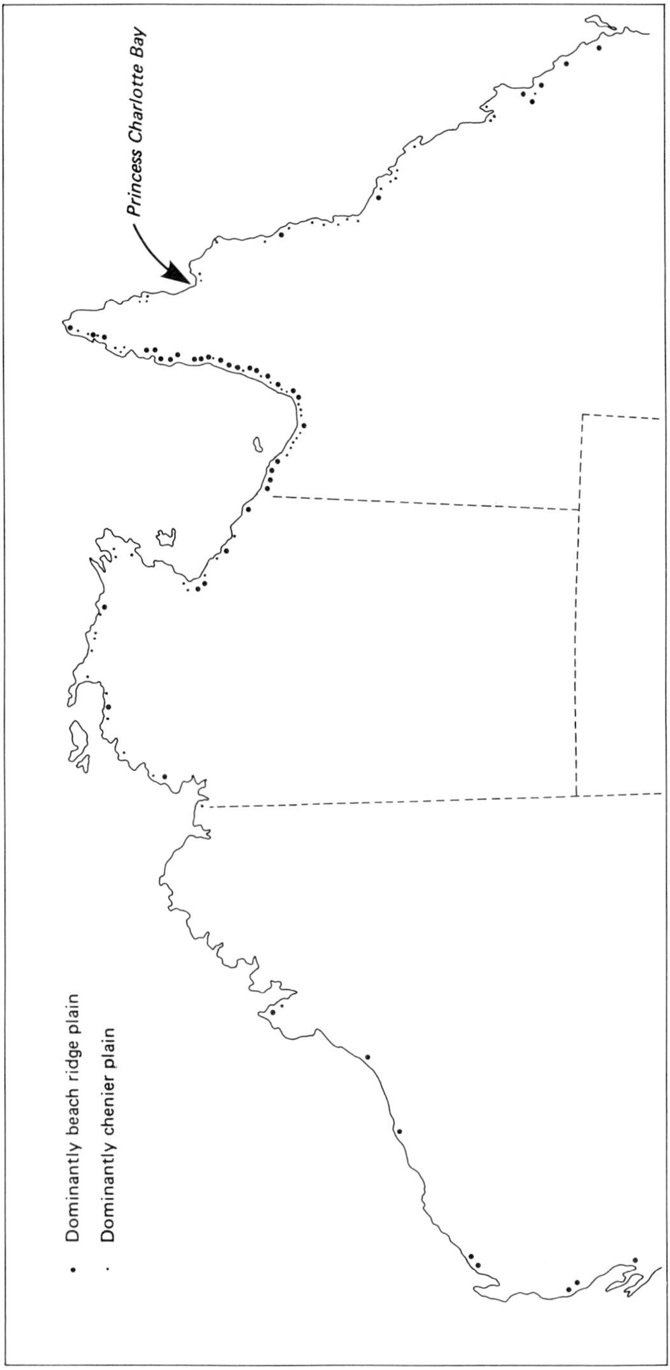

FIGURE 1. Distribution of beach ridge and chenier coasts in northern Australia, based on data from Dr. R.W. Galloway, CSIRO Canberra, with kind permission.

chenier ridges are widely flooded. Tidal channels drain the mudflats, which are inundated during the dry season by highest spring tides only, often locally rather than widely. Tidal rivers meander through the plains, and intertidal mangrove forests typically lie along their seaward margin and along the lower reaches of tidal channels and rivers.

Studies in Australia and elsewhere have shown that chenier plains are coasts of Holocene progradation, with deposits of ridges and flats generally becoming younger in the seaward direction. Progradation of chenier plains is not simply a process of smooth accretion because the ridges represent events of coarse sediment deposition separated by periods of mudflat progradation. To explain the alternation of two modes of deposition is the central problem of chenier plain geomorphology.

It is attractive to think that episodic deposition of coarse ridge-building materials is caused by unusually high energy events such as tropical cyclonic storms. This is too simple, as may be guessed from the fact that many chenier plains have far fewer ridges than the number of big storms which would be expected in their Holocene history. Different explanations have been proposed, many involving variations of sediment supply from nearby rivers. For example, Gould and MacFarlan (1959) linked chenier deposits near the Mississippi River to shifts of the position of its mouth, which altered sediment flux to the chenier system. Alternatively, there may be episodic breakdown of an offshore or longshore sediment store which alters the flux onshore. Augustinus (1978) shows that a mechanism of this type acts on the Surinam coastal plain, which is affected by periodic migration of shoals away from the Amazon River. Effects of possible sea-level fluctuations were linked by Schofield (1960) to chenier ridge formation in a New Zealand example. In the northern Australian context, variations of sediment supply were discussed by Cook and Polach (1973) and Cook and Mayo (1977) in their Broad Sound study; Rhodes (1980; 1982) argued that sediment variations were caused by Holocene climatic fluctuations. We believe that chenier systems have an internal dynamic which regulates alternation between ridge and mudflat deposition. The effects of changes of external sediment flux are not denied, such as in the Mississippi example. A case study will be described, however, where river-mouth shifts have not occurred and climatic or sea-level changes need not be invoked.

An illustration of the concept of 'internal dynamic' may be made through an analogy. Imagine a rainwater tank with a leak such that at half-full level it corresponds with the average rainfall it receives. The rain input varies with storms and droughts, but if the tank is large enough it will overflow less often than large storms occur. The occurrence of rare overflow events is regulated by the internal dynamic of storage and leak functions, in relation to the input.

The case study on chenier plain formation for which this concept is developed was undertaken in Princess Charlotte Bay, north Queensland (Fig. 2). A mangrove-fringed plain of narrow chenier ridges with broad hypersaline mudflats lies adjacent to tidal rivers at the head of this large bay. Near the rocky headlands at its northeast end the chenier plain passes into a short plain of shelly beach ridges. Research involved stratigraphic and dating studies of the history of formation of the plain, coupled with palynologic analysis of changes of mangrove communities during ridge-building events. Holocene sea-level changes relative to the region were analysed in detail from coral reef studies in the nearby northern Great Barrier Reef. Many others were involved in the project and are warmly thanked for allowing the writers to refer to their results, including Dr. J. Beaton (archaeology), Dr. R. Buckley (plant ecology), Mr. H. Polach (radiocarbon dating), and Mr. E. Wallensky and Ms. J. Hutson (coral and sea-level survey).

Morphostratigraphy of Princess Charlotte Bay

Princess Charlotte Bay, on the eastern side of Cape York Peninsula, is a wide mangrove-fringed bay protected to the east by Permian to Cretaceous rocks of Cape Melville, Bathurst Heads, and the Flinders Islands. A broad tract of Quaternary sediments inland of the Bay includes several tidal rivers which flood widely in the summer wet season. These features are shown in Figure 2, with an enlargement of the area of detailed study northeast of the Marrett River.

The study area has a mangrove frontage running 10 km from the Marrett River to its rocky terminus near Bathurst Head. The mangrove forest is several hundred metres wide and has several parallel zones of different mangrove communities. Behind the forest are broad mudflats interspersed by shore-parallel chenier ridges. Figure 2 shows these elements plus the indented rocky margin of the hills behind the chenier plain. Anticipating some later data, radiocarbon dating shows that the rocky margin was the shoreline about 6500 years ago and that the chenier plain has prograded seawards since that time.

Episodic building of chenier ridges in an otherwise muddy sedimentary environment can be explained only by linking present processes with the Holocene history. Depositional

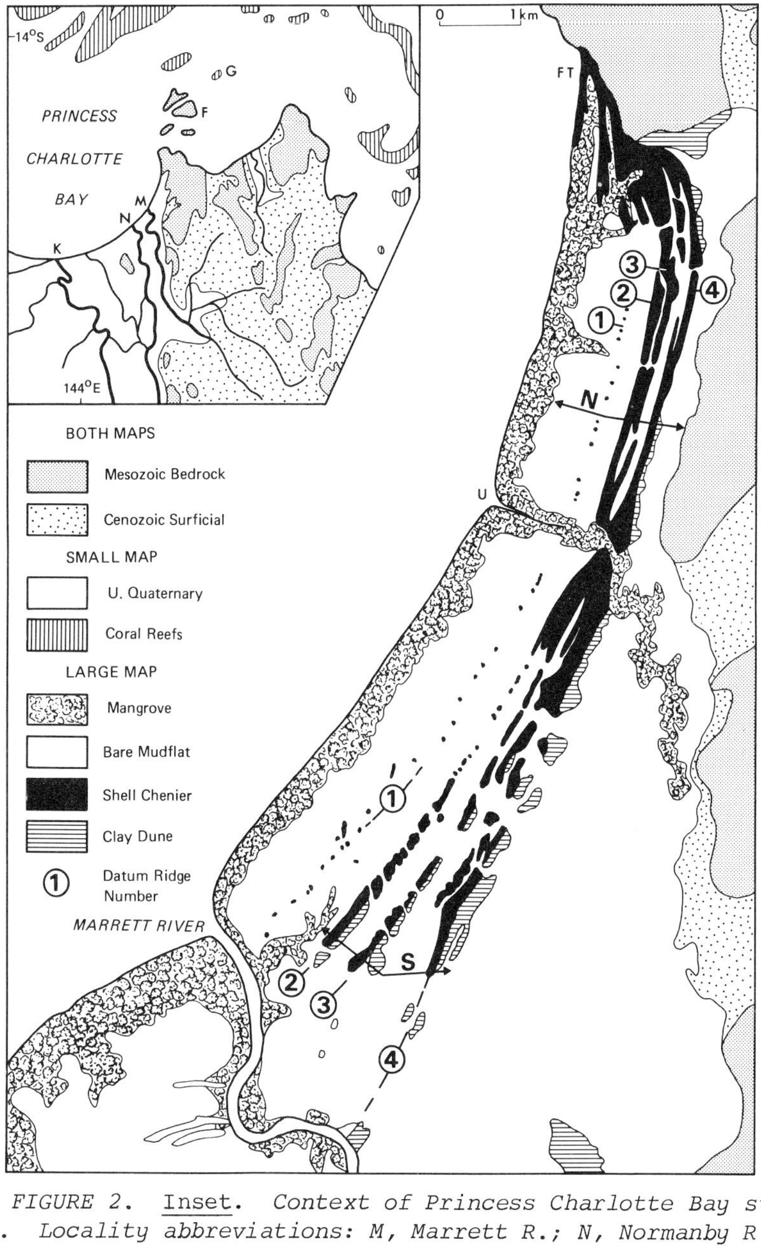

FIGURE 2. Inset. *Context of Princess Charlotte Bay study area. Locality abbreviations: M, Marrett R.; N, Normanby R.; K. Kennedy R.; F, Flinders Is.; G, King Is.* Enlargement. *Major elements of the study area. Ft, Figtree; U, Unnamed Creek; N, North transect; S, South transect. Numbers indicate datum ridges referred to in text and subsequent diagrams.*

history is discussed first because this draws attention to processes needing further examination; therefore, the initial approach is geological. Stratigraphic units of the chenier plain are continuous with modern geomorphologic features, so the analysis is couched in terms of *morphostratigraphic units*. This concept of defining mapping units in both geomorphic and stratigraphic terms was introduced to coastal studies by Thom (1967); Rhodes (1980; 1982), showed its usefulness in chenier plains.

Shallow stratigraphic sections across the Princess Charlotte Bay chenier plain are shown in Figure 3 (traverse lines indicated in Fig. 2). These are based on backhoe trenches at positions shown on the sections, plus some short stratigraphic cores. From topographically lowest to highest positions, the following morphostratigraphic units are defined. Comparable units are recognised in the Gulf of Carpentaria by Rhodes (1980; 1982).

(i) *Shallow subtidal muds.* The seabed gradient in Princess Charlotte Bay is about 1:1000 seawards from mean low-water spring tide level (MLWS), sediments are mostly muds, dominantly of clay sizes (<8μ), and sparsely quite shelly. Low ebbtide banks of silty and sandy muds near stream mouths locally are shelly. Shelly facies typically contain both whole shells and fragments of local benthic molluscs. The excavations on which Figure 3 is based did not reach this unit.

(ii) *Lower intertidal muds.* Seawards from the mangrove fringe to MLSW, mudflats exposed at low tide have a gradient around 1:1000 and are dominantly clay-sized, dark gray muds with rather sparse colonies of both surface-dwelling and burrowing molluscs and other invertebrates. Low ebb-tide banks near stream mouths locally contain more sandy and shelly facies. At the northern end of the study area the lower intertidal muds include muddy sands and sandy muds towards their seaward limit, sometimes rippled. All variants of this facies were seen in the stratigraphic trenches (Fig. 3).

(iii) *Mangrove mud and shell facies.* The coastal mangrove fringe is *c.* 300 m wide and is 1.0 to 1.3 m higher on its landward side than on its seaward side; *ie* it has a gradient of about 1:300. Two different sedimentary facies occur: dark gray muds sometimes containing sparse whole shells of mangrove-dwelling molluscs, and a shelly facies. The latter are fine to medium shell gravels in thin sheets or broad low sheets between 0.05 and 0.4 m thick. Thicker drifts resemble cheniers in being composed of clean shell gravel, whereas thin deposits characteristically are mixed shell and mud. Vertical bioturbation by crabs and other organisms is widespread. Trenches dug in the living forest floor show enmeshed mangrove fibrous roots to about 0.8 m below the surface, but the organic content otherwise was measured as rather low (typically <12%). All variants of these facies were seen in the stratigraphic trenches, where the mangrove mud and shell facies was an easily-recognised zone. The woody and fibrous root content of the Holocene deposits was lower than beneath the modern forest, suggesting some degradation after forest has abandoned a site. Comparable mangrove facies, without the shelly gravel component, occur beneath mangroves along tidal creeks.

(iv) *Upper intertidal muds.* Uniform clay-size gray muds, or sparsely shelly, occur on the flats behind the mangrove fringe and between cheniers, up to the level of mean high-water spring tide (MHWS). Salinity of these muds ranges up to 10% and thin dessicated algal mats are common in the dry season. Irregular haematitic mottles and stains within older exposures in trenches are interpreted as due to oxidation down dry-season shrinkage cracks.

(v) *Supratidal clays.* The upper 10 cm or so of the deposits of the mud flats between cheniers, in areas inundated only in the wet season, are gray to buff-coloured salty clays. These tend to form silt or fine sand-size pellets in the dry season, presumably through precipitation of weakly cementing salts in the capilliary fringe, and they visibly deflate in strong winds.

(vi) *Chenier ridges.* Long, low ridges of clean shelly gravel with crests typically 1 to 2 m above MHWS, are vegetated with grassy shrubland or mixed woodland with local groves of 'vine forest' with deciduous elements (e.g. *Bombax*). The base of the most recent chenier ridge (at the northeast end of the study area) is about 0.3 m below the level of the highest mangrove forest floor.

(vii) *Clay dunes.* Vegetated dunes of buff-coloured clay overlie the upwind side of cheniers, building to as much as 4 m above MHWS at sites where broad upper intertidal and supratidal mudflats lie to windward. Most examples currently are eroding, and they are inferred to have formed in the past from wind-transported pelletised supratidal clays.

(viii) *Human occupation deposits.* The study area contains extensive deposits related to prehistoric human occupation of the chenier ridges. There are two forms: shell middens and shell mounds. Typically, middens are 15 to 40 cm thick and vary in extent from a square metre to broad sheets over chenier crests. Mounds are approximately circular, ranging from 8 to 25 m or so in basal diameter, and 1 to 4 m thick. Both types of occupation deposit are distinguished easily from chenier shell gravels, being built of food species of shellfish (usually more than 98% of *Anadara*), and have dark gray or black interstitial soil.

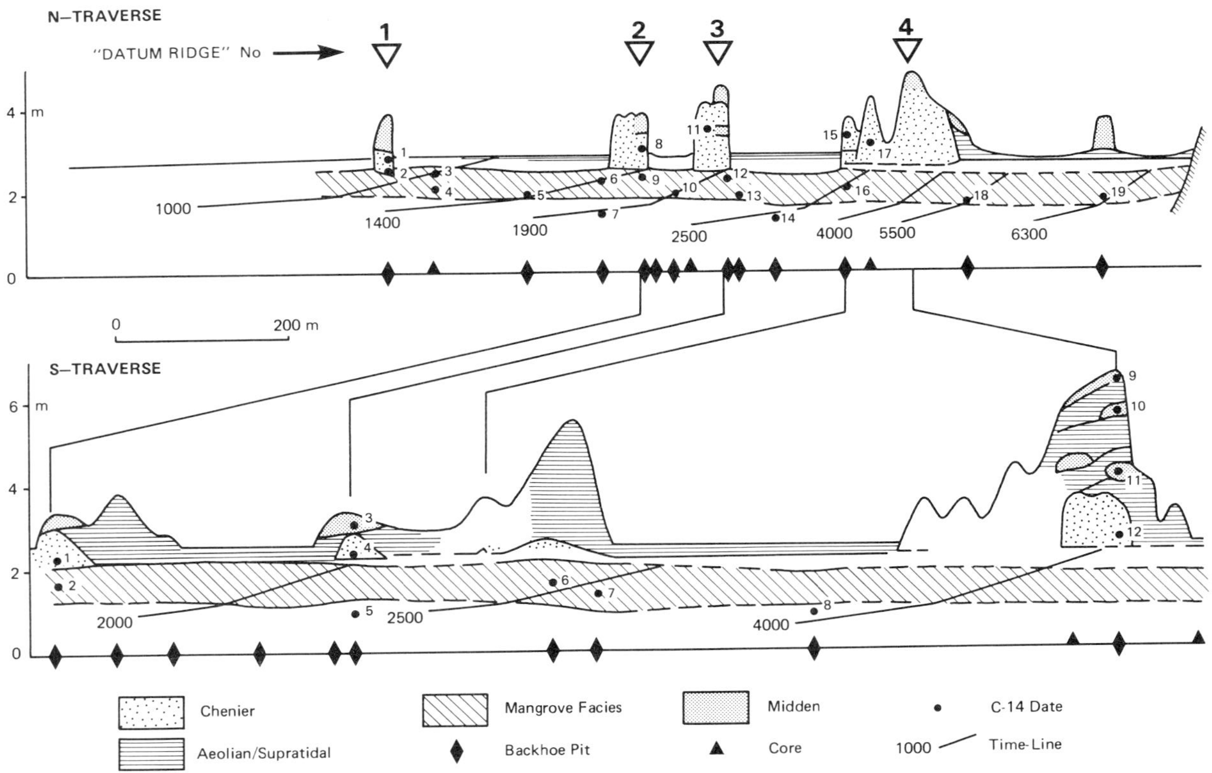

FIGURE 3. Morphostratigraphic cross sections of Princess Charlotte chenier plain. Locations of N and S traverses shown in Figure 2. (See Table 1 for ^{14}C dates).

TABLE 1. Radiocarbon results Princess Charlotte Bay

N - traverse

Sample No. (on diagram)	Lab No.	^{14}C age	Sample type
1	2955	710 ± 70	shell (c)
2	2685	510 ± 80	shell (c)
3	3091	1010 ± 110	shell
4	3261	990 ± 110	shell
5	2962	1380 ± 70	shell
6	2683	1290 ± 90	shell
7	2682	1690 ± 65	shell
8	2960	1200 ± 70	shell
9	2686	1380 ± 90	shell
10	2684	1910 ± 110	shell
11	B1752	1275 ± 85	shell
12	2687	2170 ± 80	shell (c)
13	2957	2120 ± 100	shell
14	2688	2250 ± 80	shell
15	2689	1900 ± 70	shell (c)
16	2690	2480 ± 80	shell
17	2458	2460 ± 135	shell (c)
18	2691	5580 ± 90	mangrove organic
19	2692	6270 ± 100	mangrove organic

S - traverse

Sample No. (on diagram)	Lab No.	^{14}C age	Sample type
1	2954	1670 ± 70	shell (c)
2	2961	1470 ± 60	mangrove organic
3	2327	1540 ± 115	shell (c)
4	2949	1960 ± 70	shell (c)
5	2693	2520 ± 80	shell
6	2325	2290 ± 110	shell
7	2958	2340 ± 80	shell
8	2953	3690 ± 70	shell
9	2697	640 ± 80	shell
10	2696	600 ± 80	shell
11	2695	1450 ± 110	shell
12	2694	4030 ± 90	shell (c)

Notes:

1. All numbers are ANU lab numbers except sample B1752 which is BETA Laboratory result.
2. All shell samples corrected for environmental effect.
3. (c) after shell refers to chenier ridge sample

Stratigraphic pits and cores, logged in terms of these units, are the basis of the sections in Figure 3. The next part of this paper is anticipated in Figure 3, which summarises radiocarbon dating results. Before these are discussed, the following morphostratigraphic features are noted.

Chenier ridges at Princess Charlotte Bay become lower as they become younger. This is mainly related to clay dune and occupation layer overburden, rather than crest height of the primary chenier deposit. Only 'ridge 1' (youngest) has a distinctly lower chenier crest, and this feature is not really a ridge but a discontinuous chain of islets (Fig. 2). Chenier bases (*ie,* contact between shelly ridge base and intertidal muds) have similar heights throughout. Similarly, the mangrove facies occur within the same vertical range across both sections. The total vertical range of mangrove facies, through all stratigraphic exposures, is similar to the present-day mangrove facies range, though somewhat higher in elevation. The map (Fig. 2) and sections (Fig. 3) show that the cheniers occupy a relatively narrow band within the full spread of Holocene progradation.

The shore-parallel chenier system passes abruptly northwards into a smaller beach-ridge plain. Although the chenier ridges are composites of several coalesced lesser ridges, the total number of ridges in the chenier system is much fewer than in the beach-ridge system. For convenience, the ridge systems in Figures 2 and 3 are given 'datum ridge numbers'. The composition of the beach ridges is similar to the cheniers, *ie,* 100% shell debris. Technically the beach ridges resemble cheniers in that they sit on a mud substrate but they are referred to here as beach ridges because they coalesce with no intervening muddy swales. These are traced through the transition zone in Figure 4 which shows that relatively few cheniers were deposited in the same time as many beach ridges accumulated. Examination of the chenier-beachridge transition throws light on the processes of chenier formation. The following sections show how radiocarbon dating is fundamental in this analysis.

Radiocarbon Dating

The progradational history is determined through radiocarbon dating of stratigraphic traverses shown in Figure 3 (see Table 1 for details). All units other than supratidal clays and clay dunes can be dated from shell material; the mangrove facies can also be dated through woody material. Dating was done by Mr. H.A. Polach and his staff of the ANU Radiocarbon Laboratory. All results are in years Before

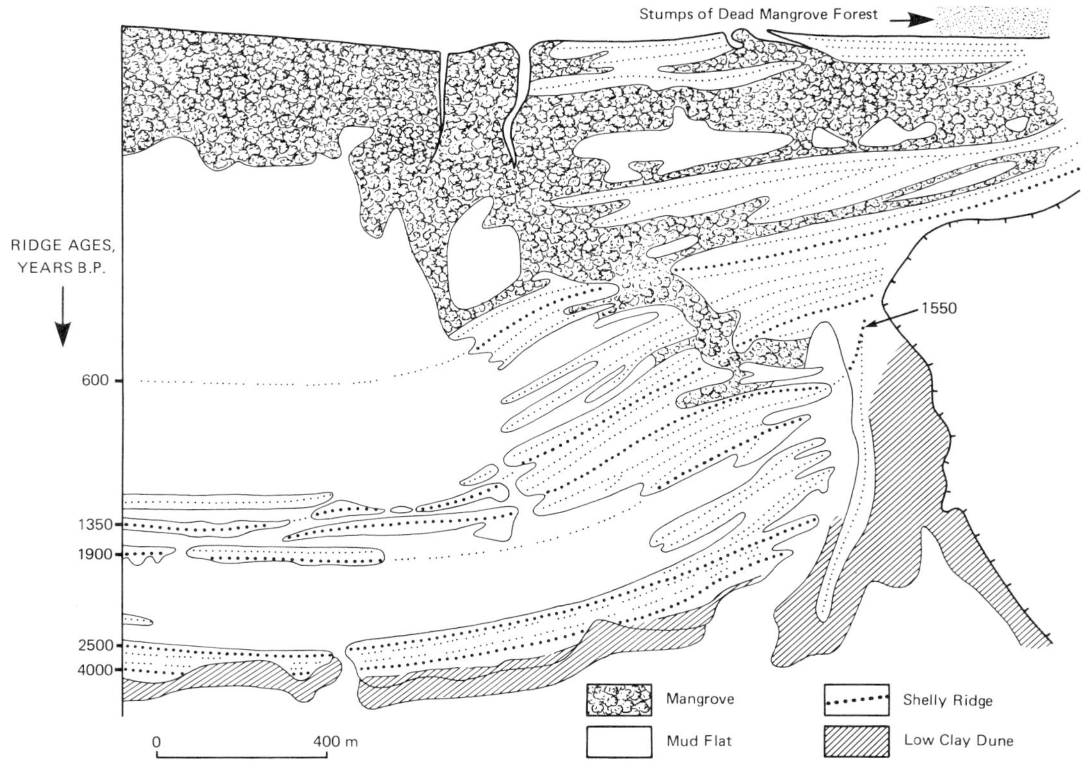

FIGURE 4. Transition from chenier to beach-ridge plain at north end of study area.

Present (B.P.), with mangrove and shell dates being placed on a common scale by (a) applying the isotopic fractionation correction (Stuiver and Polach, 1977) giving *conventional radiocarbon ages*, and (b) by subtracting the *environmental correction factor* from the shell dates (see Chappell, 1982a, for review of these matters). The environmental correction used was to subtract 450 years in accordance with Gillespie and Polach (1979). (The Gillespie-Polach result for southeast Australia appears applicable to North Queensland shells: Polach *pers. comm.*).

Radiocarbon ages have variable relationships to age of deposit from which the samples were taken. Shells from chenier deposits lived at some indeterminate time before they were swept up into the chenier ridges. For mangrove facies and lower intertidal muds, however, shells selected for dating were articulated valves from assemblages interpreted as having lived when surrounding muds were accumulating. Generally, mangrove root material will be younger than surrounding sediment. Finally, shells from human occupation layers are assumed to have been alive when harvested.

On the basis of the radiocarbon results from lower intertidal and mangrove facies shells, thin 'time lines' oblique to the stratigraphic units are drawn in Figure 3. Positions of dated samples are shown, and only one contradictory age result occurred. By 'contradictory' it is meant that age results oppose stratigraphic order of superposition outside statistical confidence limits (using the tests given by Chappell, 1982a).

The ages of datum ridges in the chenier sequence are shown in Figures 2 to 4. These ages apply to the datum ridges only, and not to the composite cheniers as a whole. They are based on large unworn shells selected from the stratigraphic pits where these cut into cheniers. These results are expected to predate deposition events, as noted above. However, the age difference is thought to be small, on the evidence of comparable ages of shells from intercalated middle and chenier layers.

Trends and Processes

Inferences can be drawn about progradation processes and episodic ridge formation, from morphostratigraphic and dating results. Some of these processes are difficult to observe in the field particularly when they occur only rarely, such as ridge-building events.

Figure 5 shows major trends over the last 6000 years. Figure 5a shows progradation of the north chenier section, measured from the bedrock shoreline. Figure 5b shows progradation of the south chenier section (location, Fig. 2), though only since 4100 B.P. This is an arbitrary starting point, near the rear of ridge 4, because the earliest Holocene shoreline is not known at this site.

Both Figures 5a and 5b indicate progradation increasing towards the present. At the north section, increase is very marked since about 2000 B.P. At the south section, progradation is rapid since the earliest date (ridge 4), and the increasing tendency is less marked. As the south section is near a major fluvial sediment source, this may be expected. However, it would be unwise to interpret Figures 5a and 5b as showing increasing fluvial sediment flux through time. Within an infilling bay, as time passes the shoreline will migrate seawards increasingly rapidly for reasons of volume geometry shown in Figure 6a. Further, as nearshore gradient diminishes, so the tendency for fine sediment deposition should increase.

There is a trend towards decreasing chenier ridge formation towards the present, correlated with increasing progradation rate. Figure 5c shows the number of chenier ridges per 1000 years, reckoned within the intervals between formation of successive datum ridges through to the present (ridgeless) mangrove shoreline. Also shown are beach ridges per 1000 years, which are discussed later. Figure 5d converts the figures to number of ridges per 100 m of progradation. There is a clear decline towards the present. This is particularly so at the north section, which is remote from the river input.

Progradation essentially is a process of mud deposition in a bay which is becoming shallower as shown in Figure 6a. Figures 5a and 5d show that chenier formation diminishes as muddy progradation increases, which suggests a link between the two. The beach ridge data from the north end of the study area are illuminating in this context.

Ridges cannot form without a supply of shell material. In the last 4000 years, three times as many beach ridges have formed as there are separate ridge elements in the chenier series. In the last 2000 years the ratio is nearer 6 to 1 (Fig. 5c). The morphology of the beach ridges, splaying to the south (Fig. 4), indicates a northern source of shell which must have been excavated by storm events to drive the debris onshore and southwards. The observed occurrence of rock-dwelling species in the beach ridges supports this. Cheniers, on the other hand, are composed of mud and sand-dwelling species. It seems that increasing muddy sedimentation in the chenier sector of Princess Charlotte Bay has inhibited the influx of shell from the chenier sources.

Two further details, based on the stratigraphic and dating study, contribute to an explanation of progradation versus

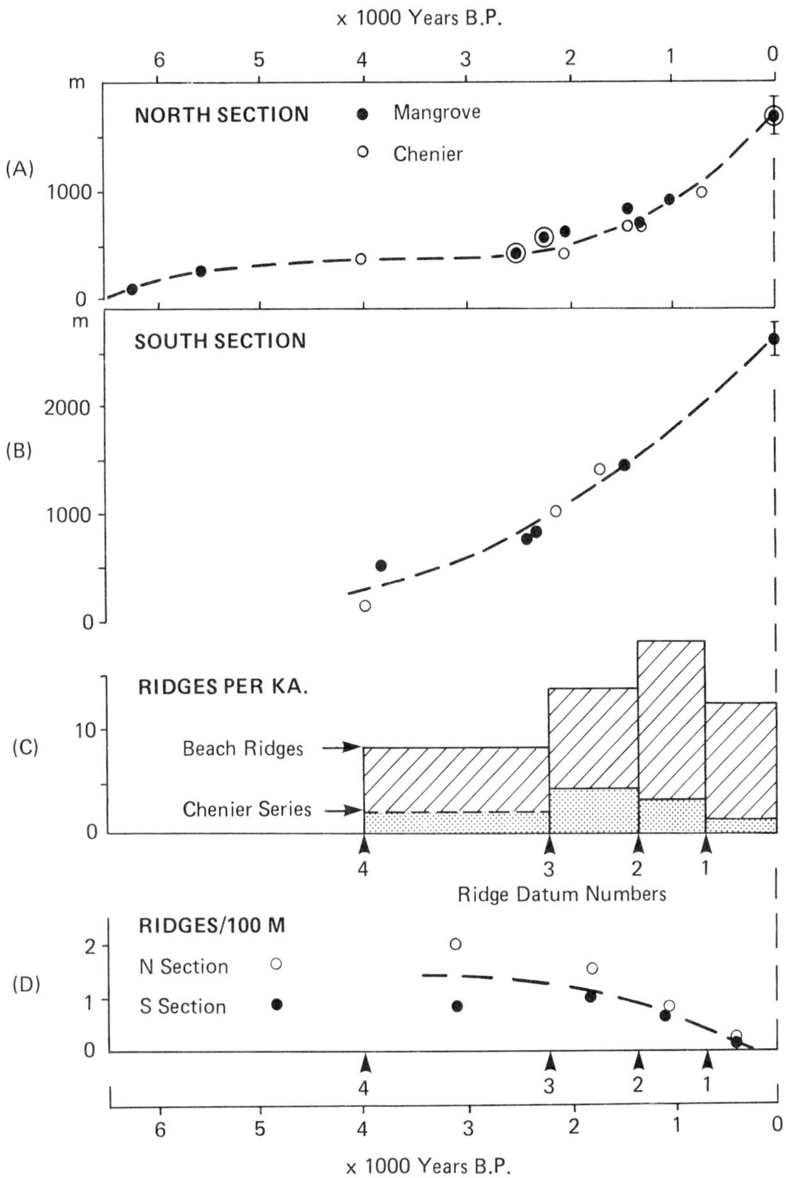
x 1000 Years B.P.
6
5
4
3
2
1
0
m
(A)
1000
0
NORTH SECTION
Mangrove
Chenier
m
SOUTH SECTION
(B)
2000
1000
0
RIDGES PER KA.
(C)
10
0
Beach Ridges
Chenier Series
4
3
2
1
Ridge Datum Numbers
RIDGES/100 M
(D)
2
1
0
N Section
S Section
4
3
2
1
6
5
4
3
2
1
0
x 1000 Years B.P.

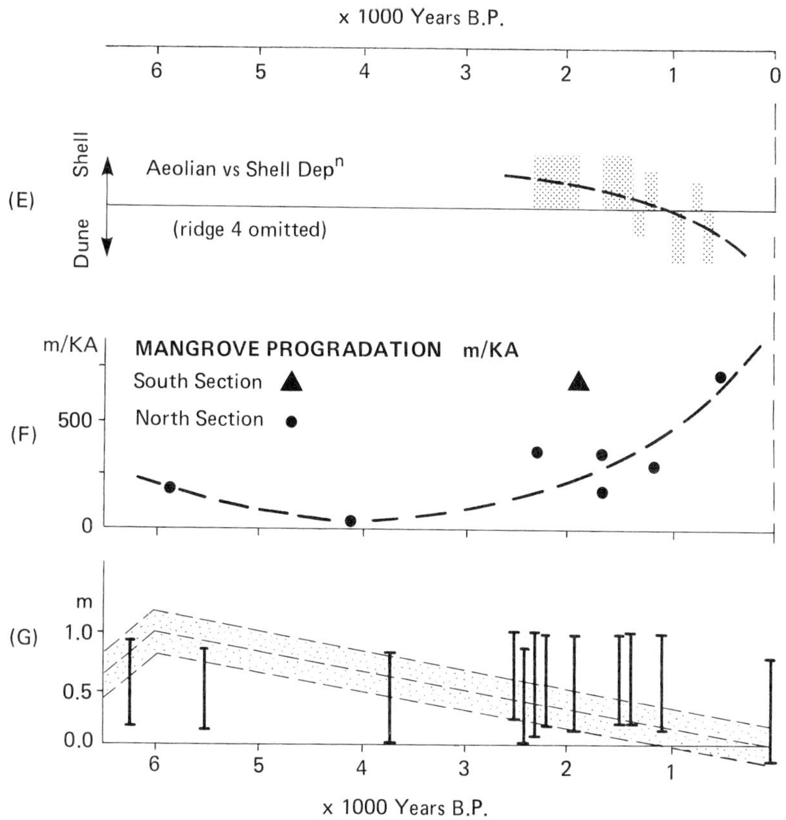

FIGURE 5. Major trends in last 6000 years. (a) and (b): Seaward advance of shoreline at N and S traverses, based on dated mangrove and chenier facies. (c): Estimated number of ridges formed per thousand years, in 4 different periods. (d): Chenier ridges per 100 m of progradation, averaged within 4 separate periods. (e): major phases of aeolian and shell deposition, with dark areas proportional to approximate volumes. Ridge 4 complex omitted due to insufficient dating. (f): Mangrove progradation, m/1000 years. (g): Sea level, northern Great Barrier Reef inner islands (speckled band shows +/- 1 std. deviation), and vertical ranges of dated mangrove facies from Figure 3. Datum for coral-based sea level curve = MLWS; mangrove datum is 1.0 m higher.

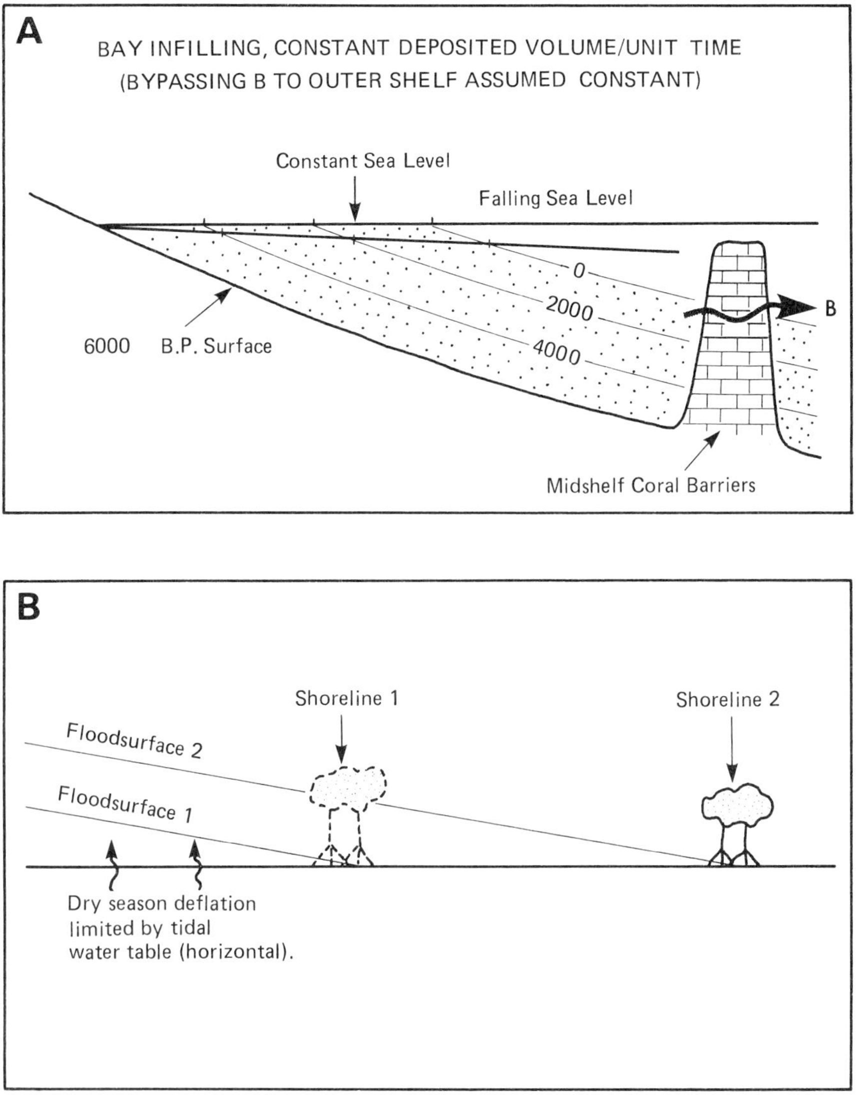

FIGURE 6. (a) Relationship between bay infilling and progradation: horizontal line, stable sea level case; sloping line, falling sea level case.

(b) Progressive increase of wet season inundation of coastal plain as coast progrades, thereby increasing flood deposition on mudflats with consequent increase of dry season aeolian deflation.

chenier formation. The first concerns a different source of evidence for muddy terrigenous sediment influx.

Clay dunes occur on the upwind sides of chenier ridges where relatively extensive supratidal flats occur to windward. This is especially true at the southern section (Fig. 3), although the same occurs to a lesser extent in the shorter northern section. Clay dunes are redeposited pelletised supratidal clays, deflated during the dry season from the mudflats. The process of wet season deposition and dry season deflation, described from the Cambridge Gulf by Thom *et al.* (1975) and from Gulf of Carpentaria chenier plains by Rhodes (1980), was seen occurring in Princess Charlotte Bay. At the present time, most clay dunes are eroding. It is inferred that for clay dunes to form, a wet season supply of mud to the sources of deflation must be substantially greater than now to ensure the supply of dune-building material.

There is good evidence of sharp episodes of clay dune formation in the past. Figure 7 shows the stratigraphy of the largest clay dunes in the area, on ridge 4 at the south section where the upwind source area is greatest. Occupation layers within the stratigraphy give rather precise dates bracketing times of dune formation. The only uncertainty associated with the evidence is related to the unknown time of pedogenesis between occupation and onset of the next phase of dune building (Fig. 7 shows pedogenic horizons). Figure 5e shows the timing of clay dune episodes in relation to chenier building events. Also indicated in Figure 5e are shell chenier and clay dune volumes from ridge 3 onwards, estimated from stratigraphic exposures. It seems clear from Figure 5e that formation of shelly chenier ridges declines as clay dune formation increases. This resembles the previous result that chenier formation declines as muddy progradation increases. Clay dunes will tend to form more easily, as progradation increases. Mud deposition over the older mud flats will increase as the coast advances because the wet season flood gradient is steeper than the dry season tidal surface, as shown in Figure 6b. Thus the quantity of deflatable clay increases with progradation.

Closer inspection suggests interesting pulsations of mud input to the system. The last phase of clay dune building was substantial but brief, as the radiocarbon dates show (about 600 years ago, Fig. 7). Previous pulses also seem to have been rather brief. It is intriguing that no chenier formed after the last 'mud pulse', and that only a minor chenier formed some 200 years after the previous pulse. These details are taken up in the later synthesis.

The final interpretive matter concerns the role of mangroves in the prograding system. Radiocarbon dates from mangrove facies in successive pits, across the sections of Figure 3,

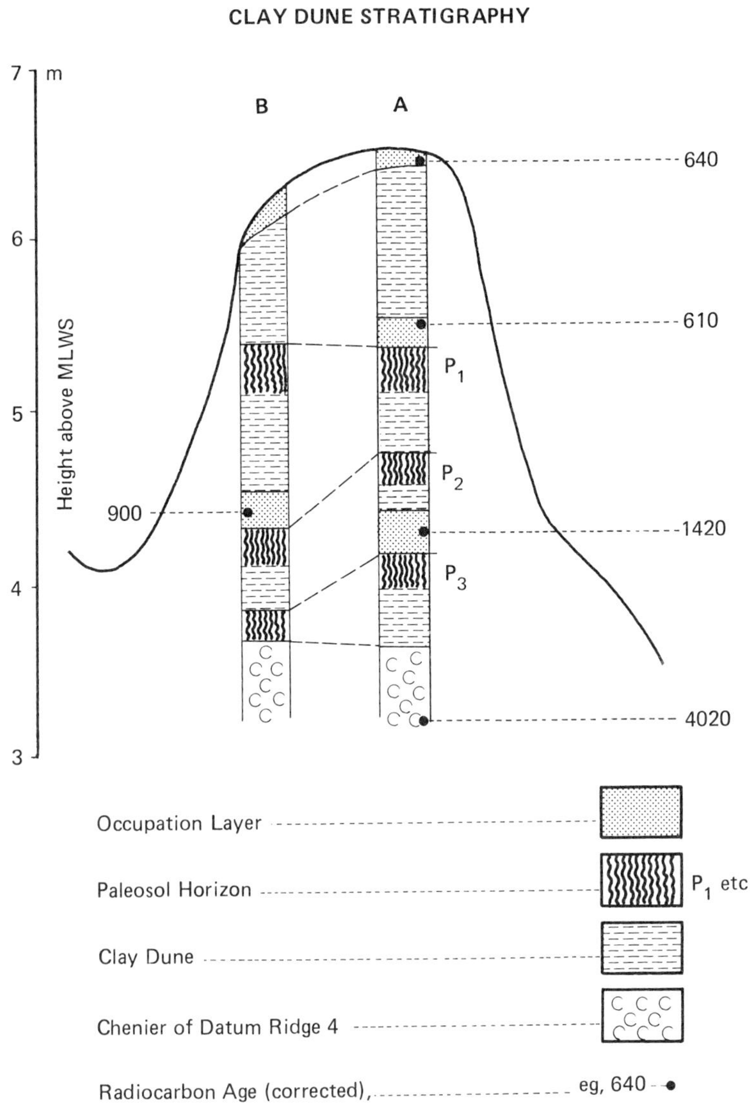

FIGURE 7. Dated stratigraphy of clay dunes. Column A on S-traverse; column B is lateral continuation on Marrett river-bank.

are converted to mangrove progradation rates in metres/1000 years in Figure 5f. The rate appears to have been lowest about 4000 years ago, and has steadily increased since then. Rate estimates are not precise because the positions of dating samples within the mangrove fringe at their times of deposition are not known. However, it is suggested that these results reflect the width of the mangrove zone through the last 6000 years. If the mangrove fringe was wide then similar mangrove ages should be found between widely spaced sites. This is the same as finding a high 'mangrove progradation rate'. Figure 5f therefore suggests that the mangrove fringe became increasingly wide over the last 4000 years. It also may have been relatively wide around 6000 years ago.

There is a long-standing question of whether mangroves are mere passengers on an intertidal landscape produced by physical processes (*eg.* Thom *et al.* 1975). It is seen in Figure 3 that the intertidal gradient is steeper in the mangroves than on the flats to either side. This suggests active sediment trapping by mangroves otherwise the gradient would be uniform across the entire muddy intertidal profile. The extent to which this is peculiar to Princess Charlotte Bay is unknown. Surveys in the Gulf of Carpentaria by Rhodes (1980), for example, suggest no such steepening of the mangrove profile. However, in our study area, trapping of shell debris is visible in low shelly bars within the present fringe. Accretion of mud appears to be active in lobate patches at the back of the fringe where youthful stands of mixed *Ceriops/Avicennia* saplings appear to be colonising landwards. Normally the rear margin is limited by high soil salinity of the upper intertidal flats (measured at up to 10%), and the forest typically is of stunted mature or dying trees. It is suggested, therefore, that in this area the mangrove fringe actively promotes progradation by trapping both shelly and muddy sediments, with the latter being greater towards the rear. Undoubtedly trapping is affected by mangrove forest density and width. Our observations at narrow fringes elsewhere, as well as of others (*eg.* Jennings and Coventry, 1973), show that chenier sediment can pass right through a less retentive fringe. Figure 5f suggests that sedimentation and increasing mangrove fringe width have supported one another over the last 4000 years.

Relative Sea-Level Changes

Sea-level changes relative to Princess Charlotte Bay are more reliably estimated from fringing coral reefs of inner islands of the northern Great Barrier Reef, than from the chenier plain stratigraphy itself. Progressive compaction of the muddy intertidal and mangrove facies is shown by an increase in bulk density of these sediments towards the rear of the plain. Compaction beneath the chenier ridges was visible in backhoe trenches cut through the ridge margins, where the chenier bases were seen to be lower where the overburden was thickest.

Studies of prograded fringing coral reefs, where coral microatolls can be good indicators of relative sea-level, were initiated by Scoffin and Stoddart (1978) in the northern Great Barrier Reef. Using mainly similar evidence, plus chenier studies from the Gulf of Carpentaria (Rhodes, 1980), Chappell and others (1982) show that Holocene sea-level change relative to north Queensland varies with location, apparently due to hydroisostatic effects. According to these results, sea level relative to Princess Charlotte Bay is expected to have fallen by a little over 1 m since 5500 B.P.

More recently, Chappell (1983) has presented sea-level evidence from fringing coral reefs at sites lying on the same 5500 B.P. isobase. Relative sea-level change is expected to be similar at all these sites. Over 40 data points suggest smoothly falling sea level from 6000 B.P. to present, with a standard error of +/- 0.2 m about the trend. This is shown in Figure 5g, and is adopted as the sea-level curve for the period of formation of the Princess Charlotte Bay chenier plain. No evidence was found for significant oscillations of sea level in the last 6000 years (Chappell, 1983). The chenier plain stratigraphy in Figure 3 is consistent with this.

The matter of progressive compaction of the sediments beneath the chenier plain deserves more comment. Where this has occurred it can falsify sea level interpretations from cheniers or analogous sequences. The nearest set of coral reef sites to Princess Charlotte Bay used by Chappell (1983) for the sea-level curve of Figure 5g is at the Flinders Islands (Fig. 2), where all observations fall directly on the sea-level curve shown in Figure 5g including several around 5000-6000 B.P. On the same diagram are plotted vertical bars representing dated mangrove facies recorded in stratigraphic pits. These are aligned with the coral-based sea-level curve using a different datum, explained in the Figure caption. The measured range of the modern mangrove facies also is shown. The effect of compaction is suggested by the difference between the falling sea-level curve and more uniform heights of dated mangrove facies. There is evidence of progressive compaction with increasing age in that bulk density of the intertidal muds increases, being about 15% greater in 5000 B.P. material than in 1000 B.P. to modern material, and permeability correspondingly decreases. The writers believe that tectonic effects are not involved in the region.

Finally, geometry of sedimentation in a shallowing bay suggests that progradation will accelerate even though the input sediment flux is constant (Fig. 6). The steady fall of sea level in the last 6000 years will enhance the accelerating tendency, as shown by intersections of the sloping line in Figure 6a. The importance of this effect cannot be estimated without knowing past values of the lower intertidal gradient, which cannot be gauged from our shallow stratigraphy. Other than this possible effect, sea-level change in the last 6000 years is thought to have had no role in the history of the chenier plain.

Chenier–Mangrove Interaction

The sediment-trapping role of the mangrove fringe was described above, and a marked tendency of shell debris to collect in lobes and low banks within the modern fringe was noted. These shelly deposits are found as layers within the mangrove facies exposed in stratigraphic pits, and may be thought of as incipient cheniers which did not develop into ridges, presumably because of inadequate flux of shell into the mangrove fringe. Two observations show that a higher shell flux can kill major components of the mangrove community:

(i) Thicker shell banks occur locally within the present fringe, analogous to the sparse chain of islets which are labelled 'datum ridge 1' in Figure 2. These banks typically lie about midway through the fringe, with *Rhizophora* communities on the seaward side and surrounded by *Bruguiera/Ceriops* communities. Where these banks are thicker than about 40 cm, one finds dead stumps of *Bruguiera/Ceriops* within them and young trees of species characteristic of a shelly habitat of higher elevation, in particular *Osbornia*. Large mature *Avicennia* grow through the banks, often with their pneumatophores buried by shell. These *Avicennia* are thought to be members of the original forest which have survived shell deposition while their Rhizophoraceae neighbours perished.

(ii) Dead stumps of a fringing *Rhizophora* forest are exposed in the mud flats immediately seawards of the northern beach-ridge plain (Fig. 5). Mature *Avicennia* grow on the seaward ridge, with pneumatophores buried by up to 70 cm of shell. The mud-floored swale behind this ridge has a mixed youthful *Osbornia/Ceriops* forest, and excavation showed stumps of *Rhizophora* 70 cm below the swale surface. This appears to be a more extreme example of death of a *Rhizophora* forest fringe, caused by rapid deposition of a shell ridge.

Without the trapping action of a sufficiently wide and dense mangrove forest fringe, there is little obstacle to landward movement of shell debris across the upper intertidal flats. Under these conditions a storm may drive a new chenier back until it meets an older ridge, as discussed by Jennings and Coventry (1973). This has occurred in the past. Tracing ridges on aerial photographs (1:10,000 scale) shows that the youngest chenier grouped with datum ridge 4 to the north section separates from it to the south, and joins datum ridge 3. Radiocarbon ages confirm this (Fig. 3).

Closer examination of chenier/mangrove is possible through pollen analysis of stratigraphic cores. Two cores taken on the north traverse are reviewed here and are linked with stratigraphic pits to produce a detailed picture of episodic mangrove growth and chenier formation. Pollen extraction and counting was done by standard methods in the palynology laboratory of the Department of Biogeography and Geomorphology, ANU.

Stratigraphic pollen material must be interpreted in terms of modern pollen assemblages in different sedimentary environments. Accordingly, surface sediment samples were collected, in pairs, from 12 sites across the north traverse. Pollen counts were found to be consistent between paired samples from each site. Four other traverses also were sampled and showed trends similar to those at the north traverse. Airborn pollen also were sampled by traps over 18 months but showed distributional patterns rather different from pollen in surface sediments, presumably due to effects of water-born transport (Grindrod, 1983). It is considered, from field observations, that flowering of all major taxa was sufficiently abundant to be representative.

It was found that the proportion of Rhizophoraceae pollen decreases landwards from the mangrove fringe. *Avicennia* differs in showing relatively high proportions near the ridges where these trees occur along ridge margins, as well as locally high values near *Avicennia* stands in the mangrove fringe. It was found that the ratio of total Rhizophoraceae count (*Rhizophora + Bruguiera + Ceriops*), against the total count of dominant non-halophytic taxa (Myrtaceae + Graminae + Cyperaceae) is a good distance-dependent index. This index is plotted in log form across the north traverse in Figure 8a, showing a smooth decrease landwards from the mangrove fringe to ridge 2. Across ridge 2 the index falls abruptly by a factor of 5, which is thought to reflect the barrier effect of ridge 2 against water-born mangrove pollen from the fringe.

Figure 8a provides a simple guide for interpreting stratigraphic pollen data. Figures 8b and 8c show values of the Rhizophoraceae/non-halophytic index, in log form, at close intervals down two cores. Locations of cores are shown at the top of Figure 8, core 2 being landward of datum ridge 1 and core 7A being directly behind datum ridge 2. Inspection of the surface sample data suggests two critical values of the index, one being a value of +0.25 near the rear of the fringe and the other being -0.5 at ridge 2 (lines M and R in Fig. 8a). Stratigraphic samples with index values >0.25 are interpreted as having been deposited within a mangrove fringe, and samples with values significantly below -0.5 may be inferred to have been deposited behind a ridge. Lines M and R are drawn along the core records in Figures 8b and 8c and the whole diagram shows how the cores are zoned in these terms.

Two other mangrove pollen variables are shown in Figure 8. In core 7A, *Avicennia* pollen count, reckoned as a percentage of total mangrove count (Rhizophoraceae + *Avicennia*) rises steeply in the upper core, above the point where the fringe appears to have moved from the core site. The *Avicennia* index is highest above 23 cm core depth, in the zone apparently deposited to the rear of a ridge. This is consistent with the observation, above, that *Avicennia* trees often form a line at ridge-base. Hence, the main phase of formation of chenier ridge 2 appears to have occurred in the interval represented by 23-33 cm depth in core 7A. In core 2, the % ratio of *Bruguiera/Ceriops* to total mangrove pollen is plotted (Fig. 8c). This rises steeply in the upper part of the core, indicating succession of *Rhizophora* forest to *Bruguiera/Ceriops* forest. This resembles the spatial succession found in the modern fringe. It is noted that the *Bruguiera/Ceriops* percentage stays high for about 8 cm of the 'upper intertidal' zone identified by the simple Rhizophoraceae/non-halophytic index. This compares with the observed lower pollen influx of *Bruguiera/Ceriops* pollen relative to *Rhizophora* in surface sediments in the mangrove fringe (Grindrod, 1983), and indicates that the simple major index used in Figure 8 is not quite unequivocal.

These results are combined with stratigraphic facies data, plus pollen counts from stratigraphic pit samples described by Grindrod (1983), in developing a model of episodic progradation in the next section.

The Internal Dynamic of Chenier Formation

Passage from the chenier system into the northern beach-ridge plain shows that chenier ridges do not form every time there is a potential ridge-building event. In the last 2000 years, six times more beach-ridges formed than chenier ridges. The governing factor must be adequacy or otherwise of the source of ridge-building shells.

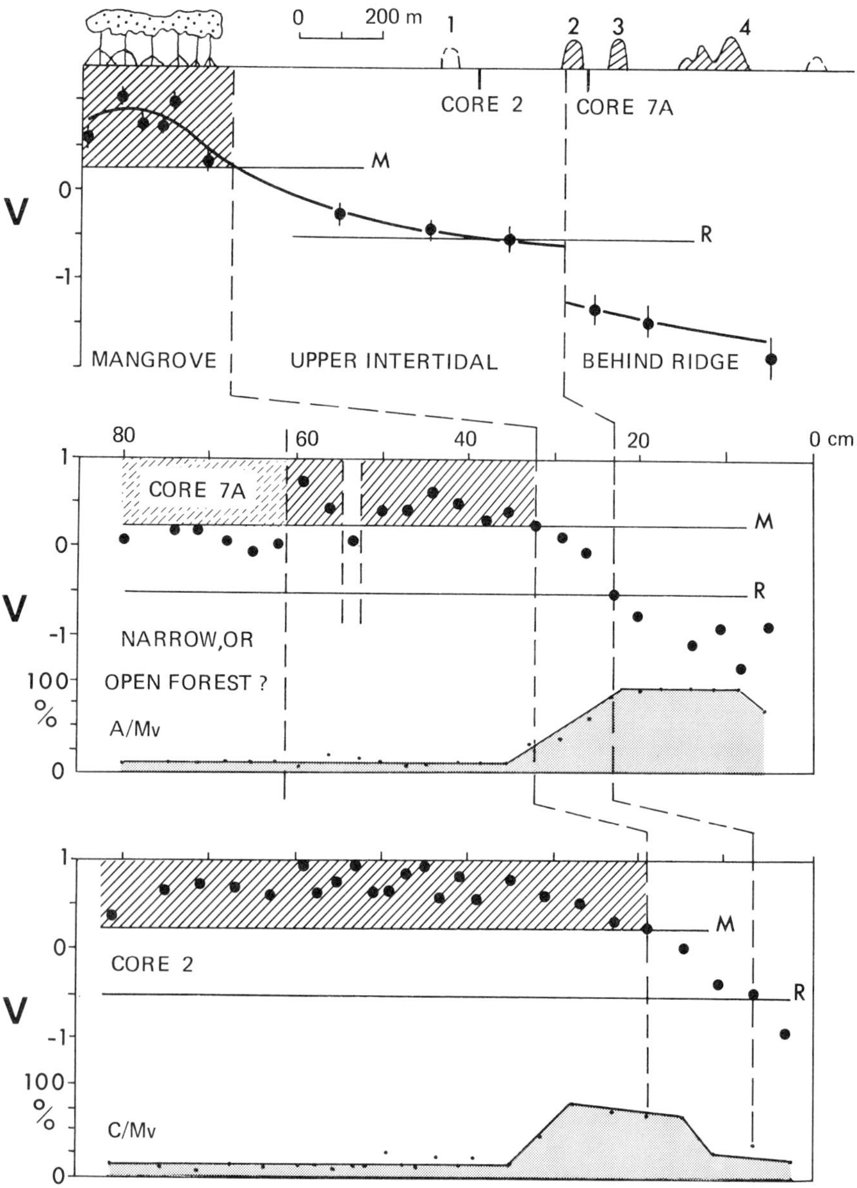
0
200 m
1
2
3
4
CORE 2
CORE 7A
M
R
V
0
-1
MANGROVE
UPPER INTERTIDAL
BEHIND RIDGE
80
60
40
20
0 cm
1
CORE 7A
NARROW,OR
OPEN FOREST ?
100
%
A/Mv
CORE 2
C/Mv

In the case of the beach ridges, over the last 1350 years there have been 16 ridges built, counting double or splayed ridges as a single event. This gives a recurrence interval of about 85 years. While this is much longer than the average recurrence interval of cyclones in the region (about 1 year; Lourenz, 1977), it is similar to the recurrence of cyclones which produce strong local geomorphologic effects. The recurrence interval of appropriately large storm surges is probably a better figure. There are no data for Princess Charlotte Bay, but the recurrence interval for storm surges at least 1 m above highest astronomical tide at Thursday Island is estimated by Hopley and Harvey (1979) to be around 80 to 100 years, which matches the beach ridge result.

Coral shingle ridges on Great Barrier Reef islands are analogous to shell beach ridges; their formation by cyclone surges is described by Hopley (1982). At King Island just outside Princess Charlotte Bay (Fig. 2), the frequency of coral shingle ridges over the last 1100 years is very similar to that of our beach-ridge plain (dating of King Island features done by the senior writer). The coral reef source of shingle is very different from intertidal/subtidal shell beds in its roughness and sedimentary matrix. Similar frequencies of ridge formation in these different environments suggests that recurrence interval of suitable storms is the limiting factor, rather than source production of ridge materials.

Much lower frequency of chenier formation, adjacent to the beach-ridge plain in the study area, indicates that supply of shell from the chenier source area is much lower than from the beach ridge source. There are two alternative explanations:

(i) Shell production is the same in both chenier and beach ridge source areas, but the shell material is continually being buried by mud and thus rendered inaccessible. Chenier formation occurs whenever muddy sediment input falls below a critical level allowing wave-winnowing and removal of mud to expose shell debris, which is then swept shorewards to form a

---

FIGURE 8. Palynologic data from surface samples on N traverse (top), core 7A (middle), and core 2 (bottom). The index $V = \log_{10}$ (Rhizophoracaea pollen count/total of Myrtaceae + Graminae + Cyperaceae counts). Heavy dots indicate measured values of V at positions or core depths shown. A/Mv and C/Mv represent Avicennia, and Ceriops/Bruguiera, percentage ratios of total mangrove count in each sample. Horizontal lines M represent approximate lower limit of the V-index in a dense mangrove forest, and R represents approximate limit of V on upper intertidal flat in front of a ridge.

chenier. Gould and MacFarlan (1959), Cook and Polach (1973), and Rhodes (1982), amongst others, attribute chenier formation to this mechanism.

(ii) Shell production usually is low in the chenier source area and becomes relatively high occasionally, for some reason. Cheniers form only when production is sufficiently high. Chappell (1982b) suggested that shell production may be low after large influxes of muddy sediment. In this case, times of increased mud input would correlate with mudflat progradation without chenier formation, which also is predicted by explanation (i). Both explanations are consistent with the trend of accelerating muddy progradation with accompanying diminution of ridge building which is found in Figure 5.

There is some evidence that shell production is affected by high mud input. The period since datum ridge 1 (600 B.P.) manifestly has been one of rapid mud progradation. Shellfish communities on the present lower intertidal and shallow subtidal bed are sparse and, notably, many of the larger bivalve species found in the cheniers are virtually absent. Traces of prehistoric human occupation also are relevant. Occupation, evidently heavily based on the bivalve *Anadara*, is judged from midden and shell mound ages to have been extensive between about 2500 and 1000 B.P. (Beaton, *pers. comm.*). Middens younger than 1000 B.P. are relatively sparse and no middens younger than 470 B.P. have been found. Live *Anadara* are very sparse today and clearly this species, at least, has all but vanished during the phase of heaviest mud sedimentation. Disappearance of *Anadara* is not attributed to human predation, otherwise the *Anadara* content should diminish in size and proportion in the youngest middens (*eg.* patterns of size diminution and species replacement in Holocene middens in California; Beaton, 1983). This is not the case in the study area.

It is concluded that increasing muddy sedimentation has progressively reduced shell production in the lower intertidal and shallow subtidal source area of the cheniers in northeast Princess Charlotte Bay. Beach ridges continued to be built from shell sources of a similar species composition near the rocky shore at the northeast end of the Bay, which were transported longshore by storm drift as shown by southward-pointing splay features of the ridges.

The respective roles of mud influx, mangrove trapping, and shell supply, can be brought together in a model of episodic progradation by examining in detail the record from 2000 to 600 B.P. Stratigraphic and radiocarbon age data for this interval, on the north traverse, are shown in detail in Figure 9. All dates shown are based on whole shells except for those in core 2 which come from shell fragments. Thicker dashed

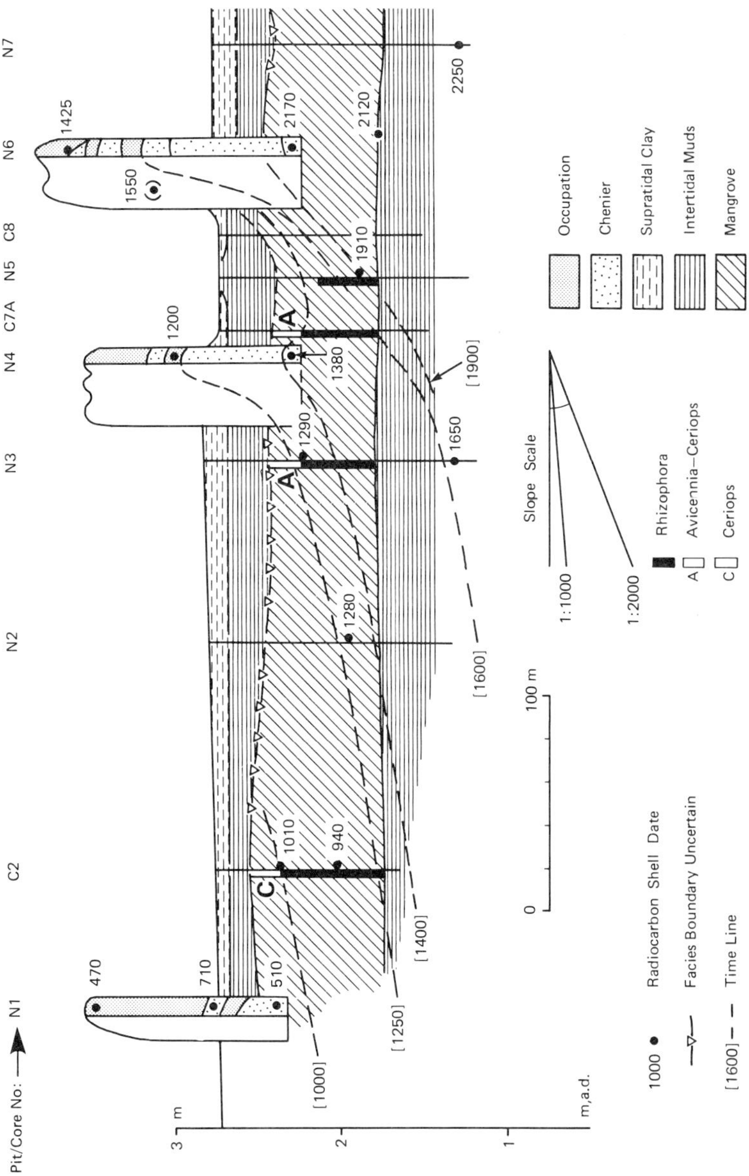

FIGURE 9. Detailed stratigraphic diagram, central part of N traverse. Note: upper limit of Rhizophora *zone not known in pit N5.*

lines oblique to facies boundaries are inferred 'time lines'. A slope scale at the bottom of Figure 9 shows present lower intertidal gradient and the gradient of steeper portions of the present mangrove fringe. Numbers of radiocarbon ages are too few to trace all time lines confidently, but the lines have been drawn to maintain parallelism as far as possible.

Despite dating uncertainties, there are two groups of apparent mud depositional rates, one 'faster' and the other 'slower'. Faster deposition occurs around 2100 B.P., from 1600 to 1400 B.P., and from 1250 to 1000 B.P. Slower rates occur between 1900 and 1600 B.P., and from 1400 to 1250 B.P.

Data of Figure 9 are combined with pollen results shown on Figure 8 (plus information from pit N3, not described here), to produce a model of alternating ridge formation and mangrove extension at nine time horizons from 2000 to 600 B.P. This is represented in Figure 10. Two different patterns are recognised. From 2000 to about 1200 B.P., 7 chenier-building events are recognised within the ridge 2 and ridge 3 complexes, while progradation was relatively small. This is termed the 'cut and recover' interval, when time-lines show alternating 'faster' and 'slower' deposition. The mangrove fringe was relatively narrow (100 to 150 m) and was episodically damaged during ridge-building events. Two such events are recorded palynologically in core 7A, for example: a brief phase of reduced mangroves around 53 cm core depth, and a major termination of *Rhizophora* forest around 30 cm (Fig. 8b). These particular events are recorded by shell depositional layers at the same stratigraphic depths in the adjacent chenier.

The second pattern begins around 1250 B.P., and is one of rapid muddy progradation under a wide mangrove fringe. During this time, pulses of muddy sediment input are inferred from the dated phases of clay dune formation labelled D1 to D7 in Figure 10 (see also Fig. 7). The only chenier feature built since 1200 B.P. is the sparse chain of islets referred to as 'datum ridge 1', at around 600 B.P. (average of two lowest dates in pit N1, Fig. 9). These islets resemble the patchy shell banks within the modern fringe. They probably drifted back through the fringe (similar drift is visible in the modern fringe), and it is not known whether the mangroves were much damaged during their formation. Figure 10 is drawn to suggest that fringe recovery and advance was rapid shortly after formation of ridge 1 islets. This pattern, from 1250 B.P. to the present, is termed the 'rapid prograding' interval.

The internal dynamic of chenier formation differs between the cut and recover mode, and the rapid prograding mode. In the former, chenier formation events appear to have occurred almost once a century, on average. There is little reason to suppose that shell production was greatly suppressed at any

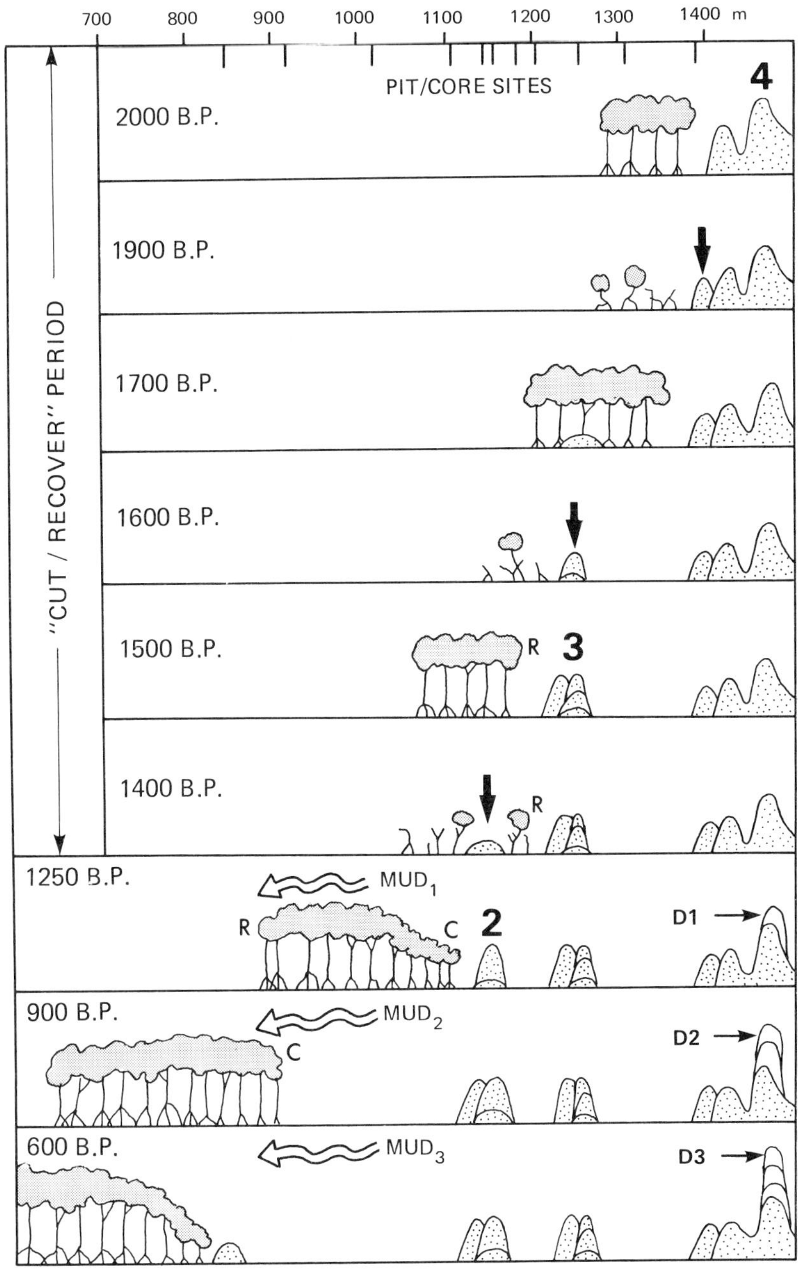

FIGURE 10. Nine stages of progradation from stratigraphic and palynologic data which are summarised in Figure 9. R = Rhizophora *forest, C =* Ceriops *forest; D1 to D3 = phases of clay dune activity.*

stage between 2000 and about 1200 B.P., either in terms of shell volumes accreted to cheniers in this interval, or from dated sequences of Aboriginal *Anadara* middens (Beaton, *pers. comm.*). Chenier building is suggested to occur when the balance between shell production, muddy sedimentation, and mud winnowing is such that shell quantities at the surface in the intertidal and subtidal zones are higher than average. At these times, sufficiently large storm surges and waves entering the Bay will build a chenier. Otherwise, when the volume of available shell is less, shell drifts within the mangrove fringe will occur. It is noted that shelly facies are common within the stratigraphic mangrove horizon from 2300 to about 1200 B.P. Whether a new chenier accretes to the face of a prior ridge, or whether it comes to rest some distance seaward, depends on the mangrove shell-trapping process. It is suggested that trapping is poor if the fringe is narrow or is sparse through prior damage. Elements of the process are outlined in Figure 11A.

There is a transition in the mode of progradation around 1200 B.P., from what may be termed the *cut and recover* pattern to a *rapid prograding* pattern. The differences between these are seen to be as follows. Major fluctuations of mud influx need not be invoked in the *cut and recover* mode. Higher than average shell availability, in source areas, depends on interaction of processes with different temporal patterns. Mud influx from rivers in Princess Charlotte Bay depends on wet season rainfall and surface sediment availability in river catchments. The latter is likely to vary with prior rainfall and fire history. Mud winnowing from shellbeds is influenced by northwest wind-wave patterns as well as passage of relatively distant storms. Shellfish productivity may not be constant, other things being equal. Hence, in this mode the formation of a chenier depends on a concatenation of prior circumstances and an appropriate local cyclone.

The *rapid prograding* mode differs in that chenier formation is much less frequent. The mangrove fringe is broad and focusses prograding sediment. Availability of chenier-building shell is constrained by mud deposition in the source areas. Since about 1200 B.P. muddy progradation has accelerated in Princess Charlotte Bay, which was linked earlier with changing geometry of sedimentation (Fig. 6a). Clay dune stratigraphy (Fig. 7) indicates pulsation of mud input which occur much less frequently than major cyclones, for reasons which are not known. However, an overall higher rate of mud sedimentation since 1200 B.P. is indicated by the evidence. Observations described earlier suggest that shell productivity in source areas is limited by high mud influx, although the effect on shell availability of continual burial cannot be discounted.

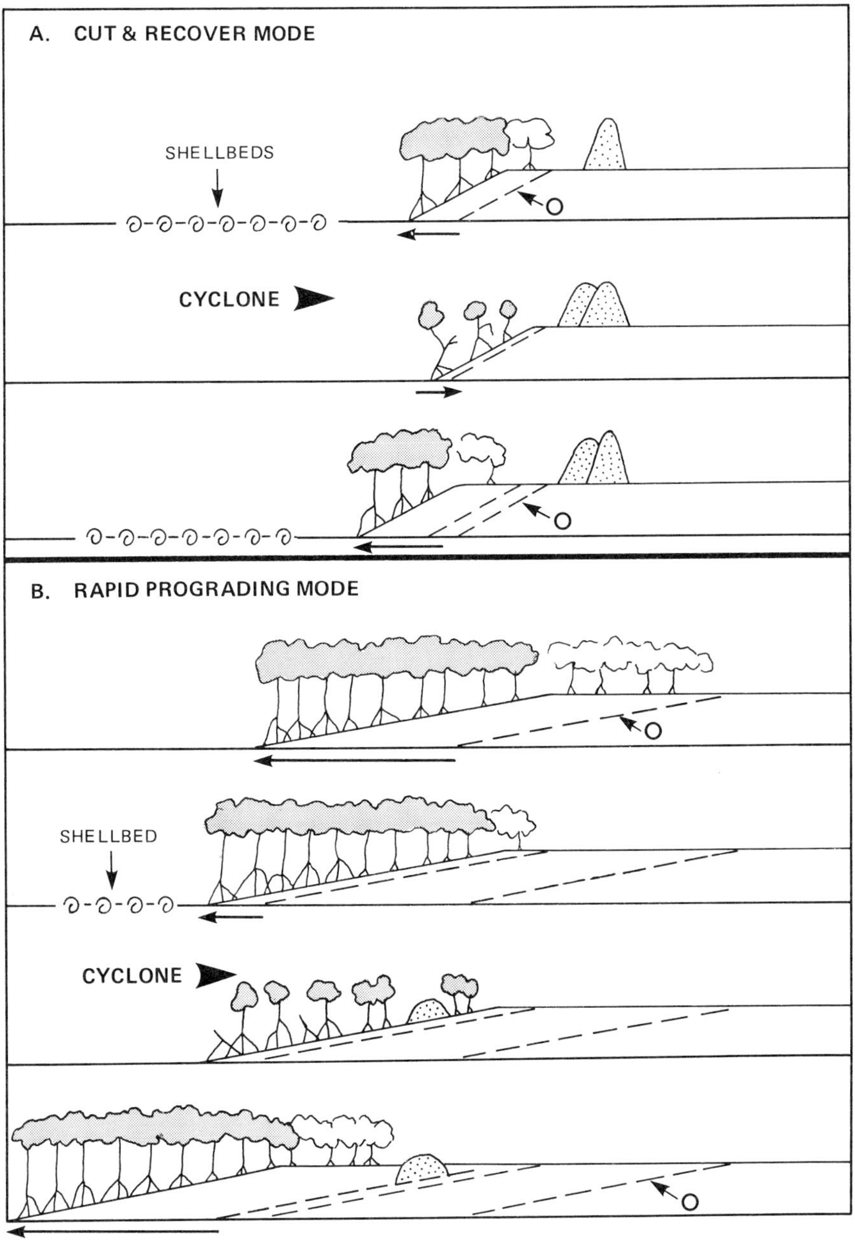

FIGURE 11. Two contrasting modes of chenier formation on prograding coast. Horizontal arrows represent relative progradation rates (see text for details).

If productivity of major ridge-building species becomes very low after major mud influxes then the dynamic of ridge building is limited by shell-bed recovery time, relative to the interval between major episodes of mud input. In this connection it is notable that the last very high mud influx occurred 600 years ago, as judged from well dated clay dune D3 (Fig. 10). All evidence suggests that shell-bed recovery has been minimal since then. Elements of the rapid prograding mode are outlined in Figure 11B.

Conclusions

Our detailed study of Princess Charlotte Bay chenier plain illustrates a suite of processes which are expected to apply broadly in similar coastal environments in northern Australia. General conclusions, relating to trends over the last 6000 years, are as follows.

(i) The trend of progradation, accelerating or otherwise, is a consequence of the geometry of the depositional bay or basin, and should not necessarily be interpreted as reflecting a like trend of terrigenous sediment influx. A trend towards change from 'cut and recover' chenier building to rapid progradation, similarly may be caused by depositional geometry, probably controlled by reef barrier growth offshore in an imperfectly closed basin.

(ii) Sea-level fluctuations appear not to have occurred in the last 6000 years and need not be invoked to explain chenier ridge formation. Sea level relative to Princess Charlotte Bay appears to have fallen smoothly by about a metre in the last 6000 years, according to evidence from nearby coral reefs.

Processes which contribute to progradation and episodic chenier building include the following.

(iii) The mangrove fringe focusses sedimentation, both in trapping storm drifts of shelly material and in promoting mud deposition. The latter appears to be more significant at the rear of the fringe. When the fringe is narrow, or is extensively damaged through onshore passage of a new ridge, the chenier may pass through it and accrete to a prior ridge some distance landward.

(iv) Shell availability in source areas is a primary factor in chenier formation. This is influenced by the rate of previous muddy sedimentation. There is evidence that when mud sedimentation exceeds a certain rate the shell productivity is significantly inhibited.

(v) In any analogous study it is necessary to identify the mode of progradation and episodic ridge building. The rapid progradation mode essentially is dominated by muddy sedimentation, fluctuations of which affect shell availability. The cut and recover mode essentially involves interaction of storms, mud deposition, and mangrove fringe. Although closely-spaced cheniers are formed this process differs from the mode of formation of a beach-ridge plain, where mud deposition has little effect.

In the light of these points we cannot identify any changes in climatic boundary conditions over the last 6000 years. The trends from 6000 B.P. to present may reflect nothing more than geometry of sedimentation in the Bay, and there is no reason to believe that the transition from 'cut and recover' to rapid progradation has a different primary cause. While three major influxes of muddy sediment are recognised in the last 1250 years, we do not attribute their absence earlier in the record to changes of catchment hydrology or forest cover. Simply, earlier influxes may not have been retained to the same extent in the intertidal and subtidal zones. We cannot speculate as to causes of these pulse-like influxes, which may represent extreme events related to concatenation of several circumstances in the catchment.

Probably the most interesting new research development in our study is the introduction of detailed pollen analysis of the intertidal stratigraphy. This has shown the interaction of episodic progradation and mangrove forest succession, during formation of the chenier plain. Its application to other stratigraphic cores would further resolve details of past processes and succession. It is time-consuming work, but we expect that a similar combination of palynologic and morphostratigraphic analysis will illuminate the dynamic history of the important mangrove coastal systems in northern Australia.

References

Augustinus, P.G.E.F., (1978). *The Changing Shoreline of Surinam*. PhD thesis, University of Utrecht, The Netherlands, 232 pp.

Beaton, J.M., (1983). The nature of Aboriginal exploitation of mollusc populations in Southern California. In: Terrell and Irwin (eds.), *Biogeographic Models in the Pacific*. University of Hawaii Press (in press).

Chappell, J., (1982a). Radiocarbon dating uncertainties and

their effects on studies of the past. In: W. Ambrose and P. Duerden, (eds.), *Archaeometry: an Australian Perspective*. Dept. Prehist., R.S. Pac. S., ANU, 322-335.

Chappell, J., (1982b). Sea levels and sediments: some features of the context of coastal archaeological sites in the tropics. *Archaeology in Oceania* 17, 69-78.

Chappell, J., (1983). Evidence for smoothly falling sea level relative to north Queensland, Australia, during the past 6000 years. *Nature*, 302, 406-408.

Chappell, J., Rhodes, E.G., Thom, B.G. and Wallensky, E., (1982). Hydro-isostasy and the sea level isobase of 5500 B.P. in North Queensland, Australia. *Marine Geology* 49, 81-90.

Cook, P.J. and Mayo, W., (1977). Sedimentology and Holocene history of a tropical estuary (Broad Sound, Queensland). *Bureau of Mineral Resources Bulletin* 170, 206 pp.

Cook, P.J. and Polach, H.A., (1973). A chenier sequence at Broad Sound, Queensland and evidence against a Holocene high sea level. *Marine Geology* 14, 253-268.

Curray, J.R. (1969). Shore zone sand bodies, barriers, cheniers, and beach ridges. In: D.J. Stanley, (ed.), *The New Concepts of Continental Margin Sedimentation* A.G.I. Short Course Lecture Notes, A.G.I. Washington D.C.

Davies, J.L. (1972). *Geographical Variation in Coastal Development*. Oliver and Boyd, Edinburgh; 204 pp.

Gillespie, R. and Polach, H.A. (1979). The suitability of marine shells for radiocarbon dating of Australian prehistory. In: R. Berger and H. Suess, (eds.), *Proceedings Ninth International Conference on Radiocarbon*, Univ. of Calif. press; 404-421.

Gould, H.R. and MacFarlan, E. (1959). Geologic history of the chenier plain southwestern Louisiana. *Transactions, Gulf Coast Assoc. Geologic Societies*, 9th Annual Meeting Houston 9, 261-270.

Grindrod, J.F., (1983). Holocene history of mangroves on a prograding tropical coast. Unpub. MSc thesis, Biogeography and Geomorphology, ANU.

Hopley, D. (1982). *The Geomorphology of the Great Barrier Reef*. Wiley, N.Y., 453 pp.

Hopley, D. and Harvey, N (1979). Regional variation in storm surge characteristics around the Australian coast: a preliminary investigation. In: R.L. Heathcote and B.G. Thom, (eds.), *Natural Hazards in Australia*, Aust. Academy of Science, Canberra, 164-185.

Jennings, J.N. and Coventry, R.J. (1973). Structure and texture of a gravelly barrier island in the Fitzroy Estuary, Western Australia, and the role of mangroves in the shore dynamics. *Marine Geology* 15, 145-167.

Lourenz, R.S. (1977). *Tropical Cyclones in thc Australian Region July 1969 to June 1975.* Bureau of Meteorology Canberra, 111 pp.

Otvos, E.G. and Price, W.A. (1979). Problems of chenier genesis and terminology - an overview. *Marine Geology* 31, 251-263.

Rhodes, E.G. (1980). *Modes of coastal progradation, Gulf of Carpentaria.* PhD Thesis, Australian National University Canberra; 357 pp.

Rhodes, E.G. (1982). Depositional model for a chenier plain, Gulf of Carpentaria, Australia. *Sedimentology* 29, 201-221.

Russell, R.J. and Howe, H.V. (1935). Cheniers of south-western Louisiana. *Geographical Review* 25, 449-461.

Schofield, J.C. (1960). Sea level fluctuations during the last 4000 years as recorded by a chenier plain, Firth of Thames, New Zealand. *New Zealand Journal Geology Geophysics* 3, 467-485.

Scoffin, T.P. and Stoddart, D.R. (1978). The nature and significance of microatolls. *Philosophical Transactions, Royal Society of London, Ser, B,* 248, 129-139.

Stuiver, M. and Polach, H.A. (1977). Discussion: reporting of ^{14}C data. *Radiocarbon,* 19(3), 355-363.

Thom, B.G. (1967). Mangrove ecology and deltaic geomorphology, Tabasco, Mexico. *Journal of Ecology* 55, 301-343.

Thom, B.G., Wright, L.D. and Coleman, J.M. (1975). Mangrove ecology and deltaic-estuarine geomorphology; Cambridge Gulf - Ord River, Western Australia. *Journal of Ecology* 63, 203-232.

11

Sand Barriers of Eastern Australia: Gippsland — A Case Study

B. G. THOM

Department of Geography
University of New South Wales
Duntroon, ACT

Introduction

Research into the properties and evolution of coastal sand barriers was pioneered in Australia by E.C.F. Bird (1961, 1965, 1973, 1978) with his work in the Gippsland region of eastern Victoria. Since his initial Ph.D research at the Australian National University in the late 1950s, there have been many other studies of barrier morphology, sediments, stratigraphy and age structure at various locations in eastern and southern Australia. It is perhaps appropriate in a book which considers recent advances in Australian coastal research, to return to the site of the much quoted early work. This study aims at providing a preliminary account of morphostratigraphical relationships in the Gippsland region based on drilling and identification of lithofacies. Use is made of barrier models developed elsewhere both in Australia and overseas.

Barrier Types and Depositional Environments

Definitions

The term barrier is commonly applied to bodies of detrital sediment, which rise above present sea level and block off or impound drainage from the hinterland. Typically, they are elongate sand bodies, parallel to the shore, and separated from the 'mainland' in whole or in part by an estuary or lagoon, swamp or marsh, or sand or mud flat. Various types of

COASTAL GEOMORPHOLOGY
IN AUSTRALIA
ISBN 0 12 687880 3

topographic association with the 'mainland' can be recognized. These include: complete detachment as 'barrier islands', perhaps forming a barrier-island chain separated by inlets; connection at one end to the mainland with an opening at the other end which is being progressively infilled by alongshore sand transport - a barrier spit; and attachment to both headlands (perhaps with a minor inlet opening) with little evidence of alongshore sand transport - a bay barrier.

There are three major geomorphic elements in any barrier system:

(i) the sand barrier complex itself, which rises from below low water to above high tide; generally, this complex is partly vegetated and consists of dune ridges landward of the subaerial beach which may or may not be inundated during high magnitude storm surge conditions; it also includes nearshore and inshore (the shoreface) sedimentary environments which are dynamically linked to the beachface by depositional processes (Wright and Thom, 1977);

(ii) the enclosed or partly enclosed body of water (lagoon or estuary) and associated backbarrier (backdune) sand or tidal flats and marsh/swamp terrains, which lie behind the barrier complex; and

(iii) the channels which cut through the barrier and connect the lagoon or estuary to the open sea; these tidal inlets are commonly associated with tidal deltaic sand bodies.

Each of the basic geomorphic elements noted above constitutes a sedimentary environment in which distinctive depositional processes operate (Reinson, 1979). The resulting sediment accumulation or facies, if preserved, possesses characteristic traits reflecting the dominance of a particular process or processes. Various facies types are identified in barrier complexes, and investigations of such in modern barrier systems in recent years have provided the basis for the interpretation of barrier rock sequences.

Vertical sequences of barrier sediment types and internal structures, based in part on modern environments and in part on buried sequences, have been proposed by Hunter *et al.*, (1979). Sediment characteristics vary with elevation above sea level and depth below sea level at any given barrier location. Many of these characteristics have been described from barrier sequences in North America and Europe (see Hayes and Kana, 1976; Kraft 1971, Kraft and John 1979; Bernard and Le Blanc, 1965; van Straaten, 1965; Davies *et al.*, 1971).

Two major factors are seen as influencing the distribution and properties of facies in a barrier sand complex. These are: (i) energy conditions (primarily wind and wave, and to a lesser extent tidal) within a given barrier environment; and (ii) mode of accumulation in response to changing sea level and sediment supply. The importance of energy conditions is demonstrated in a comparison of sections through barrier complexes involving regressive sand accumulation in areas of contrasting inner shelf steepness and energy conditions. In relatively low energy cases from the Gulf of Mexico (Matagorda and Galveston), the lower shoreface consists of fine-grained sediments which are extensively burrowed (see Davies *et al.*, 1971, Fig.6). This type of sequence has also been reported from Kiawah Island in South Carolina (Moslow and Colquhoun, 1981) and Sapelo Island in Georgia (Howard and Reineck, 1982). The lower shoreface is little more than 10 m below present sea level in these cases. In contrast, the more energetic coasts of the Netherlands and N.S.W. (Newcastle Bight) possess a coarser shelf sand at depth, a relatively steep shoreface (nearshore zone), and high dune ridges on the barrier (van Straaten, 1965; Roy and Thom, 1981; Thom 1984, in press). Finer-grained sediments are generally, but not exclusively, absent from shoreface and inner shelf environments.

Barrier Types and Facies

Figure 1 depicts the sectional geometry and generalized morphology of four barrier types (including variants) which occur on the high-energy east coast of Australia (Thom, 1974; Thom *et al.*, 1978; Roy and Thom, 1981). Similar types have been recognized in other parts of the world (see van Straaten, 1965; Dickinson *et al.*, 1970). In eastern Australia, these types reflect different modes of accumulation on an embayed rocky coast where sea level has risen rapidly (1 to 1.5 m/100 years) until 6500 years ago then remained more or less stationary during the so-called Holocene 'stillstand'. The following is a summary of major attributes of each type.

Regressive or prograded barriers. For this type abundant supplies of nearshore sand are available at some stage for onshore sand transport. Sand accumulation takes place on the beachface and is blown inland into low-relief foredune ridges where it is bound by vegetation. Over time these dune sands are subjected to leaching and soil development (see chapters in this volume by Hesp, and Thompson and Bowman). Further accumulation results in seaward migration of the beachface and the formation of a beach-ridge or strand plain. The presence of a parallel ridge and swale pattern on maps or air photos is a distinctive characteristic of this type of barrier complex (Fig. 1). In eastern Australia, heavy mineral seams are often concentrated in the upper beachface beds beneath a thin (2-4 m)

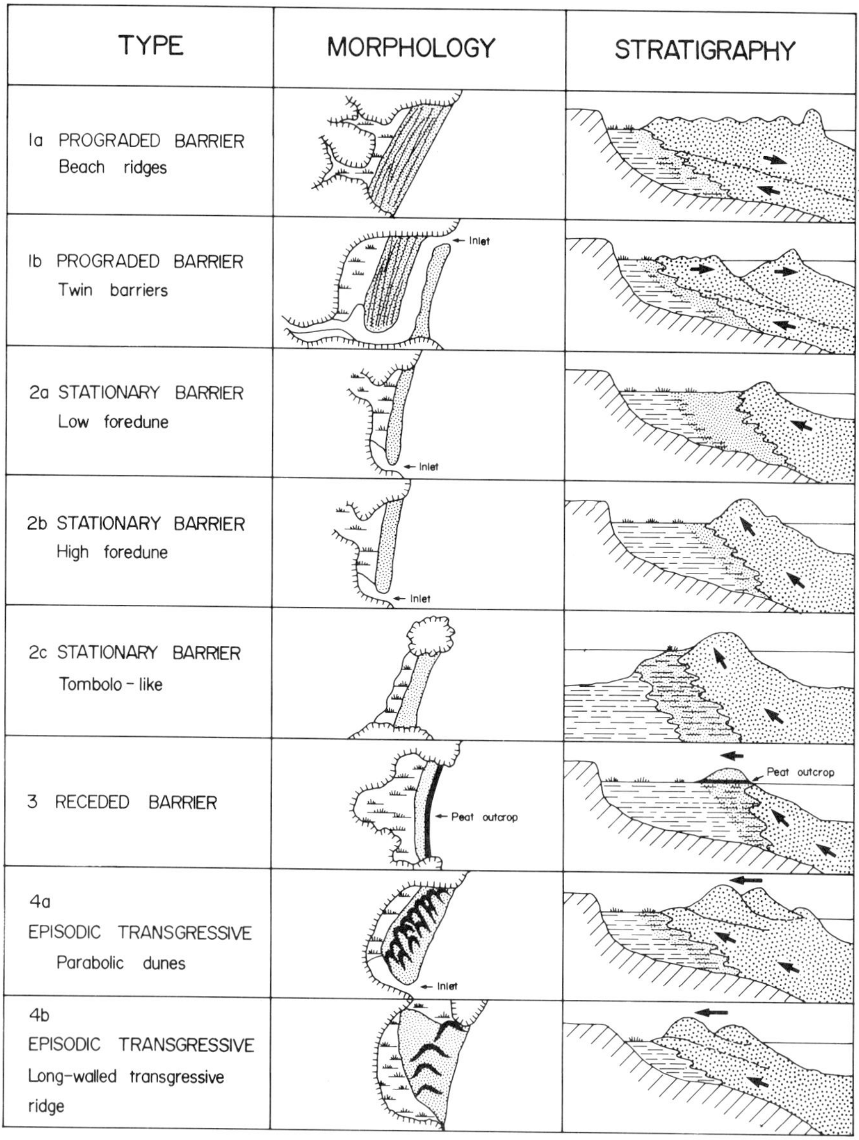

FIGURE 1. Morphostratigraphic types of bay barriers following Thom et al. *(1978).*

dune cap. Generalized vertical sequence of facies and their characteristics for this barrier type in comparison with others is shown in Figure 2.

Stationary barriers. This type is also termed aggradational or vertical build-up and has been described in Texas (e.g. Padre Island) as well as eastern Australia (see Dickinson *et al.*, 1972; Thom, 1974; Thom *et al.*, 1978; Thom, 1984, in

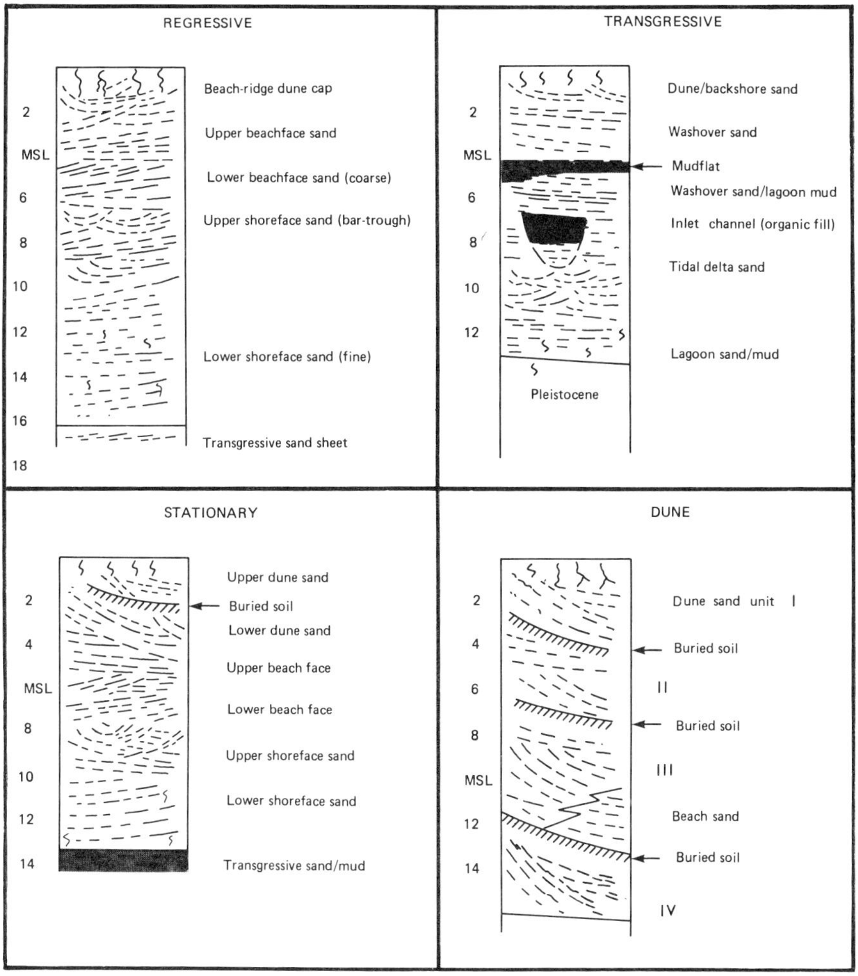

FIGURE 2. Generalized vertical sequences of sand facies in four common barrier types recognized in eastern Australia.

press). According to Hayes and Kana (1976), the sand facies is shaped like a mushroom in cross-section with the landward side expanding by washover deposition and the seaward side by limited beach-ridge accretion. In some Australian examples, the upper portion of the barrier may accrete as a result of the stacking of foredune sands into a complex hummocky dunal ridge containing buried soils. Time lines tend to slope either vertically or at a high angle beneath stationary barriers indicating limited seaward accretion and the maintenance of a close balance between conditions influencing shoreline migration over time (Fig. 1). Figure 2 depicts general facies relationships for this barrier type.

Transgressive or receded barriers. Transgressive barriers are characteristic of eroding shorelines. Kraft (1971; see also Kraft and John, 1979) has documented facies characteristics of this type on the coast of Delaware. An example from eastern Australia is that described by Thom (1984, in press) from Dee Why near Sydney. Sub-tidal and intertidal backbarrier facies dominate the section below sea level and contact between facies may be sharp and erosional. This type is generally associated with a rising sea level and/or a shorezone deficient in sand relative to energy conditions.

Episodic dune barriers. In eastern Australia there are both Holocene and Pleistocene examples of the episodic dune barrier (Thom, 1974; Roy and Thom, 1981). The best known examples are those of the dune sand islands of southern Queensland (Fraser, Moreton, Stradbroke). Stacking of sand blown inland from either an eroding or receding beachface, leads to vertical as well as horizontal accumulation of thick, cross-bedded sands (Pye, 1983). Wave-deposited sediments are restricted to a narrow shoreline fringe. A Holocene example occurs near the Myall Lake, N.S.W., in the Eurunderee embayment (see Pye and Bowman, this volume, Fig. 5). Aeolian sands may extend below present sea level (Fig. 2). This type is found in areas of very high wind energy and abundant sand supply, and may result in development of extensive sand sheets downwind of shorezone facies.

Barrier-inlet type. Migration of inlets alongshore can result in erosion and subsequent infilling of tidal-inlet channels. Modern examples occur on the Atlantic coast, U.S.A. (e.g. Shackleford Banks, N.C., Susman and Heron, 1979). The facies model is a fining-up sequence with a thinning-up trend in cross-bed set thickness (Reinson, 1979, Fig. 24). It is characterized by an erosional base with tidal-channel sand facies. It is not a common type in eastern Australia where on-offshore sand transport seems to dominate and inlets have not migrated very much in the Holocene. An example from Gippsland will be discussed below.

Composite or hybrid barriers. Clear-cut differentiation of a particular barrier or section of a barrier into one of the five types mentioned above is not always possible. Figure 1 illustrates several variations of the four basic types common in eastern Australia, prograded, stationary, receded and episodic dune. In many cases a transgressive sediment sequence may exist beneath a barrier sand body. These transgressive deposits may have formed (and later be reworked) during the Postglacial Marine Transgression (P.M.T.) (see Thom, 1984, in press, and discussion below on Gippsland cases). There are examples where beach ridges have been partially reworked into massive dune ridges (e.g. Newcastle Bight, Thom *et al.*, 1981). There are also a few cases where regressive sands have been partially reworked by a migrating barrier inlet. Such processes may be relevant to the formation of twin barriers of Holocene age (type 1b, Fig. 1, see below for discussion of a Gippsland case). The association of transgression and inlet migration is considered by Kumar and Sanders (1976) to be the origin of many linear sand bodies on inner continental shelves. Nevertheless, the classification used above does provide a basis for identification of relationships between facies types which have evolved over time into barrier sequences.

Lagoonal (Backbarrier) Facies

Behind the sand complex which forms in the shorezone of the growing barrier is the flat-lying sediment basin which is termed the backbarrier environment. In this zone sediments accumulate as a result of a number of processes:

(i) deposition by suspension (to a lesser extent bed load) from *rivers* discharging into lagoons from the hinterland; typically, these sediments form pro-delta muds and tidal flats in the lagoons or estuaries blocked by sand barriers;

(ii) *biogenic* production of organic and carbonate skeletal material below MSL and on tidal flats bordering lagoons;

(iii) *chemical* precipitation within and on the surface of lagoonal sediments;

(iv) *washover* sand deposition as extensive sheets which have been transported by storm surges across the shorezone and frontal dune ridges into the lagoonal water body;

(v) *aeolian* reworking of exposed sand surfaces to produce backdune flats; and

(vi) channel-fill deposition of *flood-tidal delta* origin.

As noted by Reinson (1979), one of the characteristic traits of backbarrier deposition is interbedding; commonly, there are sequences of mud, sand and organics each with distinctive fossil assemblages and sedimentary structures reflecting primary depositional processes and subsequent diagenesis (see also Schwartz, 1975; Reineck and Singh, 1975; Hayes and Kana, 1976). Mud content may increase with distance from the shorezone. Thus there is a marked contrast between the predominantly clean sand sequences of shorezone barrier environments (especially those from high energy coasts), and interbedded backbarrier sequences.

In many areas of the world, tropical cyclones and extra-tropical cyclones both produce storm surges which overtop low relief frontal dunes. Modern examples have been described in North Carolina (Schwartz, 1975) and Texas (Andrews, 1970; Hayes and Kana, 1976). Fan-shaped deposition is rapid often producing flat-lying beds with coarser layers (including reworked shell) at the base, grading into mixed sand and shell, then into relatively shell-free sand (Andrews, 1970). Horizontal stratification dominates resulting from flow across a subaerial surface depositing small delta-foresets into a body of standing water (Schwartz, 1975, Fig. 10). Aeolian processes may modify or bury washover deposits to various degrees (Hayes and Kana, 1976). Reinson has noted (1979; see also Kraft and John, 1979) that washover deposits form a significant portion of barrier sediment complexes, especially those which have been receding. Under transgressive conditions washover is one of the main processes by which barriers migrate landward.

Washover deposition is quite inactive today along the east coast of Australia. This was not the case during the P.M.T. and in the early stillstand period. Both in Gippsland and N.S.W., sand sheets were formed behind transgressing barriers as sea level rose across the inner continental shelf. These sands were deposited into shallow lagoons containing fine-grained sediment often rich in organics (e.g. sea-grass beds). When sea level stopped rising *c.* 6500 years ago, the initial barrier ridge could be breached and overwashed during storms, and there is evidence (see below) for continued washover sand accumulation into barrier lagoons. Locally these sands are rich in heavy minerals (e.g. Fens Embayment, near Port Stephens, N.S.W.). This process ceased at various times between *c.* 6000 and 4000 years ago as shorelines prograded and sand dunes grew to heights which exceeded maximum levels of storm surges. As a consequence, backbarrier surfaces in some areas have been reworked by wind into low relief dunes, backbarrier channels have been infilled with lagoonal muds (organic and inorganic), and subaerial surfaces have become well vegetated. Swamp vegetation producing a thin peat layer (to 1 m), often covers the backbarrier surface (e.g. in the Myall Lakes region, N.S.W.).

The character of other backbarrier sediments varies greatly from one area to another depending on embayment configuration, climate, river discharge, tidal range, vegetation and estuarine organic assemblages. No attempt will be made here to present a comprehensive description. Some aspects of the Gippsland Lakes will be discussed below (see also Davis *et al.*, 1977; Thom *et al.*, 1983).

Quiet-water deposition of clays and silts from suspension occurs in shallow to deep lagoonal/estuarine waters where a river source produces pro-delta accumulation. These sediments are often rich in brackish-water molluscs and microfauna. Various deltaic environments of deposition can be recognized in and adjacent to the subaerial delta (for an example, see Donaldson *et al.*, 1970, Guadalupe delta into San Antonio Bay behind Matagorda Island, Texas). In N.S.W., Roy *et al.*, (1980) have described embayment fills associated with quiet-water sedimentation on a relatively high relief coast (see also chapter by Roy, this volume).

The character of tidal-flat sedimentation behind barriers is influenced by tidal range and the efficiency of tidal exchange through tidal inlets. Even in areas of moderate tidal range inlets may become blocked by infill from wave-reworked sands. In general, the greater the tidal range, the thicker are tidal-flat sediments. Classical tidal-flat deposits behind barriers are described from northwest Europe and Georgia (see review in Reineck and Singh, 1975). Little detailed research has been undertaken on similar sequences in eastern Australia. In Gippsland, Davis *et al.*, (1977) and Bird (1978) have described sedimentary features found in the Lake Reeve and Boole Poole areas.

Tidal-inlet and Tidal-delta Facies

The role of tidal-inlet migration and channel infill in the formation of barrier sand bodies has already been mentioned. Tidal-channel and tidal-delta sand bodies are closely associated with each other, and show many sedimentary features in common (e.g. grain size, sedimentary structures, reworked molluscan assemblage). Tidal currents flowing through narrow openings between barrier islands, or between barrier sand and the 'mainland', transport clastics which are deposited in ebb or flood-tidal deltas. Ebb-tidal deltas form seaward of the barrier and are subject to wave reworking; on high-energy, microtidal coasts such as those of eastern Australia, such deltas may be highly modified by waves, and frequently show limited development. Hayes and Kana (1976) argue that large ebb-tidal deltas are common on mesotidal coasts. Flood-tidal deltas grow into lagoons landward of the barrier and are little affected by wave and wind-generated processes (Hayes and Kana, 1976; Reinson, 1979, Fig. 12). Large volumes of sand may be stored in these tidal deltas.

Tidal-channel deposits form by the lateral migration of inlets associated with alongshore sediment transport. This results in barrier-spit development often with recurving sand ridges on the updrift margin of the channel. Such well-sorted beachface sand lies at the top of an inlet-fill sequence. Kumar and Sanders (1974) have shown that below low water this unit overlies 3 m of bidirectional cross beds with foreset beds indicating progradation into a shallowing channel. Towards the base, deeper channel beds occur which are dominated by lenticular, bidirectional cross strata. These may overlie a lag 'gravel' on an erosional base. The occurrence of a mixed molluscan assemblage containing reworked fauna from estuarine and marine sources is characteristic of tidal-inlet environments.

Tidal deltas have been extensively studied by Hayes (1975; see Hayes and Kana, 1976). He shows how delta channel pattern is controlled by segregated zones of ebb and flood flow both in microtidal and mesotidal areas. This flow segregation gives rise to a distinct pattern of bedform distribution (Reinson, 1979). Flood-oriented sand waves occur on shoals and ebb-oriented sand waves in adjacent channels. Bidirectional megaripples are also common on shallow surfaces.

In eastern Australia, there has been limited research on tidal inlets and deltas. Reinson (1977) provided a comprehensive description of channel flow and sediment characteristics on a flood-tidal delta at Mallacoota Inlet in northeast Victoria. The mouth of the inlet changes frequently, migrating in both directions; this is a common feature of many eastern coast inlets which are not secured by breakwaters. Water depth decreases during drought, again a common feature leading to the closure of many inlets along this coast. A net movement of sand landward characterizes Mallacoota and many other inlets in eastern Australia. Megaripples and sand waves show a dominant flood-tide orientation on intertidal sand bodies (Reinson, 1977, Figs. 23, 24). Ebb-oriented bedforms are confined to lower areas.

Roy *et al.*, (1980) have examined the general setting of tidal inlets and deltas on the N.S.W. coast. They emphasize the dominance of flood-tide deposition during and subsequent to the P.M.T., leading to the accumulation of massive quartzose sand bodies. These form at the mouths of inlets which cut the so-called 'barrier-estuary' and into the mouths of drowned river valleys. The cross-sectional character of two examples of such flood-tidal deltas and inlets have recently been documented by Thom and Roy (1983). Seismic and dating studies of these deltas and inlets show landward growth of large-scale foresets particularly in the early to mid-Holocene (Roy, 1983). Except for a study of channel scour by King (1981), little work has been done on tidal inlets and deltas in Gippsland. The large, relict flood-tidal delta at Ocean Grange will be discussed below.

Gippsland Case Study

General Setting

The Gippsland Lakes area comprises one of the largest estuary and sand barrier complexes in Australia. It has been described in numerous publications (e.g. Bird, 1965, 1978; Jenkin, 1968), but there have been few studies which describe and discuss facies characteristics and stratigraphy (see Davis *et al.*, 1977).

Figure 3 shows that the area consists of three large estuaries (Lakes Wellington, Victoria and King), an ephemeral coastal lagoon (Lake Reeve), several Pleistocene sand barrier remnants (e.g. Banksia Peninsula, Raymond Island, Sperm Whale Head) and a Holocene barrier complex (Ninety Mile Beach). Several rivers flow into the lakes, such as the Mitchell, Latrobe, Tambo, forming deltas (Bird, 1978, p. 87 ff). At present, tidal waters enter the lakes through a man-made inlet at Lakes Entrance. Prior to its construction in 1889, the natural inlet in the Lakes Entrance area was frequently closed due to wave transport of sand. Less saline conditions appear to have prevailed prior to 1889 than today (Bird, 1978).

The general configuration and processes influencing shore-zone sand transport have been discussed elsewhere (Bird, 1961, 1978; Jenkin, 1968; Wright *et al.*, 1982). This section will concentrate on a general description of facies types determined by shallow surface sampling, power auger drilling and vibro coring on the barrier complex including backbarrier flats. More quantitative aspects of sediment properties will be published elsewhere.

Holocene Barrier Facies Types

Foredune/upper beachface sand. The foredune or beach-ridge complex forming the so-called Outer Barrier of the

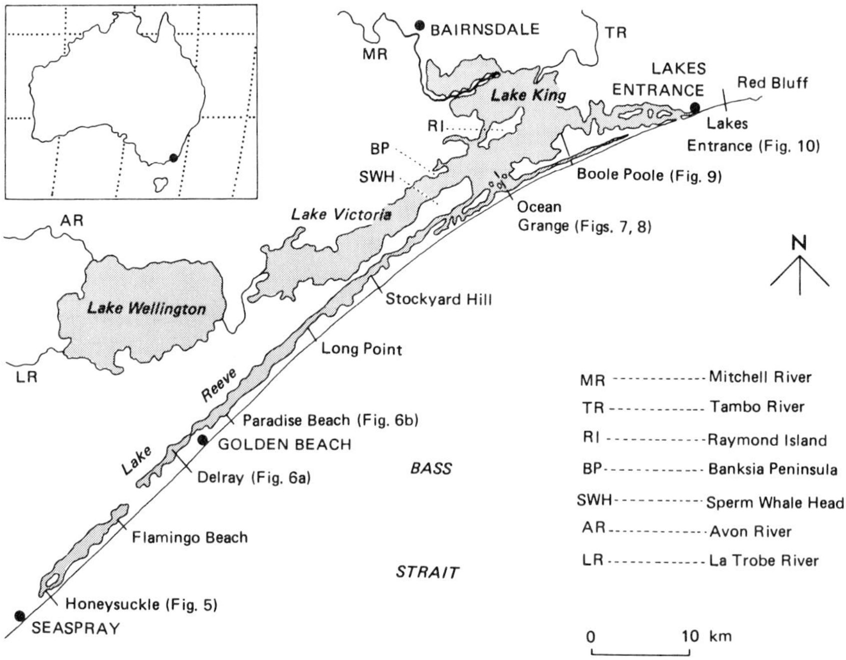

FIGURE 3. Location map of Gippsland Lakes indicating position of sections.

Gippsland Lakes varies considerably in width and height. Except in the Boole Poole area this facies occurs as a ridge 100 to 500 m wide and between 4 and 13 m high. In the southwestern area it comprises a single sharp-crested ridge which increases in bulk to the northeast. Multiple beach ridges occur northeast of Golden Beach and at Boole Poole (Fig. 3). From Ocean Grange to the east past Lakes Entrance the outer ridge is separated from the beach-ridge plain at Boole Poole and the backdune flats at Lakes Entrance by a narrow, elongate lagoon (Bunga Arm or Cunningham Arm).

The sands of the barrier ridge are typically medium/fine, well sorted, quartz rich and light yellow in colour. Very few dark minerals (rock fragments?) and heavy minerals occur in these sands. Grain size increases downwards towards the water table where medium/coarse sands are common along with shell fragments. Carbonate content also increases towards the water table. The facies is generally triangular in shape.

Podzolization has altered the colour and organic content of the upper 2 m of these sands. Soil development is minimal in the outer ridge and seaward face of the foredune where exposed to frequent blowout activity. Precipitation of calcium carbonate has taken place in areas where sand carbonate content is relatively high (approx. 10%) to form a triable calcarenite. Thin soil horizons may occur within the foredune ridge as discontinuities. Bedding is poorly developed.

Quite clearly aeolian processes have shaped the form and controlled the deposition of the upper part of this facies, most probably in association with sand-binding plants. The lower part is beach deposited, and contains somewhat coarser and more carbonate-rich sand at and just below the present water table.

Nearshore regressive shelly sand. This unit can be subdivided into depth zones. Beneath the foredune and upper beachface is a shelly sand. Frequently, the sand is quite yellow and coarse. In the upper 3 to 5 m large heavy shells such as *Glycymeris flammea* occur. The sand is mostly quartzose, but locally the carbonate content may increase to 30%. Lenses of shell are common. The zone also contains rounded carbonate pebbles. *Bankivia* and *Plebidonax* are sometimes encountered in this upper zone (= surfzone/lower beachface), as are worm specimens of *Anadara trapezia* (transported to the beachface from estuaries).

At depths of 3 to 5 m below MSL the facies often grades downwards into a more compact, medium/fine sand,fawny grey in colour and rich in carbonate fragments. A strong carbonate reaction is observed on the application of 10N HCl acid. Mixed species of shell occur in this zone including *Glycymeris*, *Bankivia*, reworked estuarine shells and fragments of Pleistocene calcarenite. Often the unit is quite well layered. Shells from this facies include fresh angular fragments and whole shells as well as brown, ironstrained and worn (rounded and burrowed) specimens. Radiocarbon dating of fresh fragments yields ages less than 7000 years to as young as 2000 years. However, discoloured worn shells when dated separately yield ages close to the limit of the method.

The nearshore shelly sand facies generally thickens to seaward beneath the barrier ridge. There are cases where this is not always apparent. To landward, the facies thins to a 'feather edge' often grading laterally into backbarrier sands.

Nearshore/beach transgressive shelly sand. A sharp discontinuity usually defines the upper contact of the shelly sand regressive facies and the shelly sand transgressive facies. The contact is marked by a sudden change to medium/coarse, shell-rich sand, orange-brown to fawny grey in colour. The unit is moderately well sorted, but frequently contains layers of coarser shell and carbonate pebbles and finer sand. A strong carbonate reaction on application of HCl acid is characteristic.

Molluscs present in this facies include a mixed assemblage of nearshore marine species (e.g. *Glycymeris*) and those commonly found on backbarrier (estuarine) subaqueous sand flats. Many of the shells are fresh and angular in appearance whilst others are worn and degraded. The former yield radiocarbon ages between 7000 and 8000 years. Species indicating reworking from older backbarrier lagoon facies (e.g. *Anadara*) may occur here.

This facies varies in shape. It is essentially a thin (1 to 4 m) unit which slopes gently seaward. This sand sheet may extend 1 to 2 km inland from the present shoreline. The upper contact has been eroded in places beneath the nearshore regressive facies. The lower contact grades downwards into a grey, finer sand with mud. Washover, beach and perhaps surfzone processes are important in the deposition of this facies.

Lagoon/washover, lower transgressive mud and sand. This unit rests on the Pleistocene contact. It is a composite unit consisting of interbedded quartzose sands, clayey sands, and sandy clays with some thin lenses of clay and peat. Typically, it is dark grey to blue grey in colour. Sands are often quite calcareous with abundant shelly fragments. Whole shell is scattered throughout and largely consists of estuarine species. Open-ocean fauna have not been observed in this unit. Sands are quite poorly sorted and possibly layered. They are mostly medium to fine. Thin fibres of sea grasses may occur in this unit.

The facies forms a sheet rising landwards. It varies in thickness from 1 to 5 m and may extend as a basal layer beneath all Holocene units from barrier front to inland margin. It is least well developed beneath the thinnest sand barriers (e.g. Boole Poole).

The sand component is most probably of washover origin accumulating behind a landwards retreating, low relief sand barrier. Radiocarbon dates are few but support the pre-stillstand origin of the unit. It can grade upward into the transgressive barrier sand, reflecting a change in dominance of wave reworking from relatively quiet-water deposition during the P.M.T.

Washover/lagoon muddy sands. Interfingering with the transgressive marine sand unit is a predominantly sand facies. It may overlie the lower transgressive mud and sand unit. Clay or sandy clay may occur as layers within the sand. The facies is commonly grey to light grey in colour.

Sands are fine to medium and moderate to poor sorted. Layering appears characteristic. Quartz sand grains dominate.

The unit is characterized by a wide diversity of estuarine shells, including *Anadara, Ostrea, Mactra, Nassarius spp.*, and *Notospisula*. Many are intact. Worn open valves of *Glycymeris* are sometimes found especially along pipeline easements crossing Lake Reeve where oil or gas pipelines are buried in this unit.

The unit either thins landward after interfingering with transgressive sands or remains uniform in thickness. Generally, the sand content diminishes landward. *Notospisula* shells seem to be more common in this direction. The facies reaches 10 m in thickness at the maximum.

Clearly this unit is of washover origin where sands and shells from the ocean side have been transported to the back-barrier environment into relatively shallow water. Muds have accumulated between major washover events. Habitat conditions were conducive to development of wide species diversity. Available radiocarbon dates suggest most deposition of this facies took place just prior or shortly after the termination of the P.M.T. There is very little evidence of shell accumulation younger than 4500 years ago.

Washover/backdune flat sands. This facies forms a thin sand cover landward of the barrier ridge. It is a surficial unit with a low relief ($<$ 2 m) dune topography. In places the sands have been moulded by lagoon waves and old inlet channels into linear ridges.

Sands are medium to fine, and moderately well sorted. They range from a dark brown to a black topsoil downwards into light yellow partially leached sand. Nodules formed from precipitation of mobilized carbonate are noted along with chalky fragments of shells at or above the water table. Near Lakes Entrance the sands are more quartzose and show the development of ground-water podzols. Organic staining is common in the B horizon. However, organic mud and peat may occur as discontinuous layers near the base of the facies.

The sands extend for distances of a kilometre or more landward of the barrier ridge. Rarely do they exceed 2-3 m in thickness. They are often cut by shallow, infilled depressions which may be former washover channels. Thin surface peat consisting of salt marsh plant detritus may form above the sand.

It is inferred that the sands were initially formed by washover processes after the termination of the P.M.T. Locally they have been reworked by winds into low-relief dunes probably in mid-Holocene times. Wave-reworking along the shores of the ephemeral Lake Reeve is currently active; Bird has recorded up to 3 new beach or 'contraction' ridges on some sections between 1957 and 1983 (1984, in press; see also Bird, 1978; Davis *et al.*, 1977).

Lagoon sands and muds. This is a heterogeneous facies, part forming to-day and part relict of previous depositional processes (e.g. washover). Davis *et al.*, (1977) state that terrigenous clay minerals are generally lacking in Lake Reeve except at western and eastern ends. They conclude that some chemical sedimentation is active producing a silty mud with a suggested composition of magnesium calcite - (Ca88 Mg12) CO_3.

The bed of Lake Reeve is often covered by an algal mat. This surface dries out in summer. Small shells of *Coxiella striata* are abundant on the surface. They are incorporated into the reworked relict sands, and into the algal mats and muds. Discontinuous laminations and fenestrae or unsupported voids do occur, but no systematic study has been made of the composition or fabrics of these surface sediments.

Muds rich in either organic or inorganic materials, also occur in the shallow subsurface. They are buried beneath thin (less 1 m) layers of washover sand. Evaporitic products do not occur in these buried muds; the muds reflect perennial water cover during deposition. They usually date in excess of 3000 years (see below).

Tidal-inlet sand and gravel. In the vicinity of Ocean Grange, extending into Lake King as a series of low sand islands and subaqueous banks dissected by shallow channels, is a relict tidal-inlet and delta complex. Davis *et al.*, (1977) suggested that these banks were part of an old tidal delta. Subsequent drilling by the Victorian Geological Survey (vibro-coring) and power augering by Geography, R.M.C., have confirmed the tidal-delta origin of the banks and islands.

Sands dominate the texture of the banks and the subsurface. Quartz is the dominant mineral. The sands tend to be medium to fine in the upper 10 m and moderately sorted. Some coarse sand layers with gravel have been noted. Organic mud layers occur as thin lenses. Shells are scattered throughout and include a mixed reworked assemblage of marine and estuarine species. Carbonate content rarely exceeds 5%. Below 10 m pebbles are quite common. Rounded and discoidal pebbles have been noted. Mixed shell species also occur in this unit.

Stratigraphic Patterns

Nine sections have been drilled to determine stratigraphic relationships within the Holocene barrier complex of Gippsland (see Fig. 3 for localities). Figure 4 shows barrier types and generalized patterns of deposition along the Gippsland coast, distinguishing between the major sediment types. A more detailed representation is provided in Figures 5 to 10. Further

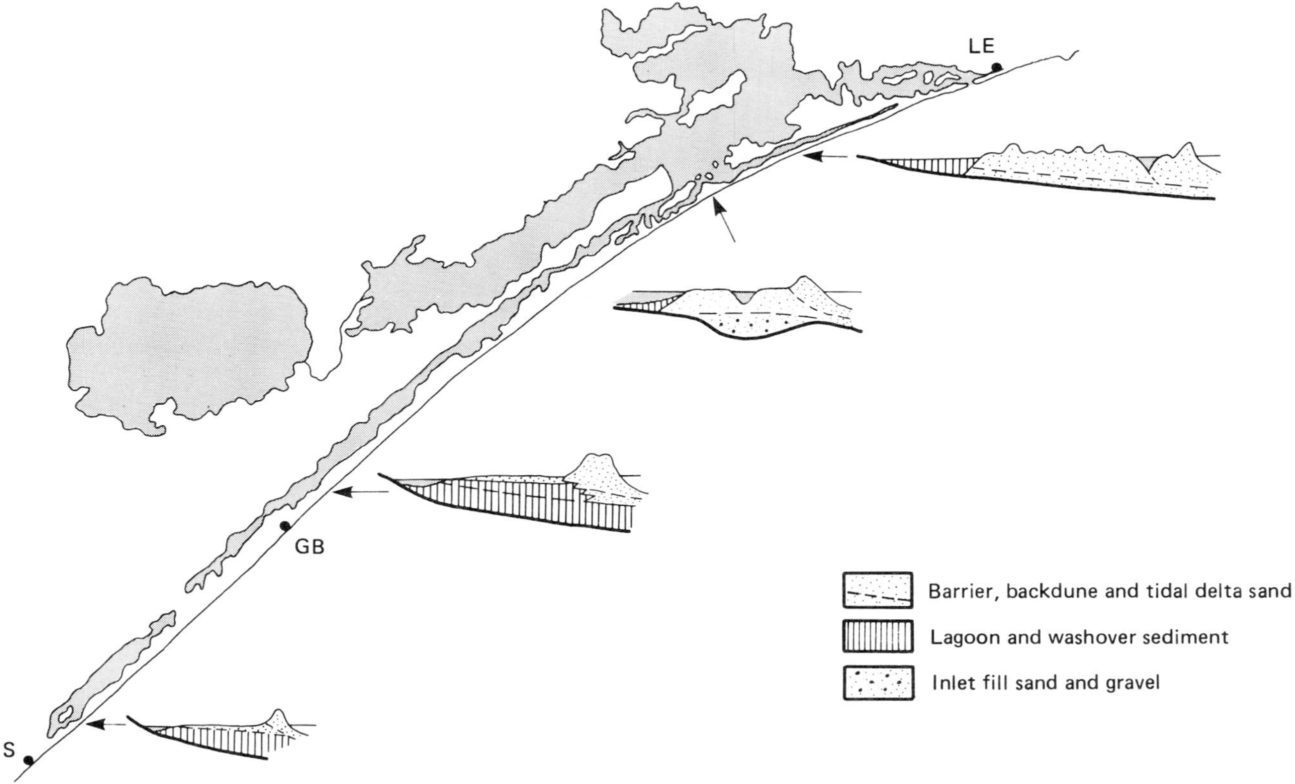

FIGURE 4. Generalized sections depicting arrangement of major facies for Holocene barriers of the Gippsland Lakes (S = Seaspray, GB = Golden Beach, LE = Lakes Entrance).

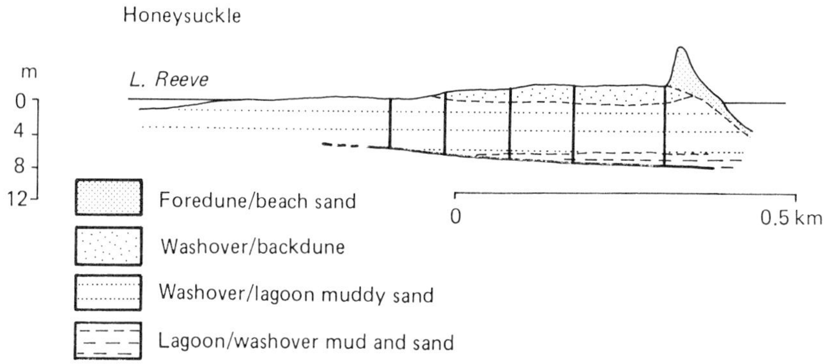

FIGURE 5. Honeysuckle section (see Fig. 3 for location), an example of a receded barrier with beach and foredune deposits transgressing backbarrier muddy sands.

information of facies properties and interrelationships, as well as age structure and problems of dating, will be published elsewhere.

Figure 5 shows the southernmost section called Honeysuckle located north of the small coastal village of Seaspray, where the barrier attaches itself to the cliffed Tertiary 'mainland'. The sand ridge is quite narrow and steep, and varies in height from 5 to 8 m increasing towards the northeast. The seaward face is often scarped by wave erosion. At Honeysuckle, the dune ridge is poorly developed in comparison with the wide backdune flat (Fig. 5). The upper washover muddy sand unit beneath this flat is rich in estuarine shells although marine shells also occur. This unit has been subdivided into upper and lower shell beds by Gill (Gill, 1970; Donner and Jungner, 1981). *Notospisula, Anadara* and *Ostrea* are common estuarine shell species which are found in the unit. They date older than 4500 years, and their present elevation above M.S.L. may reflect local sea-level change and/or lagoon closure. Figure 5 shows the occurrence of a backbarrier facies immediately beneath the foredune and upper beach sand unit. This strongly supports the morphologic impression of the development of a receded barrier at this location. A small outcrop of peat on the beach near the mouth of an inlet just south of Seaspray indicates that this barrier type extends towards the southwest.

The Delray section (Fig. 6a) is located just to the south of Golden Beach (see Fig. 3). In common with that at Flamingo (not illustrated here), this section displays many of the characteristics of a stationary barrier. The section shows a wide range of facies types. Foredune ridge and upper beach

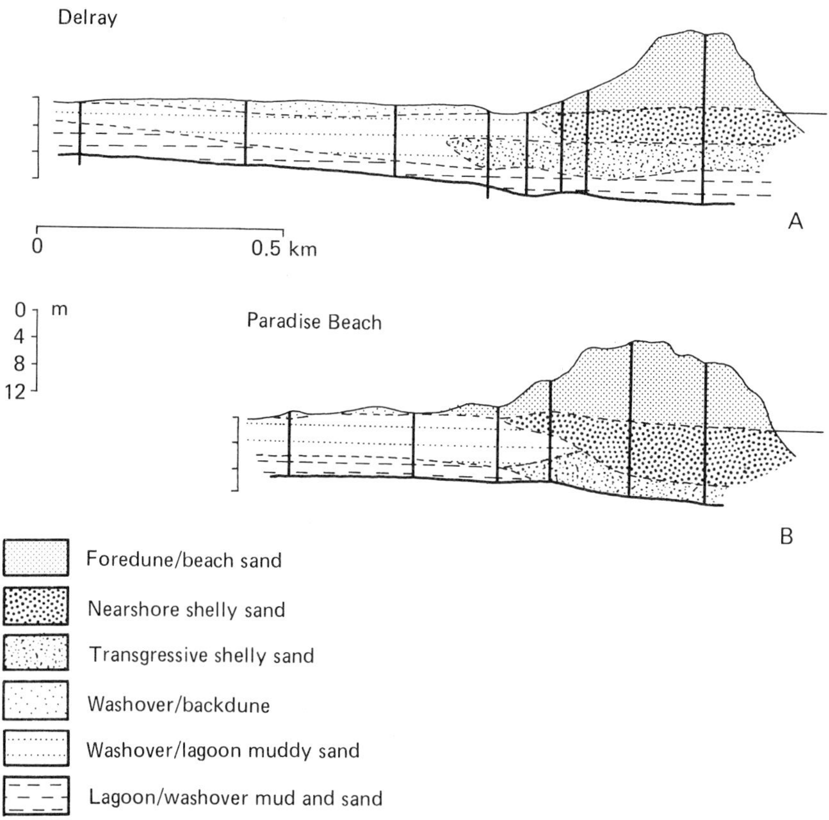

FIGURE 6. Barrier sections at Delray (A) and Paradise Beach (B) showing regressive sand accumulation. Progradation is minimal at Delray, but is expressed at Paradise Beach by a narrow set of high foredune ridges.

sands rest on a flat base and achieve a thickness of 10-12 m. Nearshore shelly sands form a thinner underlying unit and have been dated between 5000 and 6000 ^{14}C yrs BP. There is no clear evidence for extensive progradation at this site, although the occurrence of beach deposits up to 500 m inland of the present beach indicates some seawards accretion. Transgressive nearshore sands containing abundant marine molluscs interfinger landwards with a washover - lagoon muddy sand facies. Both these units are underlain by the so-called lower transgressive lagoon mud and sand facies, which appears to extend seawards of the barrier. The backdune flat consists of moderately sorted sands 1-2 m above Lake Reeve. A thin relict mudflat facies underlies these sands and is exposed in narrow channels (relict washover gutters?).

Paradise Beach (Fig. 6b) exhibits similar morphological and facies relationships to that at Delray except for three important differences: (i) although the inner part of the foredune-beach facies was deposited close to 6000 years ago, the seaward portion is younger (3000-4000 ^{14}C yrs BP); (ii) the transgressive marine sands sit directly on the Pleistocene substrate; and (iii) the backdune flat sands are continuous with the foredune sand body.

A similar depositional history is postulated for both Delray and Paradise Beach sections (and also at Flamingo, Long Point and Stockyard Hill sites located on Fig. 3) with minor local variations. As sea level rose across the inner Gippsland shelf in early Holocene a narrow barrier retreated landwards across the lagoon. This lagoon was fed by washover sediments from the seaward side; it formed a proto-Lake Reeve during the P.M.T. which extended seawards and beneath the present shoreline. About 7000 years ago, the low relief barrier reached close to the site of the present barrier ridge. Nearshore, beach and washover processes reworked sands to the present position of the marine transgressive sand facies. To landwards, the proto-Lake Reeve continued to shallow with the influx of washover sand and fines from river sources. The lagoon had continuous access to the sea through numerous tidal inlets maintaining a salinity regime conducive to the development of a diverse estuarine mollusc assemblage. Following the termination of the P.M.T., the barrier primarily grew upwards by aeolian accretion sealing most of the inlets and inhibiting washover processes. Locally, there was some seawards accretion accompanying barrier growth and consolidation (e.g. at Paradise Beach and Stockyard Hill). Over the last 3000 to 4000 years Lake Reeve has effectively ceased to act as a barrier lagoon except in the sense of being a shallow, ephemeral water body, experiencing only a very tenuous connection with the sea at the northeastern end of the lakes system.

Figure 7 illustrates the sand bank pattern in the vicinity of Ocean Grange close to where Lake Reeve joins the deeper waters of lakes Victoria and King. This area possesses many hallmarks of a relict tidal inlet and flood-tidal delta complex. The islands (e.g. Waddy, Barton) trend at right angles to the shore although they show a somewhat curved outline (Fig. 7). They are of low relief (*c.* 1 m) and are quite well vegetated. Channels between the islands feed into Bunga Arm, a barrier lagoon which stretches to the northeast. Shallow banks extend into Lake King, and are partially covered by sea grasses (e.g. Carstairs Bank).

The stratigraphy of one drill hole (No. 221) located in the centre of the tidal delta is shown in Figure 8 (see Fig. 7 for location). The base of the Holocene tends to be deeper

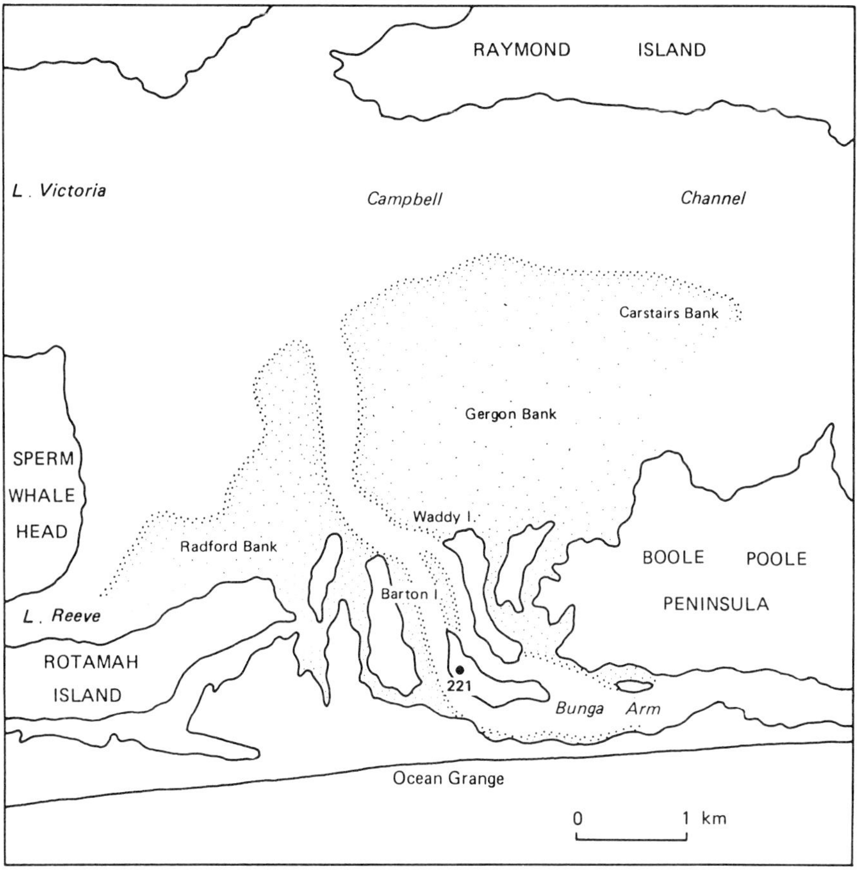

FIGURE 7. Approximate distribution of sand banks forming relict flood-tidal delta complex at Ocean Grange (location of drill hole 221, Fig. 8, is indicated).

than most other sites (-15 m to -20 m below M.S.L.). The fining upwards sequence referred to in the description of facies types predominates here. Reworked Tertiary gravels occur in the lower part of the section suggesting local concentration of tidal energy. Interpreted environments are shown on Figure 8. Preliminary dates indicate that the channel was active through the P.M.T. until about 4000 years ago. These dates are consistent with the 'extinction' of the diverse molluscan fauna along Lake Reeve. It is postulated that semi-permanent enclosure of the lakes encouraging the accumulation of organic muds commenced with the demise of the Ocean Grange tidal inlet.

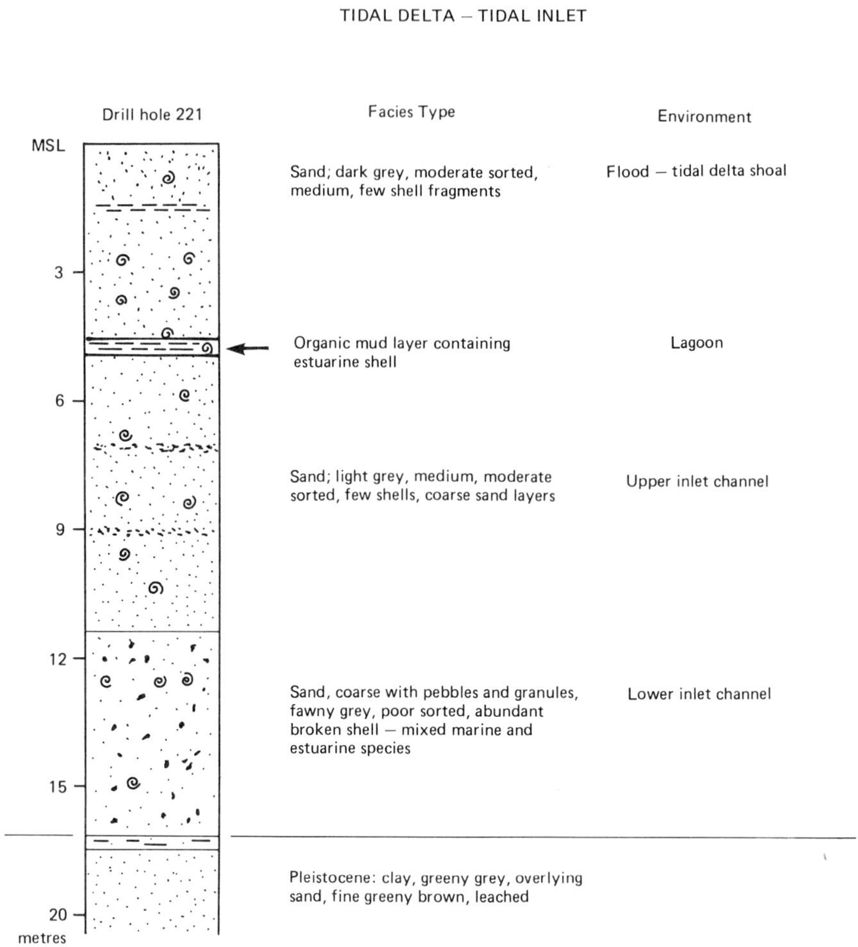

FIGURE 8. Facies types and interpreted environments of deposition from drill hole 221 Ocean Grange.

Figure 9 documents the stratigraphy at Boole Poole. This section crosses a beach-ridge plain. A thin regressive sand sheet is shown in Figure 9. Bunga Arm, an elongate coastal lagoon, lies between the beach-ridge plain and a narrow foredune ridge. This ridge is part of Ninety Mile Beach. Drilling between Boole Poole and Ocean Grange shows the presence of a gravel-rich facies, an extension of the inlet environment shown in Figure 8. The foredune is dated at less than 3000 years between Ocean Grange and Lakes Entrance (Eastern Beach). Soil development is lacking compared to strong podzolization

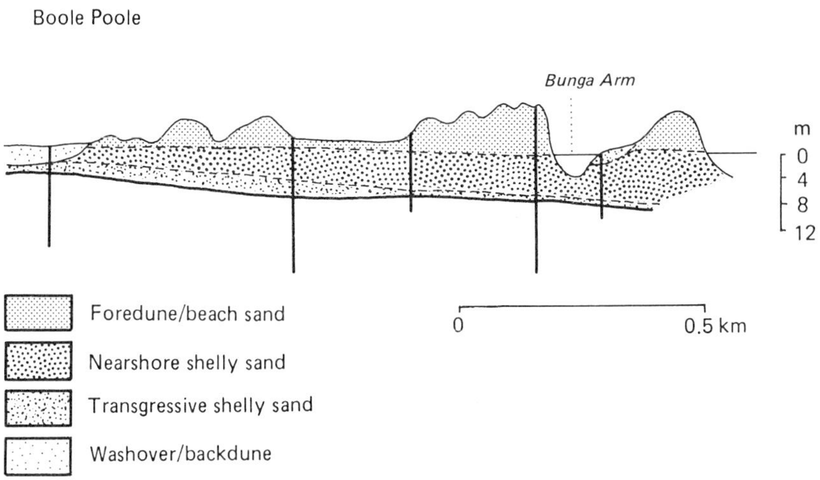

FIGURE 9. Dual Holocene barriers at Boole Poole with a broad beach-ridge plain separated from a narrow foredune by Bunga Arm, an elongate barrier lagoon.

of sands forming the beach-ridge plain on the other side of Bunga Arm. Dual-barrier development appears to be associated with the migration of the relict inlet to the northeast. Processes responsible for the division of barriers are not clear at this stage.

Eastern Beach occurs at the eastern end of the township of Lakes Entrance (Fig. 10). A dual Holocene barrier exists here, the outer part consisting of a young (dated at less than 3000 years), narrow dune ridge and backdune flat, while the inner part is formed by a broader and older sand surface. Cunningham Arm separates these two sand bodies; Bird (1978, Fig. 34) shows Cunningham Arm as once being the main channel to the Lakes Entrance inlet. The inner flat is the site of Lakes Entrance township and partly blocks off the drowned valley called North Arm. Dates from within the backbarrier sediment body suggest deposition close to 6000 years ago. Shells are quite rare, but organic muds are interbedded with moderately sorted sands. Bird (1978, Fig. 35) has indicated parallel 'dune ridges' (beach ridges) on the surface of this flat. However, these are difficult to locate given present urban development, and the stratigraphy is quite different from a regressive beach-ridge sequence, such as that observed at Boole Poole. The separation into two differently aged sand units by a lagoon implies a period of coastal erosion and redeposition. This interpretation involves maintenance and reworking of a near-shore beach sand body both at, or seawards of the present fore-

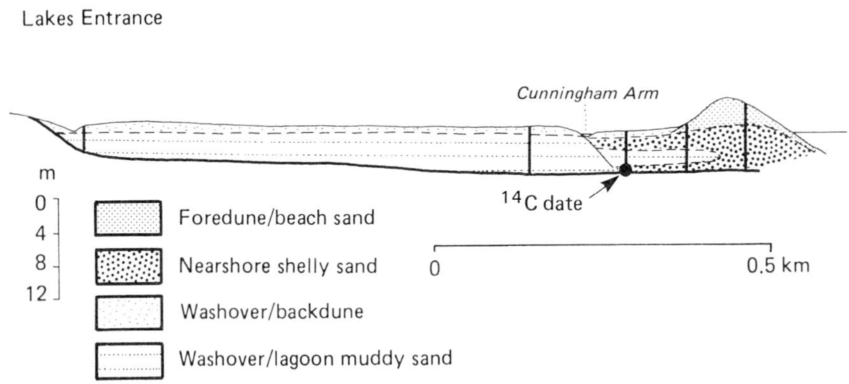

FIGURE 10. Holocene barrier complex at Eastern Beach, Lakes Entrance; truncation of mid-Holocene backbarrier sediments is indicated with re-development of barrier ridge in last 3000 years. Shelly sands date c. 2100 ^{14}C years at contact.

dune from 6000 to 3000 years ago. The present foredune ridge and the outer backdune flat have formed since 3000 years ago. This ridge is similar to the outermost barrier structure shown on the Boole Poole section (Fig. 9).

Conclusion

Coastal sand barriers of various morphostratigraphic types can be identified in southeast Australia. The generalization of Dickinson *et al.*, (1972, p. 206) as applied to modern and ancient barrier-island coastlines can also be applied to a more embayed coast.

> The internal structure of a barrier-island sand body depends on several factors, including climate, sediment supply, tidal range, and tectonism (relative sea level change). The interplay of these factors produces islands of different basic types in regard to both the shape in cross section and the vertical relations of the various types of deposits ... These types may be further modified by the effects of tidal inlets and the dynamics of sediment supply.

In central and southern N.S.W., the rocky nature of headlands and submerged bedrock ridges has restricted sediment exchange between compartments since sea level rose close to its present position *c.* 6500 years ago. The major source of sand for barrier (and backbarrier) construction appears to have been the inner shelf immediately offshore from an embayment (Roy and Thom, 1981). Wave energy has been sufficiently powerful in providing a mechanism for direct onshore sand transport during the later part of the P.M.T., and in the early 'stillstand' period. It was during the early 'stillstand' that most barriers of this coast achieved their present form. Until some sort of nearshore equilibrium adjustment between wave power, sediment supply and gradient was established, then the dominant tendency was for the beachface to be fed with sand, thereby promoting beach-ridge or episodic dune development (Thom, 1984 in press). If the gradient was too steep due to a deficiency in local supply, and yet sea level remained stable, a receded barrier would form. Stationary barrier structures develop where there is a relatively rapid adjustment of sand supply, nearshore gradient and wave energy. Any local variation in controlling processes over time can induce discontinuities and/or changes in rates of sedimentation (Thom, 1978).

The Gippsland region provides an ideal opportunity to test the validity of some of the relationships mentioned above. Four contrasting morphostratigraphic patterns found in sand barriers in Gippsland are shown in Figure 4. These patterns suggest the following mechanisms which have (and still are) influencing shorezone development. Firstly, that most sand material forming Holocene barriers has been derived from reworking inner shelf sand bodies during the P.M.T. Secondly, that since the end and perhaps during the P.M.T., there has been a predominant southwest to northeast sand transport on the beachface and in the surfzone leading to a long-term sand sink in the northeast towards Cape Howe (see Rosengren, 1981). This transport tendency is associated with persistent shoreline recession towards the southwest (the Honeysuckle receded barrier type). Thirdly, that the outer part of the Lake King embayment has served as a sand storage location in the Holocene; this is manifested by beach-ridge formation early in the 'stillstand' period (the Boole Poole and possibly Rotamah Island prograded barriers). Fourthly, that the lengthy stretch of barrier between Flamingo and Stockyard Hill quickly adjusted to conditions of wave energy and sediment supply following the termination of the P.M.T., and after an initial washover interlude, became sealed to inlet formation, wave overwash and aeolian transgression. Fifthly, that during the P.M.T., and for at least 2000 years afterwards, a major tidal inlet and associated flood-tide delta persisted at Ocean Grange. Once this inlet ceased to function ecological conditions dramatically changed in the lakes. And finally, the closure of the Ocean Grange inlet was accompanied by an alongshore extension of a barrier ridge forming Bunga and Cunningham Arms. This extension developed over a period of *c*. 2000 years from Ocean Grange in the southwest towards Red Bluff. Alongshore bypassing of sand continues today with recession in the vicinity of Seaspray being basically offset by shoreline stability elsewhere except for the occasional breaching of the narrow foredune ridge northeast of Ocean Grange.

References

Andrews, P.B. (1970). Facies and genesis of a hurricane - washover fan, St. Joseph Island, central Texas Coast. *Bureau Economic Geology, University of Texas, Report Investigation* No. 67, 147 pp.

Bernard, H.A. and Le Blanc, R.J. (1965). Resume of the Quaternary geology of the northwestern Gulf of Mexico province. *In* Wright, H.E. and Frey, D.G. (eds.), *The Quaternary of the United States*, Princeton University Press, 137-185.

Bird, E.C.F. (1961). The coastal barriers of east Gippsland, Australia. *Geogr. J.*, 127, 460-468.

Bird, E.C.F. (1973). The evolution of sandy barrier formations on the East Gippsland coast. *Proc. Roy. Soc. Victoria*, 79, 75-88.

Bird, E.C.F. (1973). Australian coastal barriers. *In* Schwartz, M.L. (ed.), *Barrier Islands*. Dowden, Hutchinson & Ross, Stroudsburg, 410-426.

Bird, E.C.F. (1978). *The Geomorphology of the Gippsland Lakes Region*. Ministry for Conservation, Victoria. Environmental Studies Series, Pub. No. 186, 158 pp.

Bird, E.C.F. (1984). Shoreline changes in the Gippsland Lakes, 1957-1983. *Proc. Roy. Soc. Victoria* (in press).

Davis, R.A.Jr., Delporte, B.C. and Marsden, M.A.H. (1977). *Morphology, Surface Facies and Sediment Distribution Gippsland Lakes, Victoria*. Report to Gippsland Environmental Study Environmental Studies Section, Ministry for Conservation, Victoria, 49 pp.

Davies, D.K., Ethridge, F.G. and Berg, R.R. (1971). Recognition of barrier environments. *Amer. Assoc. Pet. Geol., Bull.*, 55, 550-565.

Dickinson, K.A., Berryhill, H.L. and Holmes, C.W. (1972). Criteria for recognizing ancient barrier coastlines. *Society Econ. Paleontol. Mineral. Spec. Publication, No.* 16, 192-214.

Donaldson, A.C., Martin, R.H. and Kanes, W.H. (1970). Holocene Quadalupe delta of Texas Gulf coast. *Society Econ.*

Paleontol. Mineral. Spec. Publication, No. 15, 107-137.

Donner, J. and Jungner, H. (1981). Radiocarbon dating of marine shells from southeastern Australia as a means of dating relative sea-level changes. *Annales Academiae Scientiarum Fennicae*, III,131, 5-14.

Gill, E.D. (1970). Current Quaternary shoreline research in Australia. *Aust. J. Sci.*, 32, 426-430.

Hayes, M.O. (1975). Morphology of sand accumulation in estuaries. *In* Cronin, L.E. (ed.), *Estuarine Research*. Academic Press, New York, Vol. 2, 3-22.

Hayes, M.O. and Kana, T.W. (editors) (1979). *Terrigenous Clastic Depositional Environments*. Technical Report No. 11-CRD, Coastal Research Division, Dept. of Geology, University of South Carolina, Columbia, S.C., 184 pp.

Howard, J.D. and Reineck, H.E. (1981). Depositional facies of a 'high energy' beach to offshore sequence and their comparison with a 'low energy' sequence. *Amer. Assoc. Pet. Geol. Bull.*, 65, 807-830.

Hunter, R.E., Clifton, H.E. and Phillips, R.L. (1979). Depositional processes, sedimentary structures and predicted vertical sequences in barred nearshore systems, southern Oregon coast. *J. Sed. Pet.*, 49, 711-726.

Jenkin, J.J. (1968). The geomorphology and upper Cainozoic geology of south-east Gippsland, Victoria. *Mines Department, Geological Survey of Victoria, Memoir* 27, 147 pp.

King, G.M. (1981). Tidal scour in the Gippsland Lakes. *Proc. Roy. Soc. Victoria*, 92, 131-136.

Kraft, J.C. (1971). Sedimentary facies patterns and geologic history of a Holocene marine transgression. *Geol. Soc. Amer. Bull.*, 82, 2131-2158.

Kraft, J.C. and John, C.J. (1979). Lateral and vertical facies relations of transgressive barrier. *Amer. Assoc. Pet. Geol. Bull.*, 63, 2145-2163.

Kumar, N. and Sanders, J.E. (1974). Inlet sequences: a vertical succession of sedimentary structures and textures created by the lateral migration of tidal inlets. *Sedimentology*, 21, 491-532.

Kumar, N. and Sanders, J.E. (1976). Characteristics of shoreface storm deposits: modern and ancient examples. *J. Sed. Pet.*, 46, 145-162.

Moslow, T.F. and Colquhourn, D.J. (1981). Influence of sea level change on barrier island evolution. *Oceanis*, 7, 439-454.

Pye, K. (1983). Formation and history of Queensland coastal dunes. *Zeit. fur Geomorph.*, Suppl. Bd., 45, 175-204.

Reineck, H.E. and Singh, I.B. (1975). *Depositional Sedimentary Environments with Reference to Terrigenous Clastics*. Springer-Verlag, Berlin , (2nd ed.), 439 pp.

Reinson, G.E. (1977). Hydrology and sediments of a temperate estuary - Mallacoota Inlet, Victoria. *Bureau of Mineral Resources, Geology and Geophysics, Bulletin* 178, 91 pp.

Reinson, G.E. (1979). Barrier island systems. *In* Walker, R.G. (ed.), *Facies Models*. Geoscience Canada, Reprint Series 1, Geological Association of Canada, 57-74.

Rosengren, N. (1981). Dune systems on cuspate forelands East Gippsland, Victoria. *Proc. Roy. Soc. Victoria*, 92, 137-147.

Roy, P.S. (1983). Quaternary geology. *In* Herbert, C. (ed.). *Geology of the Sydney 1:100,000 Sheet 9130*, Geological Survey of New South Wales, Sydney, 41-91.

Roy, P.S., Thom, B.G. and Wright, L.D. (1980). Holocene sequences on an embayed high-energy coast: an evolutionary model. *Sedimentary Geol.*, 26, 1-19.

Roy, P.S. and Thom, B.G. (1981). Coastal Quaternary deposits of New South Wales: a model for development in the late Quaternary. *Jour. Geol. Soc. Aust.*, 28, 471-489.

Schwartz, R.K. (1975). Nature and genesis of some storm wash-over deposits. *U.S. Coastal Engineering Research Centre, Technical Memorandum*, No. 61, 69 pp.

Susman, K.R. and Heron, S.D. Jr. (1979). Evolution of a barrier island, Shackleford Banks, Carteret County, North Carolina. *Geol. Soc. Amer. Bull.*, 90, 205-215.

Thom, B.G. (1974). Coastal erosion in Eastern Australia. *Search*, 5, 198-209.

Thom, B.G. (1978). Coastal sand deposition in southeast Australia during the Holocene. *In* Davies, J.L. and Williams, M.A.J. (eds.), *Landform Evolution in Australasia*, A.N.U. Press, Canberra, 197-214.

Thom, B.G. (1984). Transgressive and regressive stratigraphies of coastal sand barriers in eastern Australia. *Marine Geol.* (in press).

Thom, B.G., Polach, H.A. and Bowman, G.M. (1978). *Holocene age structure of coastal sand barriers in New South Wales, Australia*. Geography Department, Faculty of Military Studies, University of N.S.W., Duntroon, 86 pp.

Thom, B.G., Bowman, G.M. and Roy, P.S. (1981). Late Quaternary evolution of coastal sand barriers, Port Stephens-Myall Lakes area, central New South Wales, Australia. *Quat. Res.*, 15, 345-364.

Thom, B.G., Short, A.D. and Hobday, D. (1983). *Modern Coastal Deposition in Gippsland*. Earth Resources Foundation, University of Sydney, 142 pp.

Thom, B.G. and Roy, P.S. (1983). Sea-level change in New South Wales over the past 15,000 years. *In* Hopley, D. (ed.), *Australian Sea Levels in the Last 15,000 Years: A Review*. Department of Geography, James Cook University,

Monograph Series, Occasional Paper No. 3, 64-84.

Van Stratten, L.M.J.U. (1965). Coastal barrier deposits in South and North Holland. *Mededelingen van de Geologische Stichting, Nieuwe Serie*, 17, 41-75.

Wright, L.D. and Thom, B.G. (1977). Coastal depositional landforms: a morphodynamic approach. *Progress in Physical Geography*, 2, 412-459.

Wright, L.D., Nielsen, P., Short, A.D., Coffey, F. and Green, M.O. (1982). Nearshore and surfzone morphodynamics of a storm wave environment: eastern Bass Strait, Australia. *Coastal Studies Unit, Technical Report* 82/3, University of Sydney, 154 pp.

12

Subaerial Denudation and Weathering of Vegetated Coastal Dunes in Eastern Australia

C. H. THOMPSON

Division of Soils
CSIRO
Cunningham Laboratory
St Lucia, Qld

G. M. BOWMAN

Division of Soils
CSIRO
Glen Osmond, SA

Introduction

While minor cycles of sand accretion and erosion have been observed along the east coast of Australia during recent years, virtually all of the coastal sand deposits predate European settlement. Most of our information about coastal dunes is derived from descriptions and measurements of sands that were deposited episodically during the Quaternary. Both dune shape and the mineralogy of the sands have been modified by varying periods of subaerial denudation and weathering. Therefore, there is a need to understand the development of the secondary geomorphic features that have been imposed on the primary aeolian forms, and to recognize different stages of weathering and soil development. Such changes progressively destroy the initially sharp aeolian forms and alter the composition of the sands to varying depths. They can be used to advantage in determining or confirming relative age sequences (Thompson, 1981; Thom *et al.*, 1981b), and in contributing information about the environment of the intervals between episodes of sand deposition. Most of the secondary changes in dune morphology are due to water erosion, whereas

COASTAL GEOMORPHOLOGY
IN AUSTRALIA
ISBN 0 12 687880 3

differences in weathering stage are mostly associated with exposure or internal drainage.

An appreciation of these progressive modifications must assist both description and interpretation of coastal dunes, but little attention has so far been given to these aspects (Thom, 1975). This chapter seeks to bring together existing knowledge of changes that occur in both beach ridges and high parabolic dunes in eastern Australia, once they are covered by vegetation and exposed to weathering and water erosion. Beach ridges and parabolic dunes probably represent the extremes in dune morphology along the east coast; the effects of erosion and weathering on other dune forms is expected to be intermediate between the examples given here.

Denudation

In the simplest meaning, dunes are hills of sand built by aeolian activity; they vary widely in form, elevation and extent. Aeolian activity ceases with the development of a vegetative cover, except for foredunes and areas that may receive further additions from adjacent bare sands or where the wind breaks through weak vegetation and deflates local areas within an otherwise stabilized dune. Most of the natural degradation that has occurred in vegetated dunes appears to be due to water erosion, and in discussing this we will assume that the arrival of vegetation has protected the dune from further aeolian influences.

Water Erosion

Until the arrival of colonizing vegetation, dune morphology reflects the work of aeolian processes modified by the configuration of the underlying terrain and its vegetation; any water erosion that occurs is usually erased by the continuing aeolian activity, although some evidence may be preserved in bedding planes (Bigarella, 1975). Once a dune or beach ridge is vegetated, water erosion can act without interruption from the wind and begin to reduce the new dune to local base level. Because sands generally have high infiltration rates, water erosion has rarely been considered a major factor in degradation of dunes in Australia, apart from the development of badland features in the seacliffs of some large sandmasses in southern Queensland (Coaldrake, 1962; Thompson, 1975), and the dissipation of Pleistocene beach ridges in New South Wales (Thom, 1975). However, evidence of water erosion in the form of sand and organic debris trapped upslope of obstacles, such as logs and grass tussocks, is widespread on the slopes of vegetated dunes in coastal Queensland (Thompson, 1975; Bridge *et al.*, 1984). In addition, on many steep slopes of high parabolic dunes tree roots have been exposed by water erosion, gully channels are evident, and fans of layered sands and organic debris have developed where the gullies debouch at the base of the dunes (Thompson, 1983).

Measurements on dune slopes at Cooloola, Queensland, show that water erosion is accomplished by both surface wash and raindrop splash (Bridge and Ross, 1983; Bridge *et al.*, 1984). Rates of sand movement are influenced by the degree of water repellence in the surface sands, the amount of cover (low plants and organic debris) on the dune surface, the intensity and amount of rain, and the type and extent of canopy cover. The study shows little relationship between gradient of slope and sand movement, possibly because variations in vegetative cover between dune crest and upper and lower slopes tend to mask the effect of slope at the sites sampled.

Surface Wash

The bare sands of mobile dunes and foredunes have high rates of infiltration. Hydraulic conductivities are generally greater than 600mm an hour which means that dune surfaces should accept water at rates far in excess of the rainfall intensities that have been recorded along the Australian coast. However, with the build up of litter under vegetation, an organic-enriched surface with large fungal populations develops and the surface sands become highly water-repellent. In southern Australia, water repellence in sands has been found to vary with different plant species and fungal populations (Bond and Harris, 1964; Bond, 1964).

On dune slopes, water repellence induces runoff and transport of sand and organic debris over short distances. This eventually leads to the development of gully channels with fans at their outlets (Thompson, 1983). Most of the surface wash probably occurs at the beginning of a storm when the sands are dry; its severity depends on the intensity of the rainfall and the time taken for the surface sands to become wet. Measurements under grassy forest at Cooloola show an annual average movement of 4.6kg of sand, down slope per metre width of dune, at sites of high water repellence, but only 1.5 kg of sand at sites where water repellence is negligable (Bridge *et al.*, 1984).

Water repellence is evident on very young dunes and beach ridges as soon as the colonizing vegetation has developed a litter layer. It seems to be a widespread feature in Holocene coastal sands, but is of much less importance in the Pleistocene sands. The decline in water repellence is possibly associated with changes in fungal populations in the different stages of forest succession that have been found to accompany progressive weathering and leaching of the dune sands (Walker *et al.*, 1981).

Raindrop Splash

Surface erosion induced by raindrop splash occurs in all dune systems at Cooloola under a wide range of vegetation types, from heath through woodland and grassy forest to dense rainforest (Bridge and Ross, 1983; Bridge *et al.*, 1984). Sand grains at the surface are displaced in all directions by the impact energy of raindrops and there is a nett gain of sand grains downslope due to gravity. Visual evidence of raindrop splash is provided by (i) sand grains trapped in gummy exudates and in cavities on tree trunks, (ii) in the "coat hanger" effect of sand accumulated beneath arches in fallen branches and logs, and (iii) in the "skirt" effect of sand accumulated beneath the canopies of particular low plants, such as some *Xanthorrhoea* spp.

Rates of raindrop splash are strongly influenced by the density of low plant cover such as grass and prostrate herbs or ground surface cover of litter and debris; as a consequence, rates of erosion increase markedly after fire (Bridge *et al.*, 1984).

Tree canopy has little effect on rates of erosion by raindrop splash. In rainforest and grassy forest, a large percentage of the rain intercepted by the canopy resumes its journey as much larger drops that reach about 90% of their potential terminal velocity on impact with the dune surface. The overall effect of raindrop splash is to gradually move surface sand grains downslope, thereby reducing the sharper crests of dunes to broad convex slopes and building thick colluvial mantles on the lower slopes.

Development of Secondary Geomorphic Features

Progressive down-wearing of dunes with increasing age is evident in airphoto patterns of all of the large sandmasses of eastern Queensland. Several general stages can be recognized, from the gradual development of water-scoured channels through to the formation of a drainage net (Ward and Little, 1975; Ward, 1977). The development of secondary geomorphic features induced by water erosion on large parabolic dunes of different ages has been described by Thompson (1983). Surface wash, due mainly to water repellence in the surface sands, appears to be the dominant process in the three youngest dune systems interpreted as Holocene deposits. This has led to the development of gully channels and fans on both the external and internal slopes of trailing arms and parts of dune floors. However, aeolian outlines of the dunes are still intact and trailing arms have not yet been breached by headward gully erosion, although channels 50-60 m long and fans 80-100 m in length have formed in some dunes. Marked differences in slope are evident between dune walls lacking gully development and gullied areas with channels and fans (Fig. 1). The fan deposits are clearly recognizable from their slope, morphology, bedding planes, stratified layers of organic debris, sand grains (that have been partly stripped of their sesquioxide coatings) and layered soil development.

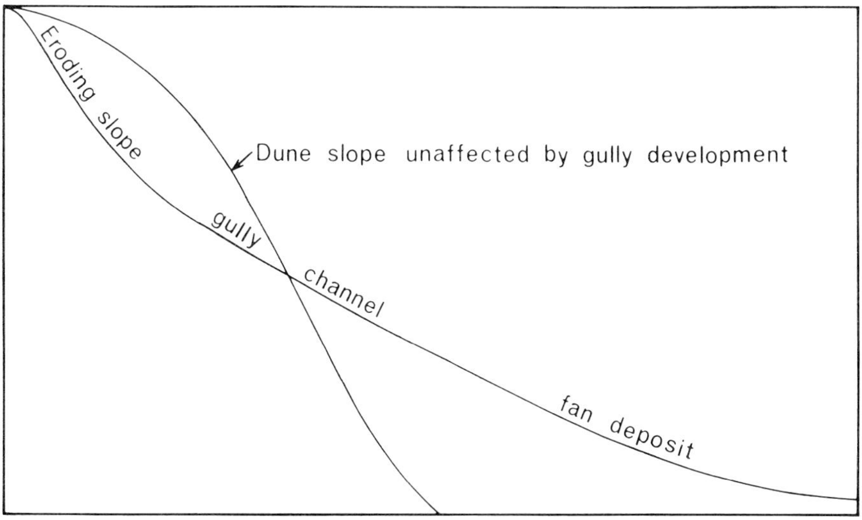

FIGURE 1. External slopes on a trailing arm of a Holocene dune at Cooloola showing changes in gradient due to gully and fan development compared with a non-gullied slope. (Data from Bridge et al., *1984).*

Raindrop splash is also active on the Holocene dunes and accumulations of quartz grains which have been stripped or partly stripped of sesquioxide coatings occur at the base of most slopes. Burrowing animals, such as bandicoots, and many soil invertebrates (e.g. funnel ants and earthworms) also contribute to the total amount of erosion on the dune slopes.

The severity of water erosion on the Holocene dunes at Cooloola may be judged from the undermining of large trees on some slopes and the exposure of tree roots on others. On slopes of maximum erosion above active gullies, these indicate a loss of up to 80 cm depth of sand from the dune slope within the life of trees estimated to be less than 150 years in age. Observations on the steep slopes of Holocene dunes in the Newcastle Bight and Myall Lakes area of New South Wales indicate slower rates of slope erosion possibly because of differences in water repellence and rainfall intensities.

The parabolic dunes of Pleistocene age are much more degraded than the Holocene systems and their morphology suggests the following sequence. In time, some of the gullies on the external slopes breach the trailing arms by headward erosion and link up with their counterparts on the internal slopes and then with those on the dune floors. This breaks the trailing arms into a series of long ridges and low saddles. The ridges have narrow crests, steep middle slopes and footslopes of colluvial and fan sands. As the reduction of the crest and upper slopes continues, the fringes of colluvial sands gradually increase in area and height. The ridges are reduced and the gradient of the side slopes decreases leading to the development of whaleback sandhills. Measured slopes from ridge crest to base at the CSIRO research sites, Cooloola, provide examples of different stages in the down-wearing process on slopes where no effects of gullying are evident (Fig. 2).

Water erosion has cut through the floors of the youngest Pleistocene dunes (Dune System 4, Thompson, 1981) at Cooloola to form deep valleys between the trailing arms. These valleys are incised in underlying older dune sands and in places extend across the sandmass to the coast. The trailing arms have been breached by headward gully erosion, but are readily recognizable in airphoto patterns as long oriented ridges with steep slopes. The apices have generally not been breached by gullying. In a few instances, the apex, the adjacent segments of trailing arms and associated areas of dune floor have escaped gully erosion. These appear to have been reduced mainly by raindrop splash and have gentle slopes and little relief between trailing arm crest and dune floor.

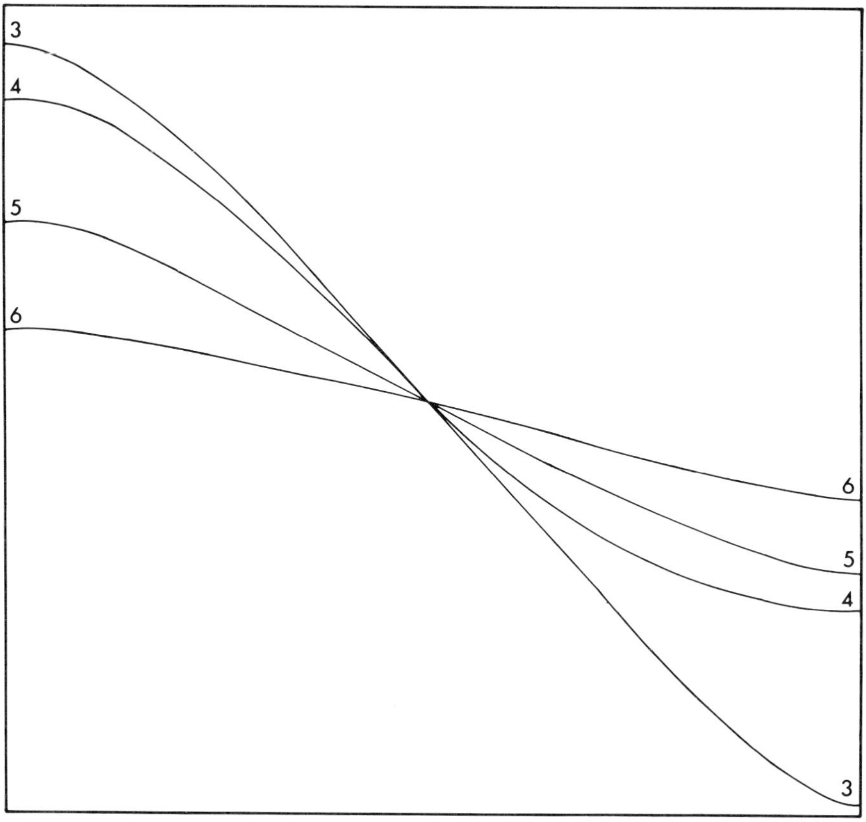

FIGURE 2. Non-gullied slopes at CSIRO research sites on dunes of different ages (Systems 3,4,5 and 6) at Cooloola showing changes with time as trailing arms are reduced to whaleback sand hills. The slopes have been plotted to a common dimension so that the midslope points coincide. (Data from Bridge et al., *1984).*

Measurements of sand movement have been made at different sites on a long external slope of one of these dunes at Cooloola (Bridge *et al.*, 1984). These show that sand is being lost from the convex dune shoulder; an almost constant rate of sand is moving down the straight middle slope and there is reduced movement on the accreting fringe. Soil examination reveals an asymmetrical depth relationship between the dune surface and the B horizon at the crest and shoulder positions, with much thinner A horizons on the dune shoulders. Such variation is only explainable by differences in erosion rates.

The next oldest Pleistocene dunes (System 5) at Cooloola has been further reduced. Extensive valley development, with local springs and flowing streams, is again evident. The ridge remnants of the trailing arms have been reduced further and the area and height of the colluvial sands is greater than in the youngest Pleistocene dunes. Water erosion has cut out all of the dune floors and the apices have been breached by gullies. The parabolic outline is only discernable in airphoto patterns from a few much-modified apices and ridge remnants. There is little evidence of active gullies at present, except for spring-fed streams. The lower slopes of these dunes appear to be entirely of colluvial sand.

The oldest Pleistocene dunes (System 6) at Cooloola are characterized by whaleback sand hills in which no evidence of the original aeolian form remains. The present landscape reflects a very long period of water erosion and the development of a drainage net due to discharge of waters from both above and below organic pan B horizons that have developed in the broad valleys. The broad convex crests and extensive colluvial slopes point to raindrop splash as the dominant process. Surface wash appears to be minimal, but there is widespread evidence of raindrop splash in 'coat hanger' and 'skirt' accumulations, which are associated with fallen branches and low vegetation.

In summary, gullying and surface wash were important processes in the early reduction of Pleistocene dunes, but raindrop splash appears to be the dominant process under the present environment. Some localized surface wash still occurs and the activities of fauna and soil invertebrates contribute to the present sand movement.

No measurements of water erosion have been made on beach ridges in eastern Australia. Simple water-drop tests during examinations of soils have shown that the surface sands of most Holocene beach ridges and foredunes are strongly water repellent. Erosion rills and miniature fans have been observed on some beach ridges, but the generally low gradients and short slopes seem to preclude gully development. However, surface run-off carrying both organic debris and sand does occur on most beach ridges during storms, and has been observed at Fraser Island and Caloundra, Queensland, and at Disaster Bay and Moruya, New South Wales.

Raindrop splash is considered to be the more important process moving sand from the crest of ridges to the swales, gradually reducing the height of the beach ridges and extending their width. Because the surface sands are partly weathered (i.e. stripped of sesquioxide coatings) this movement also affects the depth to B horizon. Deeper profiles occur on the slopes as compared with those on the crests of beach ridges, as observed at Woy Woy, New South Wales (Burges and Drover, 1953), but the centre of the swales seems to be little affected (Bowman, 1979). As the process continues, it also gradually alters the relationship between the ground surface of the beach ridges and the water-table, particularly where the swales are narrow. Such changes appear to be very slow; little morphological change is evident in ground surface and water-table relationships over the past 6000 years in dated beach-ridge sequences in New South Wales (Bowman, 1979; Thom *et al.*, 1981a).

The effects of long-term erosion in reducing beach ridges is clearly evident in the Pleistocene deposits. On airphoto patterns, Pleistocene beach ridges are much more worn and subdued than their Holocene equivalents. Frequently, they have lobed or convoluted margins with adjoining swales whereas these junctions in Holocene beach ridges are usually very even (Shepherd, 1970). In the Myall Lakes area, the crests of Pleistocene beach ridges have been reduced to 30-40 cm above adjacent swales and rarely exceed 2m, whereas the Holocene ridges are often 3m or more above the adjacent swales (Langford-Smith and Thom, 1969). The reduction of the Pleistocene beach ridges and the infilling of the swales with about a metre of peaty sand has in places led to the amalgamation of groups of primary beach ridges into much less distinct composite secondary forms (Thom, 1975). In some areas, for example Eurunderee and Newcastle Bight, New South Wales, a network of transverse surface drainage lines has developed, linking adjacent swales and dividing the beach-ridge plain into rectangular blocks (Thom *et al.*, 1984, in press). Also, gullies have been incised into the margins of Pleistocene beach ridges during periods of low sea level and these have since been infilled with fluvial or marine sediments following the Holocene rise in sea level (Thom, 1965).

Weathering

Common Constituents of Dunes

The sands of the coastal dunes and beach ridges are dominantly quartz; they contain few weatherable minerals and, on average, much less than 2% of marine shell fragments. The mineralogy of the sands has been described in varying detail from many areas, mainly in the search for heavy minerals (Beasley, 1950; Gardner, 1955; Whitworth, 1956; Connah, 1961), but also in attempts to determine the source and host rocks of the sands (Hails, 1969; Roy and Crawford, 1977) and their weathering characteristics (Little *et al.*, 1978; Bowman, 1979). These studies show that although some twenty different minerals occur in the coastal sands, quartz makes up about 98% and the rest consists mainly of the heavy minerals ilmenite, rutile, zircon and monazite, and some feldspar. Minerals such as tourmaline, magnetite, leucoxene and garnet may be common in the heavy mineral suite in some localities. Some contributions from particular rock types exposed at the coast are detectable locally, but the amounts are usually too low to affect weathering trends. Fragments of marine shells are a highly variable component; some areas of foredunes and the youngest beach ridges contain up to 10%, but the overall average is less than 1%. Shell fragments are rare in the large parabolic dunes and are restricted to a trace in some young dunes near the coast and elsewhere to a surface scatter of shells presumably carried in by man.

Sands of the beaches and unweathered zones in beach ridges and parabolic dunes are various shades of yellowish brown. Microscopic examination shows that the colour is due to thin sesquioxide coatings that appear as a yellowish tinge on grain surfaces and brown to reddish brown accumulations in niches and pits. Electron microprobe analysis of coatings on the grain surfaces (K. Norrish pers. comm.), and chemical analysis of coatings removed in nitric acid digests (Jehne and Thompson unpub. data), show that they consist mainly of aluminium, iron and silica with which is associated small amounts of magnesium, potassium, phosphorus, sulphur and calcium and traces of seven other elements. It appears that the coatings have been inherited from a previous environment of subaerial weathering and have survived water and/or wind transport during their most recent episode of deposition (Ly, 1976). Most of these sands have probably been derived from the destruction of older sand deposits by the sea, and by wind erosion as the floors of new parabolic dunes have cut deeper into underlying weathered dunes.

Shell Fragments

In eastern Australia, the sands deposited in foredunes or parabolic dunes adjacent to an active beach source invariably receive some contribution of biogenic carbonate, usually in the form of fine shell fragments. The initial shell content of wind-deposited sand is considerably less than that of beach sediments because of the general absence of whole shells and large pieces of shell. Hence, coastal parabolic dunes and the aeolian sand that caps beach-ridge plains usually have an initial shell content of much less than 10% (Bird, 1961, 1971), whereas relict beach facies may contain as much as 70% in localized lenses (Roy and Thom, 1981).

Leaching of shell fragments begins when the sands are colonized by vegetation to provide litter for decay and a source of organic acids. Studies of leaching rates of shell carbonate from English sand dunes (Salisbury, 1922, 1925) have shown that virtually all of the carbonate was removed from near the surface in 300 years. The rates of leaching of carbonate have also been examined in a radiocarbon dated sequence of dunes along the shores of Lake Michigan, U.S.A. (Olsen, 1958). In this area, about 75% of the initial carbonate content of 1.4% $CaCO_3$ in the surface 10cm was lost in 300 years. It was estimated that about 1000 years was needed to remove the carbonate to a depth of two metres. At Woy Woy, the time needed to leach carbonate from the surface 10cm of beach ridges was estimated to be 200 years (Burges and Drover, 1953). However, recent studies in which radiocarbon dates were obtained for the same beach ridges indicate that this is an underestimate and that a period of 1000-1500 years is more likely (Bowman, 1979). As previously noted (Salisbury, 1922), the rates of leaching of carbonate are governed by the initial size of the shell fragments and will therefore vary locally.

The distribution of calcium ions with depth has recently been used to assess the distribution of shell carbonate in some Holocene beach-ridge systems of eastern Australia (Bowman, 1979). Below the organic-enriched surface horizons, the concentrations of calcium are generally low throughout the soil profiles except on the seaward side of the youngest deposits where there is an age-dependent leaching front that increases in depth with distances from the sea (Fig. 3).

Feldspars

Feldspars are generally present as minor amounts in the coastal sands. Reported concentrations rarely exceed 5% and there are some indications of significant differences in concentrations between parabolic dunes, beach ridges and fluviatile and marine sands (Grimes, 1982). Feldspars are often much more abundant in local areas, such as coarse sands adjacent to river mouths (Bird, 1971, 1978) and in parts of estuaries (Roy and Crawford, 1977), compartmented coastal embayments (Bird, 1967), and in nearshore marine sands (Marshall, 1980; Grimes, 1982).

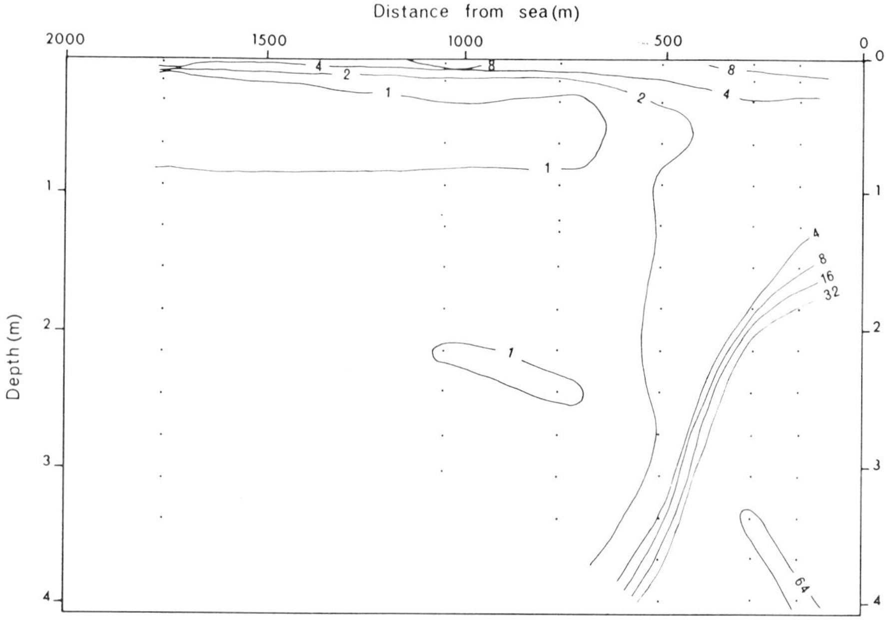

FIGURE 3. Distance/depth plot across the Moruya beach-ridge plain, N.S.W., showing the distribution of acid extractable calcium above the watertable. Note the progressive deepening of the leaching front with increasing distance from the sea (and age). Soil profiles beyond 700 metres distance have been completely leached of shell carbonate. Dots represent analysed sampling points. Calcium expressed as micromoles per gram of oven-dry soil.

Studies of surface and drill-hole samples from parabolic dunes and beach ridges in the Wide Bay area, Queensland, show that the potassium-feldspar content of the dunes is less than 1% whereas the beach ridges contain up to 2% (Grimes, 1982). Grain counts of subsurface samples from four beach-ridge plains in New South Wales, one in north Queensland and one in Tasmania show the mean values for feldspar grains from all areas to range from 2.3 to 4.9%; orthoclase is generally the most abundant feldspar present, but the ratio of potassium to plagioclase feldspar varies from 1:0.67 to 1:6.4 (Bowman, 1979).

No detailed studies of feldspar weathering have been reported in the coastal sands of eastern Australia although rates of feldspar weathering have been used to deduce approximate ages of coastal sediments overseas (Gardner, 1967). From limited examinations, orthoclase appears to be the common feldspar in the dune systems at Cooloola, and the amounts decline as weathering proceeds to greater depths in the older systems. Total potassium, as determined by X-ray fluorescence analysis, also declines for comparative depths with increasing age of dune system and this has been interpreted as reflecting the progressive weathering of potassium feldspars (Thompson, 1981).

Sesquioxide Coatings

As the newly-deposited sands acquire a vegetative cover, they are exposed to both leaching and biological processes. Together, these processes gradually remove the sesquioxide coatings from the grains near the dune surface. The constituents are incorporated in the biological cycle or deposited at lower depths or removed from the dune via ground waters. Mechanisms of mobilization and precipitation of iron and aluminium compounds in freely-drained dune sands probably involve several different chemical and biological processes (Petersen, 1976). The overall effect with time is to strip the sesquioxide coatings from the sand grains at progressively greater depths in the dunes, leading to thick accumulations of clean white quartz sands overlying coloured sands at depth.

Field evidence indicates that under a cover of vegetation, the sesquioxide coatings are lost from the upper layers of freely-drained coastal sands irrespective of the initial colour of the coatings. Both yellowish and red sands weather to clean white quartz grains and there is no evidence in the subtropical dunes of reddening with age as has been reported from tropical Queensland (Pye, 1981). Red beds and iron pans are known in the Pleistocene sands (Coaldrake, 1962; Thompson and Moore, 1984), but both appear to be associated with particular dune sand deposits.

Quartz

The resistance of quartz grains to physical breakage, the ultimate size to which the grains break naturally, and the susceptibility of different grain types to solution weathering all appear to be largely dependent on microfractures or strains and stresses that individual grains have inherited from their environment of crystallization (Moss and Green, 1975). Hence, different types of grains with varying evidence of fractures and weathering occur in most sand deposits. Also, because the microfractures within grains separate sheets of quartz that are mostly 2-20μ across, quartz is common in the silt size fraction of soils and strongly resistant to further diminution of grain size.

Four different types of quartz grains, described under the scanning electron microscope as clear and unetched, milky, saccharoidal and microgranular occur in the different dune systems on Fraser Island (Little *et al.*, 1978). The saccharoidal and microgranular grains are highly weathered with strongly etched surfaces and are generally fragile and easily broken by slight pressure; some etching is also evident on the milky grains. Although the proportion of highly weathered grains to total quartz is small in all samples, it increases with age of dune system. A similar increase in weathered sand grains with age of dune systems also occurs in samples from Cooloola (M.P. Green pers. comm.) and in beach-ridge systems in New South Wales (Bowman, 1979).

A further interesting feature in samples from old dunes on Fraser Island is the increase in the proportion of weathered saccharoidal grains in BC horizons as against A horizons. Little *et al.*, (1978) suggested that this may be due to either the destruction of these grains and their removal from the A horizons or that the moister environment of the BC horizons may be more favourable to their development. No further work has been done to resolve this question. However, the increase in silt fraction, which consists mainly of quartz, to about 7% in the top of the B horizon of soils on the Pleistocene dune systems at Cooloola, points to the gradual physical removal of silt size quartz from the A horizons. It accumulates where further vertical movement is restricted by B horizon development. Studies of *podzols* on Holocene beach ridges in New South Wales also indicate that the weathering of quartz grains is most intense in the A2 horizons, but there is no evidence of grain destruction over the 6000 year period (Bowman, 1979).

Clay Minerals

Coastal dune sands in eastern Australia have very low contents of minerals with the potential to weather to clay. However, there is a gradual increase in clay in zones of accumulation from less than 1% in the youngest Holocene dunes to about 15% in the oldest Pleistocene dunes. Both silt and clay occur in relatively thin zones of accumulation in the upper part of the soil B horizons with negligible amounts above and below. Presumably this represents the total clay derived from weathering of the overlying sand column plus that derived from atmospheric dust. It appears that the dust contributions to the Holocene and younger Pleistocene dunes are negligible, but there may be some contribution to the older Pleistocene dunes, such as the Teewah Sands which are slightly clayey (Coaldrake, 1962).

The composition of the clays has been determined for few samples. These show chlorite-vermiculite and kaolin as the main clay minerals in the young dunes and kaolin and quartz in samples from the old dunes, which is in keeping with the weathering of other constituents. No data are available for beach-ridges.

Formation of Soils

The coastal sands contain few weatherable minerals and local variations in their concentration are generally too small to affect soil weathering trends. They may, however, delay the initiation of the main process, for instance, in the case of included shell fragments or the slightly higher iron concentrations and greater amounts of haematite in some coloured sands, e.g. Teewah Sands (Coaldrake, 1962). Other variations in mineralogy between dune systems do not seem to have a significant affect on soil weathering trends, and therefore the dune sands constitute a simple and relatively uniform parent material. While leaching of some bases probably occurs ahead of plant colonization of the sands, soil development begins with the arrival of plants and associated microorganisms.

Soil profiles in coastal sands develop only under a cover of vegetation. They are the result of mobilization and precipitation of sesquioxide coatings derived from the sand grains and organic compounds from the vegetation and soil organisms. At freely-drained sites, the process leads to the development of prominent soil horizons: A1 horizons of clean quartz sand darkened and mixed with decomposing organic debris; bleached A2 horizons of clean quartz sands; streaky brown, black and yellowish brown Bhir horizons with accumulated organic, iron and aluminium compounds; and C horizons of yellowish brown unweathered sand (Brewer and Thompson, 1980). Soils with these characteristics are classed as *podzols* (Stace *et al.*, 1968; Thompson and Hubble, 1980).

The rate at which *podzol* profiles develop in vegetated sands is largely determined by the amount of water that enters the soil and the degree of erosion or sedimentation at each site. These vary on each of the geomorphic components that make up both beach ridges and dunes with the result that the stage of soil development differs for each component (Thompson, 1983).

Soil Development on Holocene Dunes

Field examinations at freely-drained sites in age sequences of both dunes and beach ridges show that with time the *podzol* profiles continue to increase their vertical dimensions (Burges and Drover, 1953; Bowman, 1979; Thompson and Hubble, 1980; Thompson, 1983). Soil development in Holocene sands shows that as weathering proceeds both A1 and Bhir horizons become progressively thicker and more intensely developed; there is also a gradual increase in the thickness of bleached A2 and in the irregularity of its boundary with the underlying Bhir horizon (Figs. 4 and 5).

Studies of soils on dated beach-ridge sequences in New South Wales (Bowman, 1979) show that as the soils become older the shell component is gradually removed and the pH becomes progressively lower. The thickness of A1, A2 and Bhir horizons increase progressively from 1000-6500 ^{14}C years (Fig. 4). There is a marked increase in both sharpness and irregularity of the A2/Bhir boundary in soils older than about 4500 years. *Podzols* about 6000 years in age are on average more acid, about two units of pH lower, than soils 1000 years old. Increasing profile development is also reflected in element concentrations, with accumulations of iron, aluminium and manganese in the B horizons of older soils and depletion in A horizons (Fig. 6). In contrast, cations such as calcium and magnesium are lost from the profiles as they age, except for minor accumulations within organic matter at the surface (Fig. 3).

As a result of water erosion, the shallowest soils are usually found on the ridge crests of Holocene beach-ridge plains. There are deeper profiles on the lower slopes (Burges and Drover, 1953) and in the swales. In some areas, the swales show no evidence of either erosion or deposition (Bowman, 1979). Erosion in the Holocene parabolic dunes has led to the shallowest soil development on the steep external slopes. Soil depth is greater on the internal slopes, crests of trailing arms and dune floors; maximum development occurs on dune floors where erosion is minimal and there may be minor additions of sediments (Thompson, 1983; Thompson and Moore, 1984). Very shallow soils are also associated with high, steeply-sloping knolls on the trailing arms, and layered soils occupy the young alluvial fans and colluvial slopes.

Soil Development on Pleistocene Dunes

Most of the soils on the freely-drained Pleistocene dunes are giant *podzols* in which depth to Bhir horizon is greater than 3 m and as much as 18 m in places. The shallowest

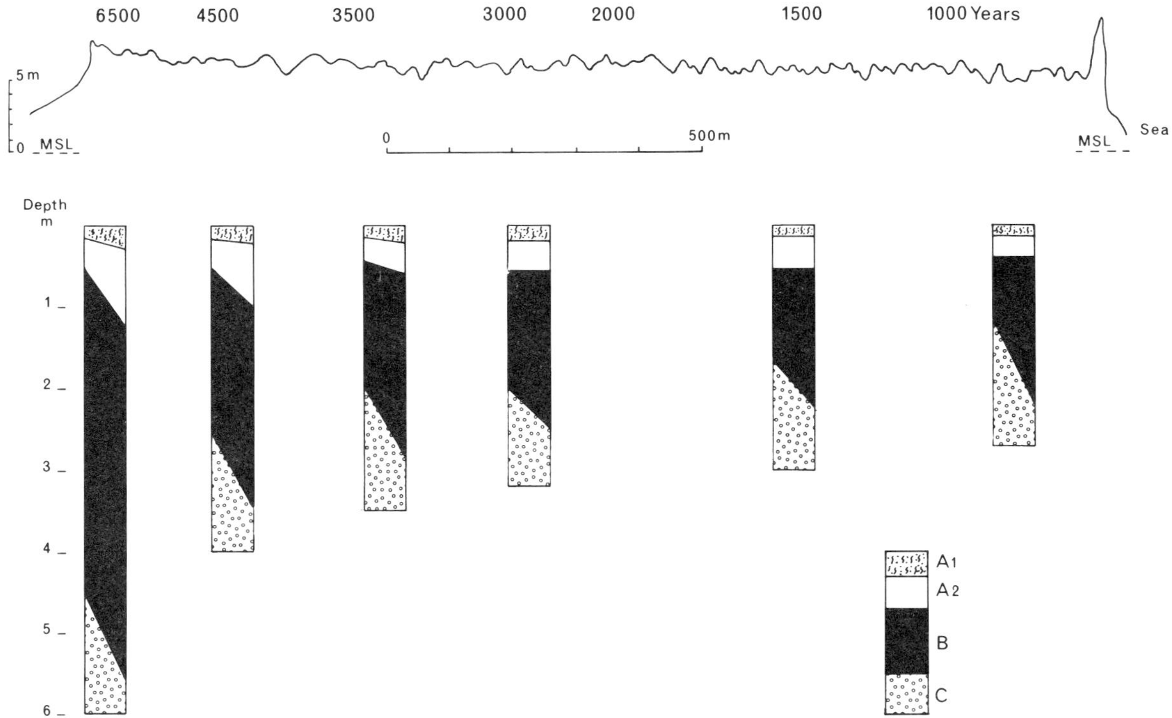

FIGURE 4. Cross-section of beach-ridge plain at Disaster Bay, showing increase in the development of podzol profiles with time. Ages in calendar years B.P.

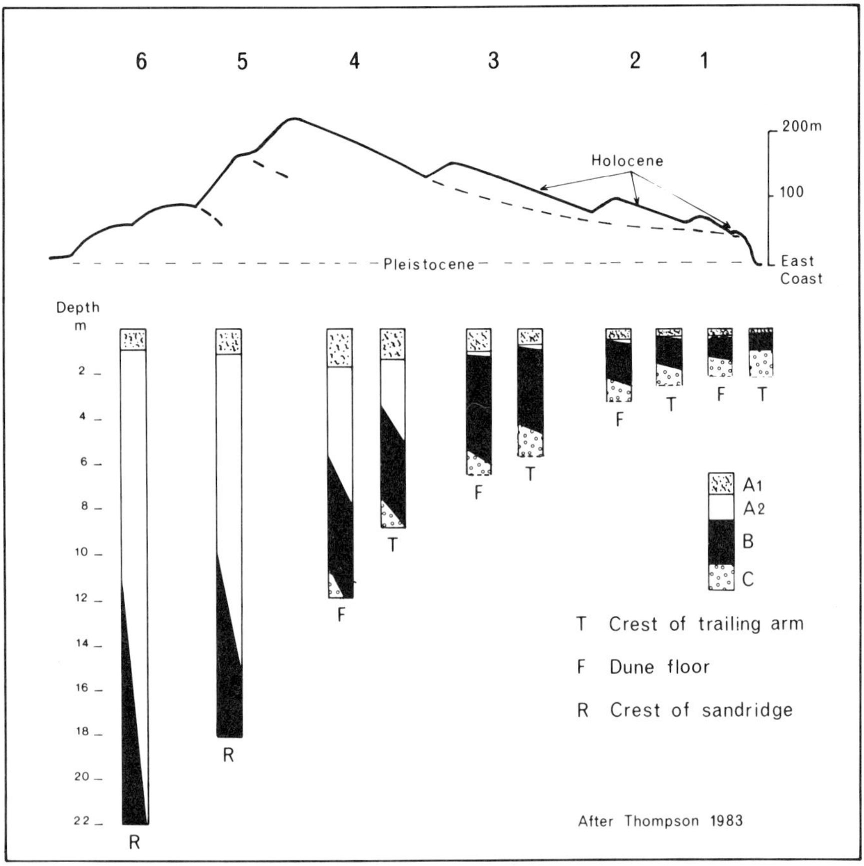

FIGURE 5. Schematic cross-section of parabolic dune systems at Cooloola showing increase in development of podzol profiles with time. Diagonal lines indicate variations in depth to underlying horizon. (From Thompson, 1983).

profiles occur at the sites of maximum erosion on steep slopes and on the convex shoulders of the dunes just below the crests. Deeper profiles occur on dune crests and the remnants of dune floors. Over-deep profiles, relative to soil development elsewhere on the dune, occur on fan and colluvial sediments at the base of the dunes. Where the highly-coloured Teewah Sands have been exposed by erosion, the soil profiles are often quite shallow, probably because the greater amounts of haematite are more resistant to mobilization than yellow-brown iron compounds and soil development is therefore slower.

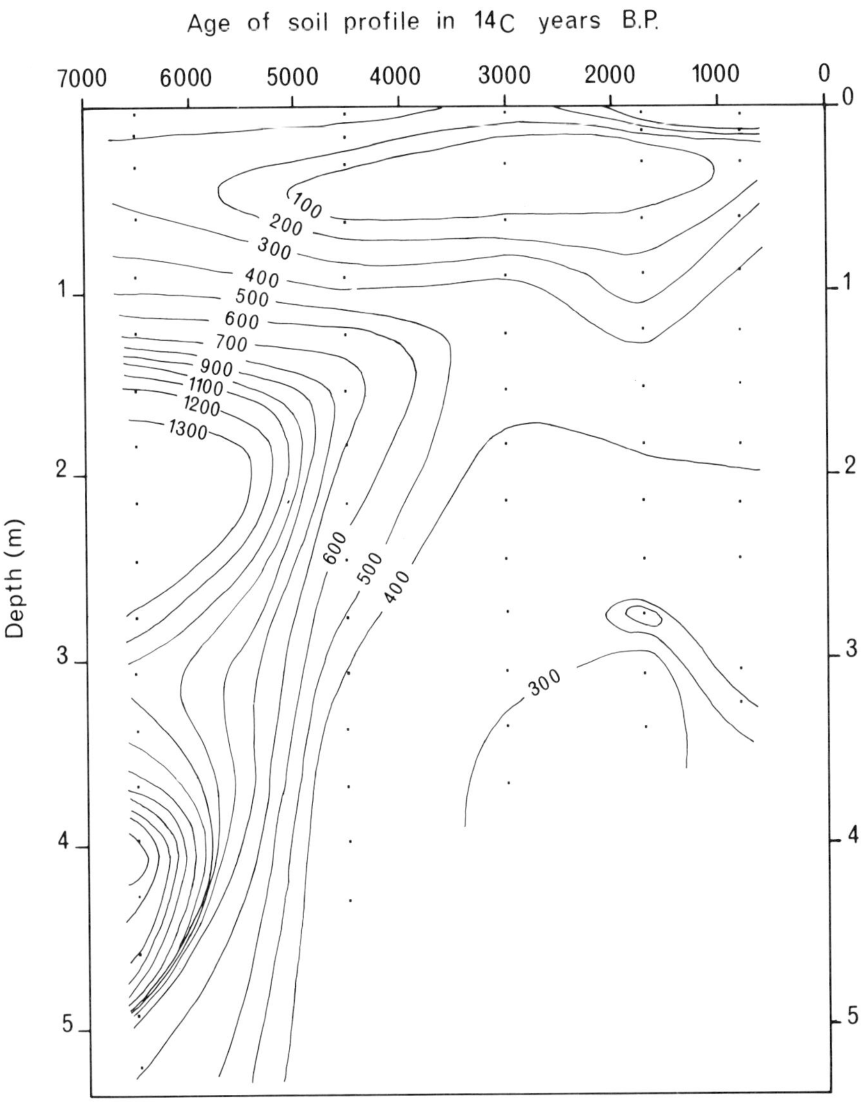

FIGURE 6. Time/depth plot of the Disaster Bay (N.S.W.) beach-ridge plain showing the distribution of acid extractable iron above the water-table. The diagram shows progressive podzol profile development with increasing age, particularly the development of a marked Bhir horizon in profiles > 4500 ^{14}C yrs BP.

Long periods of weathering lead to a marked increase in the thickness of bleached A2 relative to that of both A1 and Bhir horizons in Pleistocene dunes (Fig. 5). The irregularity at the top of the Bhir horizons increases as tongues of bleached A2 increase in diameter and penetrate deeper. These eventually develop into irregular columns of bleached sand encased in precipitated organic and sesquioxide compounds that penetrate the B horizon to depths of several metres, forming pipey B horizons (Thompson and Hubble, 1980; Thompson, 1983). The development of pipey B horizons is believed to be due to variations in the vertical distribution of water and leachates from the surface causing uneven remobilization of organic compounds and sesquioxide from the top of the B horizon.

Influence of Water-Tables

The vertical dimensions of soil profile development in the coastal sands are constrained by the depth to water-tables. The thickness of sand in the apex and trailing arms of the large parabolic dunes usually exceeds 30 m. At these sites, both topography and thickness of unconsolidated sand ensure that soil development can extend to this depth without interference from the water-table. Soil development in the oldest dunes at Cooloola, has been found to approach this depth (Thompson, 1981; Thompson and Moore, 1984). By contrast, many of the Holocene beach-ridge plains have water-tables within 4 m of the swale surface and within 7 m of beach-ridge crests.

The water-table sets a maximum limit to the depth at which soil development can occur. It also sets a maximum depth to which *podzols* can develop before iron is lost and there is a change in processes inducing precipitation of organic compounds and, as a result, in the type of B horizon developed (Farmer *et al.*, 1983). In these situations, the B horizon forms at the top of the water-table. It is dominated by organic and aluminium compounds and contains virtually no iron. In soil science it is designated as a humic or humus B horizon (Bh), and has been referred to in geological literature as 'humate' (Swanson and Palacas, 1965). These horizons are usually dark grey or black, relatively uniform laterally in both consistence and chemical composition and generally have even boundaries with the overlying A2 horizons. Both the kind of B horizon and its relationship to the water-table distinguished these soils from the freely-drained *podzols*; in Australia they have been classed as *humus podzols* (Stace *et al.*, 1968).

Where water-tables occur within a metre or so of the surface in the older Holocene sands, the *humus podzols* have soft Bh horizons of very dark brown to black organic sand. In the Pleistocene dunes, the Bh horizons are usually hard and cemented although the upper few centimetres may be soft or friable. These cemented layers in soil profiles have been referred to as 'coffee rock' (Byers *et al.*, 1938; Stace *et al.*, 1968), but where they are exposed along the beach they are generally know as 'sandrock' (Ball, 1924; Ward *et al.*, 1979). Cemented Bh horizons have very low permeability which leads to the perching of seasonal water-tables. The perched waters coloured by organic compounds are known as 'black waters', and issue from drainage lines through these soils (Reeve and Fergus, 1982). In some areas, the cemented Bh horizons are very thick and are probably the result of fluctuations in the water-tables associated with changes in sea level (Thom *et al.*, 1981b).

Field evidence indicates that the ageing process in *podzols* on freely-drained dunes in eastern Australia increases the dimensions of these profiles, eventually leading to the development of giant forms. However, the reverse occurs in many *humus podzols* and depths to Bh horizons either remain static or gradually decrease, provided there is no change in the water-table levels. In the northern hemisphere, the depth to B horizon is believed to remain relatively constant in *podzols* formed in an age sequence on drift, extending from 2000 to 10,000 years ago. (Franzmeier and Whiteside, 1963).

Influence of Vegetation on Development of Al Horizon

Although the degree of soil development and successional changes in vegetation on coastal dunes are inter-related (Walker *et al.*, 1981), the floristics and structure of the vegetation probably also influence the rate of *podzol* development at freely-drained sites. This aspect has not been examined in detail in Australia. However, a close relationship is evident between the thickness of Al horizons and amount of vegetation biomass which presumably controls the amount of organic debris returned to the soil annually (Walker *et al.*, 1981; Thompson, 1983). This relationship is not surprising because the A horizons of these *podzols* are almost entirely clean quartz sands and are deficient in elements, such as calcium, iron and aluminium, that stablize the decomposition products of organic debris in the surface horizons of other soils. The mean residence time of organic matter in the Al horizons of these *podzols* is therefore comparatively short and the amount of organic matter is closely related to the existing biomass and environmental decomposition rates; for instance, decomposition of organic matter appears to be slower in temperate than in subtropical climates. The relationship between thickness of A1 and an index of biomass in the *podzol* chronosequence at Cooloola, shows an increase in both to a maximum in Dune System 4 (the youngest Pleistocene dune) and a decline in the older systems (Fig. 7).

Influence of Other Factors

Minor variations in mineralogy between dune systems retard the initial rate of *podzol* development at some sites. Erosion also prevents the development of deep soils while sedimentation allows over-deep profiles to form. In each case, the soil weathering trend to *podzol* development continues. However, when depth of soil development reaches the water table, a different kind of B horizon is precipitated and *humus podzols* form. At freely-drained sites, variations in climate accelerate or retard rates of weathering. Major changes in climate cannot affect the soil weathering trend provided that sufficient vegetation cover remains to prevent large scale wind erosion and there are no massive inputs of colloids from atmospheric dust. Consequently, differences in depth of soil development due to variations in past climates have little effect on the kind of horizons formed but may give a distorted impression of the interval between successive dune systems.

Atmospheric accession of major ions from sea spray, as measured at Cooloola, are markedly greater for the first kilometre inland from the coast and then relatively constant for 20 km inland (Walker *et al.*, 1981). The effects of ions,

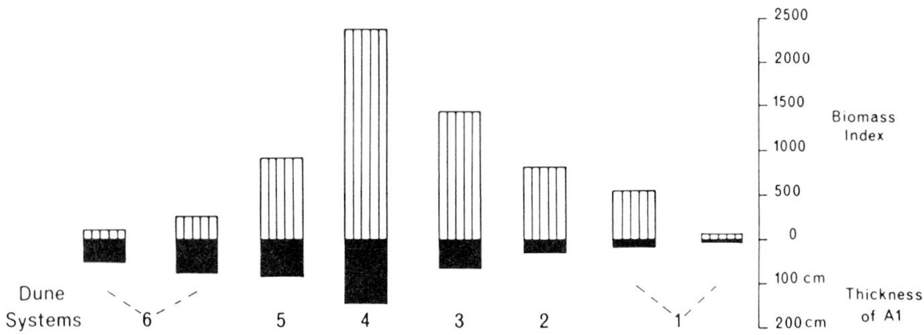

FIGURE 7. Relationships between biomass and thickness of A1 horizons in podzols on the sand dune systems at Cooloola. (From Thompson, 1983).

particularly the sodium ion, from this source on rates of soil development are unknown, but do not appear to change the soil weathering trend.

Conclusions

Water erosion is occurring on all of the vegetated coastal dunes in eastern Australia. Its severity varies with degree of water repellence in the surface sands, amount of cover on the dune surface, tree canopy cover, amount and intensity of rainfall and topographic position. There are very few sites in the coastal dunes that are not losing or gaining sand through surface wash or raindrop splash. Surface wash associated with water repellence is a major process in the Holocene dunes leading to gully and fan development; it is much less evident in Pleistocene dunes. Raindrop splash moves sand on all dune surfaces that are not protected by dense ground cover and organic debris; it moves sand to lower topographic positions and forms colluvial deposits on the footslopes.

The swales on young beach-ridge plains are the least affected by water erosion, and at these sites raindrop splash is probably just moving surface sand grains backwards and forwards. Water erosion gradually dissipates beach ridges moving sand into the swales and destroying the initial aeolian form and relief. Parabolic dunes are similarly affected; gully erosion cuts through the dune floors and trailing arms forming oriented high ridges which are progressively reduced to whale-back sandhills. Several stages in dune denudation can be recognized in progressively older dune systems. These provide additional evidence in establishing the relative age of a series of dune systems.

Soil development on the dunes begins with the arrival of plant cover. Because the coastal sands are dominantly quartz with few weatherable minerals, weathering is largely restricted to the mobilization of iron and aluminium compounds, from the coatings on sand grains, and their precipitation at greater depth. This, together with organic compounds from the vegetation, leads to the development of *podzols*. Depth and degree of soil development within a dune system are different on each geomorphic component because of the variation in soil water and amounts of erosion or sedimentation at each site. Both depth and degree of soil development also increase with age of dune system. Several stages of *podzol* development are evident from rudimentary profiles (< 1m thick) on young dunes to giant forms (> 20m thick) on the oldest Pleistocene dunes.

Comparisons of soil development on comparable geomorphic components in different dune systems provide additional evidence in establishing relative age sequences.

Holocene beach-ridge plains have relatively low rates of erosion, a lateral extent that allows a large number of sampling points across the depositional sequence, and are frequently underlain by marine shells suitable for dating. They provide suitable sites for chronosequence studies of soil development that has occurred over the past 7000 years. However, they usually contain greater amounts of marine shell fragments and more feldspar than parabolic dunes, both of which may retard *podzol* development. Also, beach-ridge plains receive high amounts of atmospheric accession over a greater proportion of their area than the large sandmasses of parabolic dunes. Most areas of Pleistocene beach ridges have water-tables near the surface that prevent the development of thick A2 horizons. Fluctuation in water-tables in response to sea-level changes may lead to the formation of a thick humus B horizon.

Holocene parabolic dunes are more eroded, have fewer recognizable steps in their depositional sequence than beach ridges and contain little datable material. Site selection for soil chronosequence studies must be made with considerable care to ensure that comparable geomorphic components are sampled on each dune. The considerable thickness of dune sand, generally in excess of 30 m on the trailing arms and dune floors near the apex, ensures that soil weathering can extend profile depth without interference from water-tables. This is important where the chronosequence includes Pleistocene dunes so that studies may be made of long-term changes in soils and the biological systems that they support.

The systems of both beach ridges and parabolic dunes that have been deposited episodically along the Australian coast during the Quaternary allow the use of either or both in investigations of temporal changes in soils, vegetation and soil biota. Each type of deposit has advantages and disadvantages; together they complement each other in the record of Quaternary coastal events.

References

Ball, L.C. (1924). Report on oil prospecting near Tewantin. *Qld. Govt. Min. J.* 25: 354-362.

Beasley, A.W. (1950). Heavy mineral beach sands of southern Queensland. II. Physical and mineralogical composition, mineral descriptions and origin of the heavy minerals.

Proc. R. Soc. Qd. 61: 59-104.
Bigarella, J.J. (1975). Structures developed by dissipation of dune and beach ridge deposits. *Catena* 2: 107-152.
Bird, E.C.F. (1961). The coastal barriers of East Gippsland, Australia. *Geogr. J.* 127: 460-468.
Bird, E.C.F. (1967). Depositional features in estuaries and lagoons on the south coast of New South Wales. *Austn. Geog. Studies* 5: 113-124.
Bird, E.C.F. (1971). The origin of beach sediment on the north Queensland coast. *Earth Sci. J.* 5: 95-105.
Bird, E.C.F. (1978). The nature and source of beach materials on the Australian coast. *In* Davies, J.L. and Williams, M.A.J. (eds.). *Landform Evolution in Australasia,* ANU Press, Canberra, 144-157.
Bond, R.D. (1964). The influence of the microflora on the physical properties of soils. II. Field studies on water-repellent sands. *Aust. J. Soil. Res.* 2: 123-131.
Bond, R.D. and Harris, J.R. (1964). The influence of the microflora on the physical properties of soils. I. Effects associated with filamentous algae and fungi. *Aust. J. Soil Res.* 2: 111-122.
Bowman, G.M. (1979). *Development of Podzol Soils in East Australian Coastal Sand Barriers.* PhD thesis, Dept. of Geography, Univ. Sydney.
Brewer, R. and Thompson, C.H. (1980). Morphology of two sub-tropical podzols formed on a siliceous dune in coastal Queensland. *In* Joseph, K.T. (ed.) *Proc. Conf. Class. Manag. Trop. Soils,* Malaysian Soc. Soil Science, 116-123. Science.
Bridge, B.J. and Ross, P.J. (1983). Water erosion in vegetated sand dunes at Cooloola, south-east Queensland. *Z. Geomorph. Suppl.* 45: 227-244.
Bridge, B.J., Ross, P.J. and Thompson, C.H. (1984). Studies in landscape dynamics in the Cooloola - Noosa River area, Queensland. 3. *CSIRO Aust. Div. Soils Divl. Rep.* No. 75 (in press).
Burges, A. and Drover, D.P. (1953). The rate of podzol development in sands of the Woy Woy district, N.S.W. *Aust. J. Bot.* 1:83-94.
Byers, H.E., Kellogg, C.E., Anderson, M.S. and Thorp, J. (1938). Formation of soil. In *Soils and Men.* USDA Year-book, U.S. Gov. Print. Office, Washington, D.C., 948-978.
Coaldrake, J.C. (1962). The coastal sand dunes of southern Queensland. *Proc. R. Soc. Qd.* 72: 101-116.
Connah, H.T. (1961). Beach sand heavy mineral deposits of Queensland. *Geol. Surv. Qd. Dept. Mines Pub. No.* 302.
Farmer, V.C., Skjemstad, J.O. and Thompson, C.H. (1983).

Genesis of humus B horizons in hydromorphic humus podzols. *Nature* 304: 342-344.

Franzmeier, D.P. and Whiteside, E.P. (1963). A chronosequence of podzols in northern Michigan. I. Ecology and description of podzols. *Mich. Agr. Exp. Sta. Quart. Bul.* 46: 1-20.

Gardner, D.E. (1955). Beach sand heavy mineral deposits of eastern Australia. *Comm. Bur. Min. Res. Aust. Bull.* No. 28.

Gardner, R.A. (1967). *Sequence of podzolic soils along the coast of northern California.* PhD thesis, Univ. Calif., Berkeley.

Grimes, K.G. (1982). Stratigraphic drilling report - GSQ Sandy Cape 1-3R. *Qd. Govt. Min. J.* 80: 79-82.

Hails, J.R. (1969). The nature and occurrence of heavy minerals in three coastal areas of New South Wales. *J. & Proc. Roy. Soc. N.S.W.* 102: 21-39.

Langford-Smith, T. and Thom, B.G. (1969). New South Wales coastal morphology. *J. Geol. Soc. Aust.* 16: 572-580.

Little, I.P., Armitage, T.M. and Gilkes, R.J. (1978). Weathering of quartz in dune sands under subtropical conditions in eastern Australia. *Geoderma* 20: 225-237.

Ly, C.K. (1976). *Depositional and mineralogical studies of Quaternary sedimentation in the Newcastle - Port Stephens area of New South Wales.* PhD thesis, Univ. Newcastle, NSW.

Marshall, J.F. (1980). Continental shelf sediments: southern Queensland and northern New South Wales. *Bur. Min. Res. Geol. Geophys. Aust. Bull.* 207.

Moss, A.J. and Green, P. (1975). Sand and silt grains: predetermination of their formation and properties by microfractures in quartz. *J. Geol. Soc. Aust.* 22: 485-495.

Olsen, J.S. (1958). Rates of succession and soil changes on southern Lake Michigan sand dunes. *Bot. Gaz.* 125-170.

Petersen, L. (1976). *Podzols and Podzolization.* DSR Forlag, Copenhagen.

Pye, K. (1981). Rate of dune reddening in a humid tropical climate. *Nature* 290: 582-584.

Reeve, R. and Fergus, I.F. (1984). Black and white waters and their possible relationship to the podzolization process. *Aust. J. Soil Res.* 21: 59-66.

Roy, P.S. and Crawford, E.A. (1977). Significance of sediment distribution in major coastal rivers, northern New South Wales. *Third Aust. Conf. Coast Ocean Engin.* Inst. Eng., Australia, 177-184.

Roy, P.S. and Thom, B.G. (1981). Late Quaternary marine deposition in New South Wales and southern Queensland. *J. Geol. Soc. Aust.* 28: 471-489.

Salisbury, E.J. (1922). The soils of Blakeney Point: a study of soil reaction and succession in relation to the plant covering. *Ann. Bot.* 36: 391-431.

Salisbury, E.J. (1925). Note on the edaphic succession in some dune soils with special reference to the time factor. *J. Ecol.* 13: 322-328.

Shepherd, M.J. (1970). *Coastal Geomorphology of the Myall Lakes Area, New South Wales.* PhD thesis, Dept. of Geography, Univ. Sydney.

Stace, H.C.T., Hubble, G.D., Brewer, R., Northcote, K.H., Sleeman, J.R., Mulcahy, M.J. and Hallsworth, E.G. (1968). *A Handbook of Australian Soils.* Rellim Tech. Publs., Adelaide.

Swanson, V.E. and Palacas, J.G. (1965). Humate in coastal sands of north-west Florida. *U.S. Geol. Surv. Bull.* No. 1214-B.

Thom, B.G. (1965). Late Quaternary coastal morphology of the Port Stephens - Myall Lakes area, N.S.W. *J. & Proc. R. Soc. N.S.W.* 98: 23-36.

Thom, B.G. (1975). Modification of coastal and deltaic terrain subsequent to deposition. *Z. Geomorph. Suppl.* 22: 145-170.

Thom, B.G., Bowman, G.M., Gillespie, R., Temple, R. and Barbetti, M. (1981a). Radiocarbon dating of Holocene beach-ridge sequences in southeast Australia. *Monograph No. 11,* Geography Dept. Univ. N.S.W. Duntroon.

Thom, B.G., Bowman, G.M. and Roy, P.S. (1981b). Late Quaternary evolution of coastal sand barriers, Port Stephens - Myall Lakes area, central New South Wales, Australia. *Quat. Res.* 15: 345-364.

Thom, B.G., Shepherd, M.S., Ly, C.K., Roy, P.S. and Bowman, G.M. (1984). Quaternary geology and geomorphology of the Port Stephens - Myall Lakes Region. *Dept. Biogeog. Geomorph.,* Monograph No. 6, ANU, Canberra (in press).

Thompson, C.H. (1975). Coastal areas of southern Queensland some land use conflicts. *Proc. R. Soc. Qd.* 86: 109-120.

Thompson, C.H. (1981). Podzol chronosequences on coastal dunes of eastern Australia. *Nature* 291: 59-61.

Thompson, C.H. (1983). Development and weathering of large parabolic dune systems along the subtropical coast of eastern Australia. *Z. Geomorph. Suppl.* 45: 205-225.

Thompson, C.H. and Hubble, G.D. (1980). Sub-tropical podzols (spodosols and related soils) of coastal eastern Australia. *In* Joseph, K.T. (ed.) *Proc. Conf. Class. Manag. Trop. Soils.* Malaysian Soc. Soil Soc., 203-213.

Thompson, C.H. and Moore, A.W. (1984). Studies in landscape dynamics in the Cooloola - Noosa River area, Queensland. 1. Introduction, soil landscapes and research approach. *CSIRO Aust. Div. Soils Divl. Rep.* No. 73 (in press).

Walker, J., Thompson, C.H., Fergus, I.F. and Tunstall, B.R. (1981). Plant succession and soil development in coastal sand dunes of eastern Australia. *In* West, D.C., Shugart, H.H. and Botkin, D.B. (eds.). *Forest Succession: Concepts and Application,* Springer-Verlag, New York, 107-131.

Ward, W.T. (1977). Sand movement on Fraser Island: a response to changing climates. *Univ. Qld. Dept. Antrop. Pap.* 8, 113-126.

Ward, W.T. and Little, I.P. (1975). Times of coastal sand accumulation in south-east Queensland. *Proc. Ecol. Soc. Aust.* 9: 313-317.

Ward, W.T., Little, I.P. and Thompson, C.H. (1979). Stratigraphy of two sandrocks at Rainbow Beach, Queensland, Australia, and a note on humate composition. *Paleogeogr., Paleoclimat., Paleoecol.* 26: 305-316.

Whitworth, H.F. (1956). The zircon-rutile deposits on the beaches of the east coast of Australia with special reference to their mode of occurrence and the origin of the minerals. *Dept. Mines N.S.W. Tech. Rep.* 4: 7-60.

13

Quaternary Stratigraphy and Sea-Level Variation in the Southeast of South Australia

D. A. SCHWEBEL

Esso Exploration and Production Australia Ltd
Sydney, NSW

Introduction

The broad coastal plain of southeast South Australia has preserved across its width a unique record of Quaternary sea-level change. Slow tectonic uplift, low coastal plain gradient and eustatic sea-level oscillations have combined to produce a series of subparallel depositional shorelines hundreds of kilometres long. Immediate post-depositional cementation of these substantially calcareous deposits has preserved the depositional geometry and stratigraphic relationships.

Quaternary shorelines have been deposited on a coastal plain which stretches from Western Victoria to the Murray River (Fig. 1). Twenty one stratigraphically separable Quaternary coastal barrier deposits occur in thirteen geomorphologically distinct 'ranges' across the coastal plain at its widest point between Robe and Naracoorte. To the south and north of this line of section many of the ranges coalesce and lose their integrity (Fig. 1). Where best developed, these ranges are up to 40 m higher than the surrounding coastal plain. Although they may be up to 10 km wide, they are generally less than 5 km wide.

The potential of these deposits to contribute to the understanding of Quaternary sea-level change has been recognised by many workers (Hossfeld, 1950; Sprigg, 1952; Cook *et al.*, 1977; Schwebel, 1978). Early studies were hampered by poor outcrop of what are essentially surficial deposits, absence of widespread subsurface control, and the unavailability of dating

COASTAL GEOMORPHOLOGY
IN AUSTRALIA
ISBN 0 12 687880 3

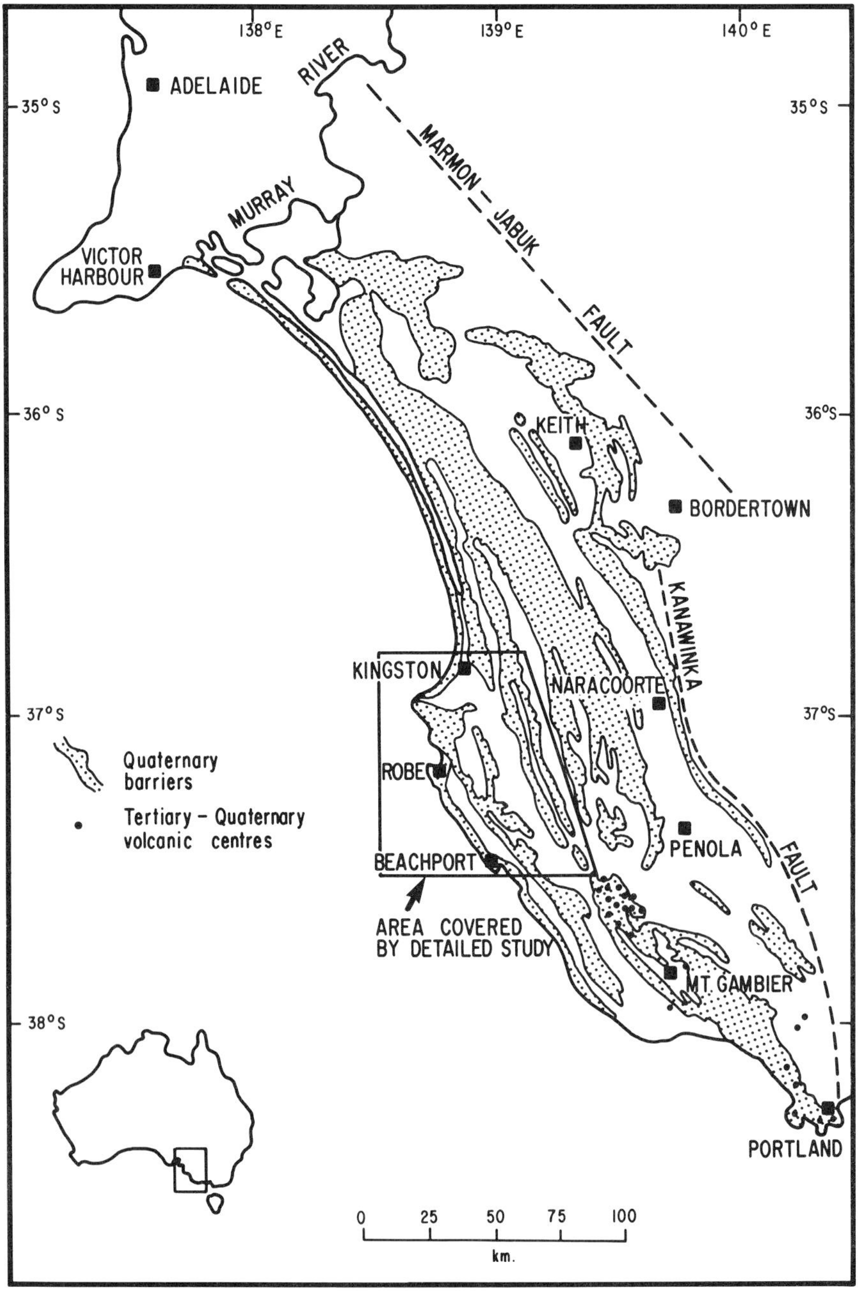

FIGURE 1. Distribution of Quaternary shoreline deposits across the coastal plain of southeast South Australia.

techniques capable of accurately determining the age of the sediments forming the stranded shorelines. Geomorphology and subjective geological reasoning were the only criteria on which to base a chronology of depositional events.

Shallow stratigraphic drilling by the Bureau of Mineral Resources during 1974-1975 provided extensive subsurface control across the widest part of the coastal plain. This drilling revealed complexities in the Quaternary sequence which require substantial alteration to existing theories of the relationship between the stratigraphy and eustatic sea-level change. Cook *et al.*, (1977) described the gross stratigraphic relationships revealed in 36 continuously cored drill holes. The detailed relationships, both between and within the stratigraphic units present across the westerly half of the coastal plain, are described more comprehensively by Schwebel (1978). It is in this area that the following detailed investigation of late Pleistocene sea-level change has taken place. An attempt has also been made to extend the study into the area where the stratigraphic understanding is not so complete.

Stratigraphic Units

A predominantly carbonate sequence was deposited across the coastal plain during the Quaternary. The stratigraphy is divisible on the basis of lithology and depositional environment into three mapable units which were deposited episodically in association with interglacial high sea-levels.

The subdivision into units on the basis of depositional environment is broadly comparable to that established by Cook *et al.*, (1977). The three units which Schwebel (1978) described in detail are:

(1) *A Beach-Dune Skeletal Grainstone Unit* consisting of grainstones deposited as the beach and dune components of barrier shorelines at and near the peak of glacio-eustatic sea-level transgressions.

(2) *A Lagoonal Skeletal Grainstone, Packstone and Wackestone Unit* consisting of sediments deposited under sheltered marine conditions created on the leeward side of those shorelines which formed as barrier islands when the peak of each transgression was reached.

(3) *A Lacustrine Calcitic and Dolomitic Mudstone Unit* consisting of calcareous mudstones deposited from saline lakes occurring in low-lying areas between the stranded coastal barriers wherever the groundwater table emerged at the surface.

The three units are not uniformly distributed across the coastal plain. The Beach-Dune Unit occurs as shoestrings of sand sub-parallel to the present coastline which separate extensive sheets of the finer grained sediments of the Lagoonal Unit and the Lacustrine Unit. The distribution of the three units in the area where the detailed stratigraphic study has taken place is shown in Figures 2 and 3.

Deposition of the Beach-Dune Unit occurred at and near the peak of eustatic sea-level oscillations. If the area leeward of the Beach-Dune Unit barrier was flooded, then the Lagoonal Unit was also deposited at this time. The Lacustrine Unit appears to have been deposited initially as part of a regressive sequence at the stabilised interglacial sea-level and then continued to accumulate in areas of outcropping water table as the sea-level retreated.

Beach-Dune Skeletal Grainstone Unit

The Beach-Dune Unit has been described by Sprigg (1952), Borlakoff (1963), Firman (1973) and Cook *et al*., (1977). It was deposited as mainland barrier shorelines and barrier islands which have been preserved as conspicuous topographic ranges on an otherwise featureless coastal plain (Schwebel, 1978). Ditches excavated to facilitate the drainage of winter rains expose excellent cross sections through the more prominent topographic ranges. Morphologic and stratigraphic relationships and the distribution of the Beach-Dune Unit are shown on Figures 2 and 3

The Beach-Dune Unit unconformably overlies Tertiary marine sediments. The upper surface of the Pleistocene elements of this unit are generally calcreted and exposed along the crest of the ranges. The Holocene element of the Unit has not been calcreted. On the flanks of the ranges, the Beach-Dune Unit is overlain by onlapping sediments of the inter-range Lacustrine and Lagoonal Units.

Morphologically, the ranges appear to be 'simple' and the product of one depositional event. While this is generally the case, the Woakwine, Robe and West Naracoorte Ranges have 'complex' internal structure and are the product of more than one episode of deposition. Within these 'complex' barriers significant unconformities separate deposits of the Beach-Dune Unit of different ages. These unconformities are identified by conglomerates and shell beds in the case of marine erosion and by soil profiles and calcretes where there has been extended periods of exposure and non-deposition (Fig. 4).

Limestones of the Beach-Dune Unit are beach and wind deposited sand-sized sediments which range in carbonate content from 3% to 80%. They are analogous to aeolianites described

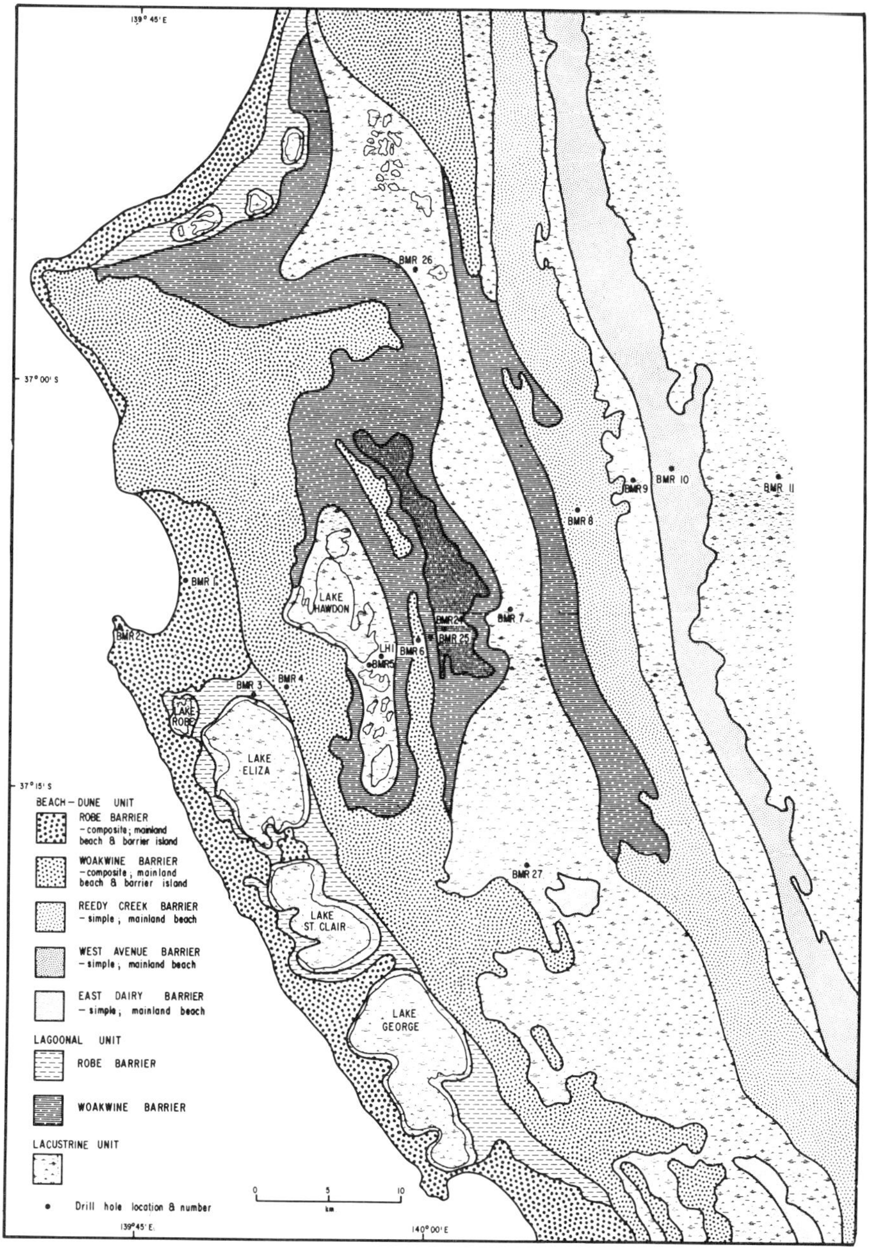

FIGURE 2. Distribution of depositional shoreline units in the vicinity of Robe.

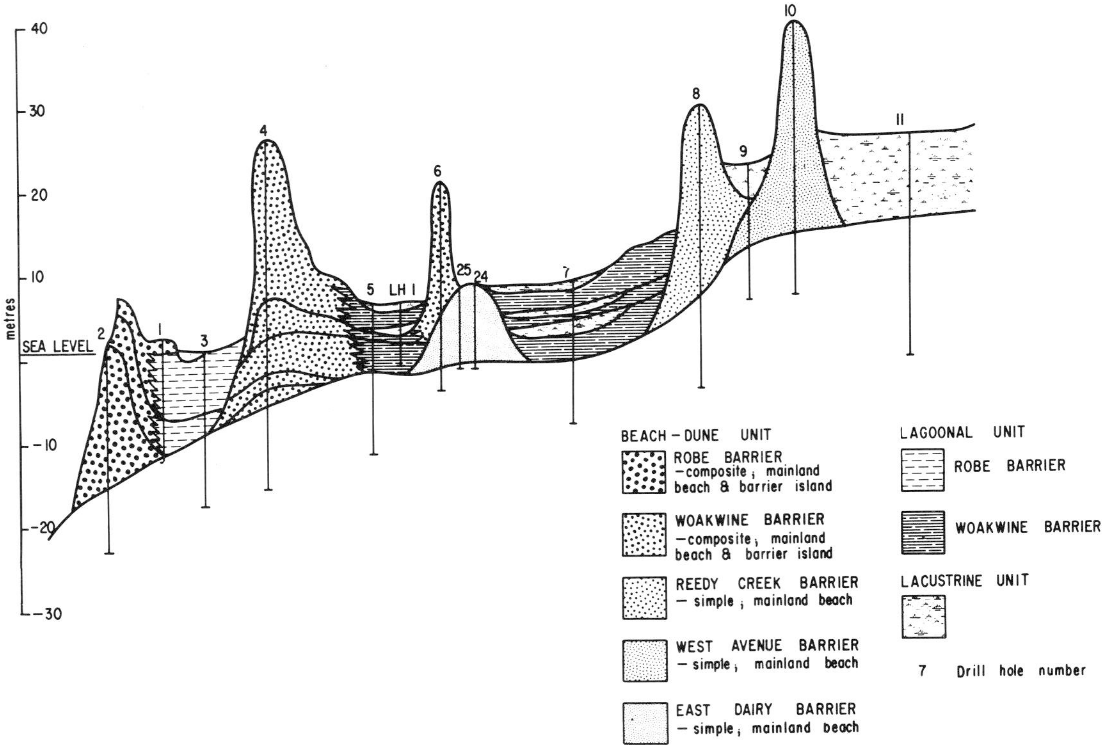

FIGURE 3. Generalized section across coastal plain east of Robe (see Fig. 2 for location of drill holes).

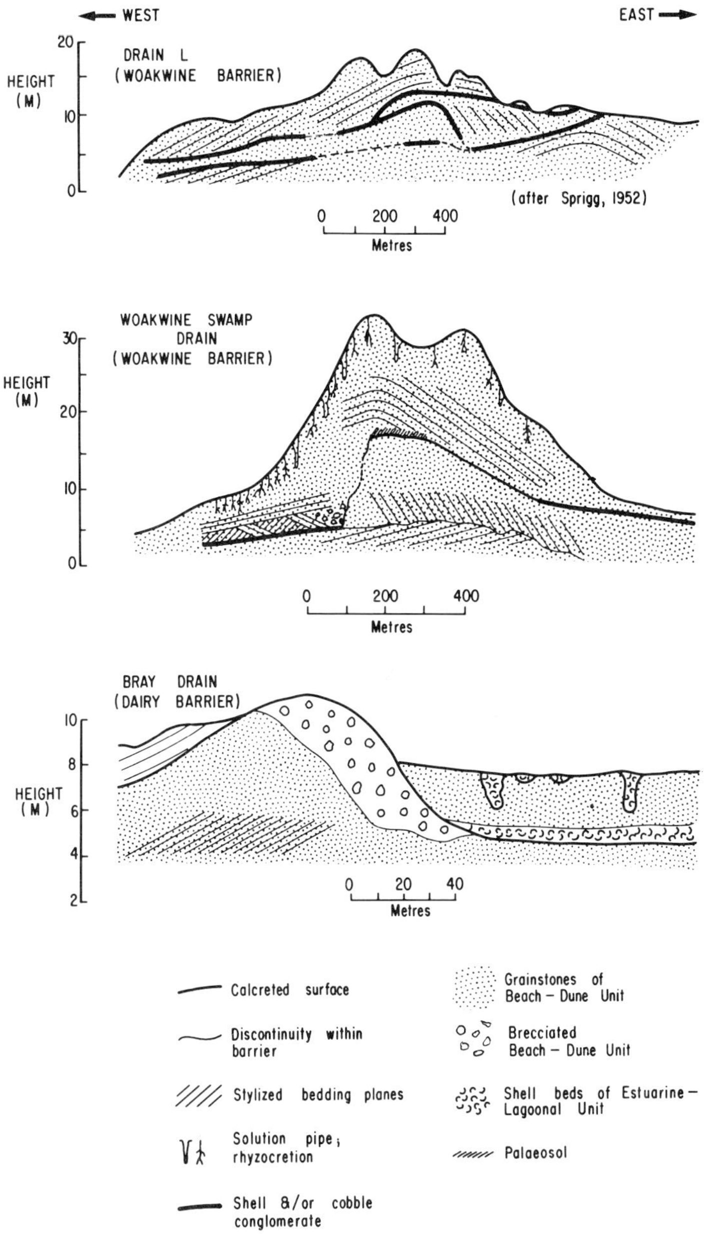

FIGURE 4. Cross-sections through complex barrier shorelines

from many parts of the world (Bretz, 1960; Fairbridge and Teichert, 1953; Ward, 1975; Yaalon and Larome, 1977). The carbonate fraction is made up principally of molluscan and calcareous algal debris and lithoclasts of older limestone. The non-carbonate fraction is almost entirely quartz, with minor quantities of K-feldspar and traces of plagioclase and heavy minerals, primarily opaque minerals, amphiboles and zircons (Colwell, 1977).

A calcrete of variable form and thickness veneers all the Pleistocene elements of the unit (Blackburn *et al.*, 1965). Calcrete development has prevented redistribution of the sediment and in each range has preserved near-perfect coastal barrier morphology of foredunes, blowouts and beach ridges. The one stratigraphically significant episode of calcretization proposed by Firman (1973) is not consistent with this observation. As there is no indication of the development of other soil profiles on the Beach-Dune Unit, it is most likely that calcrete development has occurred either continuously through the Pleistocene or only during interglacial times immediately after each episode of shoreline deposition.

Grainstones of the Beach-Dune Unit, apart from those which have been calcretized, have poorly developed calcite cements. These cements were deposited in both vadose and phreatic groundwater environments. Grain contact, meniscus and gravitational cements are typical of the vadose zone (Ward, 1975). Pore-filling cements are ubiquitous in the phreatic zone. In each 'simple' barrier, the zones of maximum cementation coincide with the top of the phreatic zone and with the surficial calcrete. Unconformities are also generally sites of maximum calcite cementation, either associated with calcretisation or pore-filling cementation (as seen in 'complex' barriers).

The original carbonate mineralogy of the Beach-Dune Unit is typical of bioclastic limestones deposited under temperate latitude conditions (Nelson, 1978). Carbonate minerals in the Holocene sediments are biogenically derived low Mg-calcite and high Mg-calcite (60-80%) and aragonite (20-40%). The principal sources of the minerals are molluscs (calcite and aragonite) and calcareous red algae (high Mg-calcite). In some areas there are traces of early diagenetic dolomite.

Pleistocene elements of the unit have low Mg-calcite, high Mg-calcite, aragonite and dolomite present as the principal carbonate mineral types. The proportion and distribution of these minerals were confirmed by thin-section petrography, x-ray diffraction and scanning electron microscopy. In successively older barrier shoreline sediments, progressive diagenesis reduces the quantity of aragonite and high Mg-calcite present in the framework to the extent that neither mineral occurs in barriers older than the Reedy Creek (Table 1). The

TABLE 1. Age estimates, measured palaeosealevel heights and uplift corrected palaeosealevel heights for each barrier shoreline which has been identified to the west of the West Avenue Barrier.

STRATIGRAPHIC EVENT	AGE ('000's years)	0^{18} ESTIMATE	HEIGHT (m) MEASURED	UPLIFT CORRECTED
Robe 1	4.3 ± 0.1	4	0	0
Robe 11	-	83	-6 to +2	-13 to -5
Robe 111	-	93	-15 to -8	-23 to -16
Woakwine 1	100 ± 30 125 ± 20	120	+12 to +17	+2 to +7
Woakwine 11	-	203	0 to +9	-16 to -8
Woakwine 111	-	218	-3 to ?	-20 to ?
Reedy Creek	-	248	+12 to +17	-8 to -3
East Dairy	-	309	-2 to ?	-27 to ?
West Avenue	-	347	+13 to +14	-12 to -10
Woakwine 1V	-	384	-5 to ?	-36 to ?
Woakwine V	-	?	-7 to ?	?

proportion of dolomite present in the Pleistocene sediments appears to be directly related to diagenetic processes associated with the mobilisation of magnesium from the metastable high Mg-calcite. There is no identifiable trend of increasing dolomite content with increasing age.

Lagoonal Skeletal Grainstone, Packstone and Wackestone Unit

The Lagoonal Unit was deposited under marginal marine conditions where lagoons formed in flooded interdune corridors in the lee of barriers of the Beach-Dune Unit which were deposited at that particular high sea level. The Unit has been observed only to the west of the Reedy Creek Barrier and is observed to be directly related to multiple episodes of deposition along the Robe and Woakwine shoreline trends (Figs. 2 and 3).

The Lagoonal Unit ranges in age from Pleistocene to Holocene. It occurs as thin sheets of sediment which have complex stratigraphic relationships. It unconformably overlies either Beach-Dune Unit, Lacustrine Unit or Tertiary sediments and is in part overlain by the Lacustrine Unit. The Lagoonal Unit is a common surficial sediment and can be shown to be a lateral facies equivalent of the Beach-Dune Unit deposited at the same high stand of sea level (Schwebel, 1978). The degree of induration is largely dependent on age and proximity of surficial calcrete development. Recent sediments are unconsolidated while their Pleistocene counterparts range from unconsolidated to strongly indurated.

The grainstones of the Unit have similar composition to the sediments of the Beach-Dune Unit. They are quartz/molluscan grainstones which were deposited around the margins of the lagoon and are representative of low energy beach and nearshore environments. The framework most commonly consists of detrital quartz and either whole or fragmentary bivalves and gastropods.

The packstones were deposited in lower energy conditions where calcareous and terrigenous mud was not winnowed from the framework of the sediment. In the more sheltered and deepest parts of the lagoon the wackestones were deposited. In these fine-grained sediments the coarser grainsize fraction consists almost entirely of skeletal material derived from organisms living in the low energy environments in which deposition took place. These are primarily molluscs, ostracods and foraminifera. Minor quantities of detrital quartz are also present.

The textural subdivision into which the sediments of the Lagoonal Unit fall exercises fundamental control over their carbonate mineralogy. The grainstones and packstones of this Unit have a mineralogy which is controlled by the skeletal components of the framework, age of the sediment and the degree of diagenesis. The mineralogy varies in a similar manner to the way in which it does for the Beach-Dune Unit. That is, the grainstones and packstones of successively older units progressively undergo diagenetic change from a low Mg-calcite-aragonite-high Mg-calcite composition to a low Mg-calcite-dolomite composition.

The wackestones also have a mixed calcite-aragonite-dolomite mineralogy. However, there is no corresponding systematic compositional variation correlatable with age. The source of the carbonate minerals is generally biogenic. However, in the finer fractions there is an increase in the proportion of aragonite at the expense of the other minerals. Scanning electron microscope study of the finest component of these sediments did not identify crystalline minerals - only 'amorphous' material is present (Schwebel, 1978). This has been interpreted to indicate that there is a contribution of aragonite to the sediment from an inorganic source. That is, it is suggested that the aragonite was precipitated directly from lagoonal waters in much the same way the aragonite sediments of the Coorong Lagoon are currently being formed (Von der Borch, 1965). These aragonitic sediments are essential to the understanding of the relationship between the stratigraphy and Pleistocene sea-level change since they are suitable for uranium series age determination (see below).

Lacustrine Calcitic and Dolomitic Mudstone Unit

The Lacustrine Unit is present as surficial deposits in all the interrange areas across the coastal plain (Figs. 2 and 3). It occurs as the unconsolidated surface sediment of Holocene saline lakes. The indurated sediments of Pleistocene lakes crop out in interrange areas to the east of the Woakwine Range. In the same area they underlie most modern lakes.

The Lacustrine Unit may unconformably overlie Tertiary sediments. It conformably overlies the Lagoonal Unit where the two combine to form an upward-fining regressive sequence. The unit may also onlap and disconformably overlie the Beach-Dune Unit and the older indurated deposits of the Lacustrine Unit.

Sediments of the Lacustrine Unit are primarily mudstones with occasional wackestones and rare packstones. Sand size grains are generally of biogenic origin and are derived from molluscs (principally gastropods), foraminifera and ostracods. Occasional quartz grains, probably of aeolian origin, are also present. The calcareous mud has a grainsize generally less than 2 microns and is commonly aggregated into pellets.

There is a systematic spatial distribution of the three lithologies of the Lacustrine Unit in which the mud content is controlled by the energy conditions prevalent in the different parts of the lake. Coarser grained, better-sorted sediments occur around the leeward margins of the lakes where wind-generated waves have winnowed finer-grained material from the sediment. Gastropod coquinas are common beach deposits along these leeward lake shores. Pelletal mudstones tend to occur on those shallow parts of the lakes where currents and waves associated with seiching periodically remove the fine, disaggregated mud from the lake surface (Noye, 1967). In these areas laminated pelletal mudstones occur. Amorphous mudstones (the yoghurt muds described by Von der Borch, 1977) appear to be representative of the more central lacustrine conditions where surface water persists through severe seasonal evaporative periods.

The Coorong Lagoon and ephemeral lake chain are considered to be the present-day analogue of the Pleistocene parts of the Lacustrine Unit. However, the complex mixtures of hydromagnesite, magnesite, protodolomite, aragonite, high Mg-calcite and low Mg-calcite present in the modern environments (Von der Borch, 1965) have not been identified in the Lacustrine Unit. Only low Mg-calcite, high Mg-calcite and dolomite have been identified in the Lacustrine Unit. This may be a real difference between the two areas or it could simply be a reflection of the sparce sample control available from the drill holes.

As the Lacustrine Unit is generally the last deposited sediment in the inter-range areas, it has in many cases been exposed for long periods of time to pedogenic processes. However, because of its fine grained nature and proximity to the water table no deep soil profiles have developed. This is particularly so since the mudstones of the Unit have often undergone early diagenetic alteration and induration which has produced a stable low Mg-calcite-dolomite mineralogy less likely to be affected by chemical processes of calcretisation. Physical fracturing and recementing of the rock is a common feature of the calcreted Lacustrine Unit.

Stratigraphic Relationships

Detailed stratigraphic drilling has shown the Quaternary sequence to have been deposited in response to the complex interaction of slow tectonic uplift and eustatic sea-level change on a broad, gently dipping coastal plain. As a consequence the resultant overall regressive sequence has been deposited in a series of transgressions and regressions. Sediments preserved as linear shorelines (and where present, associated lagoonal deposits) represent deposition at or near the peak of each of these transgressions, and provide a discontinuous record of the depositional environments active throughout the Quaternary.

The Robe and Woakwine Barriers have complex internal structure and are the result of three and five separate high sea-level stands, respectively. The deposits of each high sea-level is separated from overlying deposits by an erosional surface along which conglomerates, soils, calcretes and strong induration generally occur. Lagoonal Unit deposits on the inland side of these barriers are lateral equivalents of the upper two and three barrier shoreline deposits, respectively. The West Dairy Barrier is not considered to be the product of a separate high stand of sea-level, but rather to be the equivalent of the last episode of Woakwine Barrier deposition. It occurs only as beach deposits around the eastern margin of the Woakwine Barrier lagoon.

All of the other barriers (except the West Naracoorte Barrier) have been shown by stratigraphic drilling to be simple shoreline deposits. No deposits of Lagoonal Unit have been identified to the east of the Reedy Creek Barrier. The West Naracoorte Barrier accumulated on the downthrown side of the major northwest-southeast trending Kanawinka Fault which marks the inland limit of the coastal plain. Successive episodes of deposition may have occurred where there was local steepening of the coastal plain gradient as the fault moved episodically during the early Pleistocene. Stratigraphic drilling identified beach sediments at the base of the oldest component of the barrier. The remainder of the sediment pile in the West Naracoorte Barrier consists of aeolian accumulations separated by thick soils.

Geochronology

The sequence consisting of Beach-Dune Unit, Lagoonal Unit and Lacustrine Unit has long been regarded of Quaternary age (Sprigg, 1952, 1958, Hossfeld, 1950, Firman, 1969, 1973; Cook *et al.*, 1977). Refinement of this Quaternary age estimate was not possible until the results of uranium series age dating, radiocarbon dating and magnetostratigraphy became available.

All of the Beach-Dune Unit to the west of and including the West Naracoorte Barrier has normal magnetic polarity, and is interpreted to have been deposited since the Brunhes-Matuyama magnetic reversal 690,000 years ago (Cook *et al.*, 1977). The Lagoonal Unit stratigraphically equivalent to the youngest part of the Woakwine Barrier was deposited within the dating range of the uranium series method and age estimates of 100,000 ± 30,000 and 125,000 ± 20,000 years respectively were obtained from aragonitic muds and molluscs of the Unit (Schwebel, 1978) (Table 1). Both of these results are suggestive of a Last Interglacial age (Veeh, 1966). The remainder of the Lacustrine Unit and all of the Beach-Dune Unit is either outside the range of the Th^{230}/U^{234} method or devoid of material suitable for dati

Holocene sedimentation has occurred along the modern coastline, and uncemented sediments of the final stage of Robe Barrier deposition are within the range of radiocarbon dating and are estimated to have been deposited since sea-level peaked after the last transgression. The maximum measured carbon-14 age of molluscs collected from these sediments is 4330 ± 100 years (Blackburn, 1966). To the west of Robe in Guichen Bay, Thom *et al.*, (1981) have dated Holocene shells from the prograded bay barrier.

Lacustrine Unit sediments are of both Pleistocene and Holocene age (Von der Borch, 1977; Cook *et al.*, 1977). The unit exhibits no simple spatial variation through time and appears to be independent of barrier shoreline deposition. There is probably a general trend towards increasing age with greater distance from the present coastline. However, the unit may be deposited at any time and in any place where the water table intersects the land surface. Von der Borch (1977) has described Holocene lakes in the vicinity of Naracoorte which are actively precipitating dolomite.

Age estimates of the sediments and the preservation of their spatial relationships indicate that calcrete development was an active process throughout the Pleistocene. Uncemented sediments and unaltered sediments are generally of Holocene age.

Sea-Level Changes

Age of barrier shoreline deposition, rate of tectonic uplift and present height above sea level are the essential variables from which estimates of Pleistocene palaeosealevels can be made. The ages of the barrier shorelines that are able to be dated by conventional means have been outlined above. The age of the other deposits which occur across the western half of the coastal plain have been estimated by comparing the detailed stratigraphic sequences with the deep sea oxygen isotope record (Broecker and Van Donk, 1970; Shackleton and Opdyke, 1973, 1976). The measured and estimated ages of these barrier shorelines are presented in Table 1. There is far more uncertainty surrounding the sequence of depositional events to the east of the West Avenue Barrier. Since the stratigraphic drilling in this area did not reveal the presence of any Lagoonal Unit from which inference could be drawn about multiple shoreline deposition along the one lineament, it has been assumed that these older barriers increase in age towards the east in a uniform manner.

The rate of tectonic uplift has been calculated by two separate methods and relies on the independently derived age estimates for the West Naracoorte and Woakwine Barriers. Using the presently measured heights of the inferred palaeosealevel (see criteria outlined in Table 2), the 690,000 years age estimate for the oldest element of the West Naracoorte Barrier, and the 125,000 years age estimate for the youngest element of the Woakwine Barrier, uplift rates of between 0.05 mm/year and 0.09 mm/year have been calculated.

The present heights of the palaeosealevels which were responsible for the deposition of the barriers to the west of the West Avenue Barrier are shown in Table 2. These are based on a number of stratigraphic criteria which are also outlined in Table 2. Combination of the information contained in Tables 1 and 2 and an average uplift rate of 0.07 mm/year leads to the development of a palaeosealevel curve which is applicable to the southeast of South Australia. This curve is shown in Figure 5. The most important aspect of the curve is that it agrees well with the existing interpreted history of sea-level change for the period up to 125,000 years ago (Broecker *et al.*, 1968; Matthews, 1973; Bloom *et al.*, 1974; Chappell, 1974). For the period of 125,000 to 400,000 years the inferred sea level is no greater than that of the present time.

For the barrier shorelines interpreted to be older than 400,000 years the sequence of events and the frequency and magnitude of the sea-level changes is far more speculative (Chappell and Veeh, 1978). While a relative sequence of events can be reconstructed from the stratigraphy, the chronology relies entirely on correlation of the depositional shorelines to the peaks on the deep sea oxygen isotope record. Consequently, the sea-level curve so constructed mirrors the periodicity of the oxygen isotope record. However, within the constraints imposed by the estimated uplift rate, the stratigraphy does allow estimation of the palaeosealevels at which each shoreline was deposited. It may be significant that despite all of the assumptions inherent in the method, for the period from 400,000 to 700,000 years ago palaeosealevel is estimated to be never more than about 10 metres higher than the present sea level.

TABLE 2. Criteria used to establish range of palaeosealevel heights responsible for deposition of each barrier shoreline.

STRATIGRAPHIC EVENT	MEASURED HEIGHT ABOVE MSL (m)		CRITERION BY WHICH SEA-LEVEL HEIGHT DEFINED
Robe 1		0	Modern Sea-level
Robe 11	Maximum	2	Height of palaeosol between Robe 11 and Robe 111 Barriers.
	Minimum	-6	Presence of Lagoonal Unit in lee of Robe 11 Barrier.
Robe 111	Maximum	-8	Absence of Lagoonal Unit in lee of Robe 111 Barrier.
	Minimum	-15	Base of Beach-Dune Unit in Robe 11 Barrier.
Woakwine 1	Maximum	17	Palaeosol between Woakwine 1 and 11 Barriers.
	Minimum	12	Marine erosion surface at base of Woakwine 1 barrier.
Woakwine 11	Maximum	9	Beach aeolian transition in Woakwine 11 Barrier.
	Minimum	0	Marine erosion surface at base of Woakwine 11 Barrier.
Woakwine 111	Maximum	-	Indeterminable.
	Minimum	-3	Marine erosion surface at base of Woakwine 111 Barrier.
Reedy Creek	Maximum	17	Absence of Lagoonal Unit in lee of Reedy Creek Barrier.
	Minimum	12	Base of Beach-Dune Unit in Reedy Creek Barrier.
East Dairy	Maximum	-	Indeterminable.
	Minimum	-2	Base of Beach-Dune Unit in East Dairy Barrier.
West Avenue	Maximum	14	Absence of Lagoonal Unit in lee of West Avenue Barrier.
	Minimum	13	Base of Beach-Dune Unit in West Avenue Barrier.
Woakwine IV	Maximum	-	Indeterminable.
	Minimum	-5	Marine erosion surface between Woakwine IV and V Barrier.
Woakwine V	Maximum	-	Indeterminable.
	Minimum	-7	Base of Beach-Dune Unit in Woakwine V Barrier.

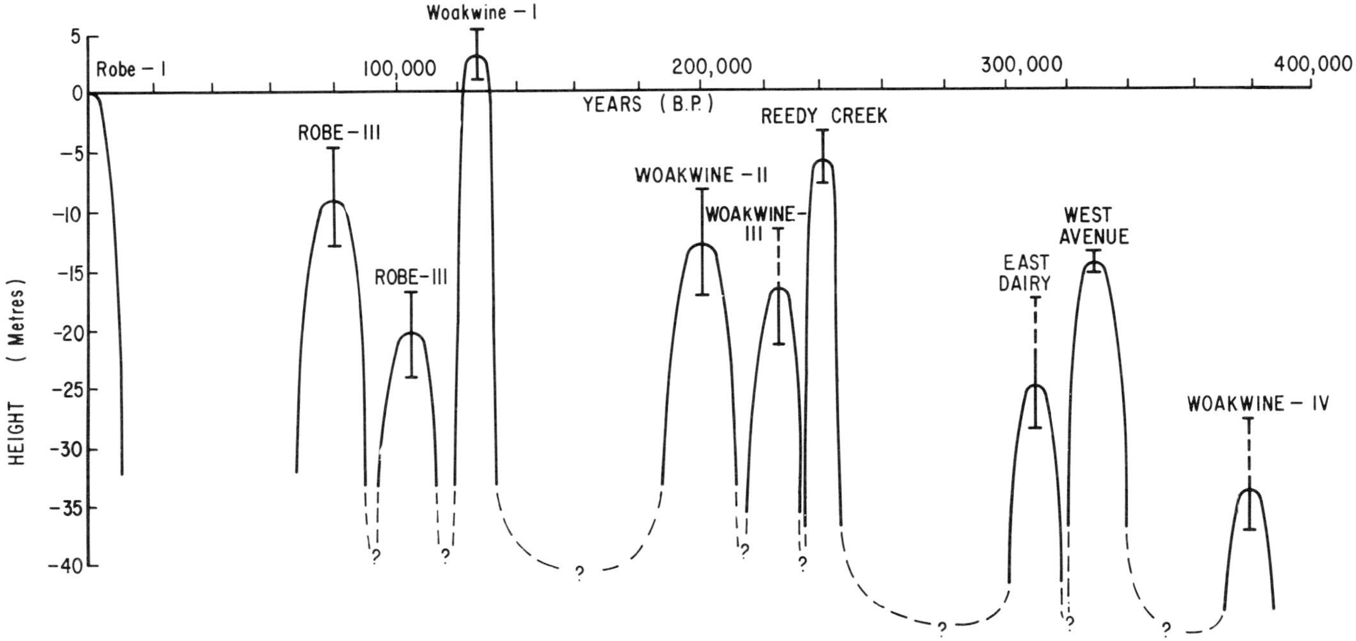

FIGURE 5. Reconstructed sea-level curve for southeast South Australia for the period 400,000 years ago to the present.

Discussion

The stratigraphy of the southeast of South Australia has preserved a history of sea-level change which when reconstructed presents a reasonable picture of the magnitude of eustatic sea-level oscillations up to 400,000 years. The most recent portion of the reconstruction, up to 125,000 years, is in good agreement with the established sea-level curves from the upper Pleistocene to the Holocene. The deep sea oxygen isotope record is the only reference with which to compare the remainder of the interpreted sea-level history.

The record of oxygen isotope variation has provided a means of estimating the absolute ages for the temporally uncontrolled portion of the stratigraphy from 125,000 to 400,000 years ago (and also to 700,000 years). Consequently, the sea-level curve derived herein and the oxygen isotope curve will have identical periodicity of maxima and minima.

The good correspondence between the sea-level and the oxygen isotope record over the last 125,000 years would suggest that the earlier record for both phenomena should be comparable. The major factor which lends credence to the sea-level curve generated from the stratigraphy of the southeast of South Australia is that, the Last Interglacial apart, the absolute heights of the palaeosealevels are not significantly greater than the present sea-level.

An essential adjunct to the derivation of a palaeosea-level curve is the determination of an uplift rate for the coastal plain. This has been done using two independent methods of determining the ages of different barrier shoreline deposits. The results indicate an uplift rate of between 0.05 and 0.09 mm/year.

There is no detailed understanding of the uplift mechanism, but coincidence of the area of maximum uplift with sites of Tertiary volcanism has been suggested by Sprigg (1952) and Cook *et al.*, (1977) as evidence of a causal relationship.

The detailed investigation of the Quaternary stratigraphy has done much to explain the mechanisms of shoreline emplacement along the southern Australian coastline. The most common Quaternary shoreline deposit is of the simple mainland beach type. This consists of a single accumulation of Beach-Dune Unit which was deposited on a gently sloping coastal plain at the peak of the transgressing postglacial sea-level. Complex barrier shorelines which contain sediments of both the Beach-Dune Unit and the Lagoonal Unit appear to occur only where the peak of the transgressing postglacial sea-level coincides with a local steepening of gradient along a topographic irregularity on the coastal plain. This type of situation is normally encountered where an older, generally eroded, barrier shoreline coincides with the peak of the transgression.

The Beach-Dune Unit is then deposited along the pre-existing lineament and the Lagoonal Unit in the flooded area in the lee of the main barrier deposit. This appears to have been the case for the only two barriers having complex internal structure. The Robe and Woakwine Barriers are the only Beach-Dune Unit deposits to have Lagoonal Unit sediments preserved on their inland sides. It is in these particular sediments that material suitable for dating by the uranium series method is preserved.

The whole stratigraphy has been preserved in its near perfect condition because of the rapid post-depositional induration of the sediments. This has been a direct consequence of their carbonate composition. Stabilization has not simply been effected by normal vadose and phreatic zone cementation of grainstones. Calcrete formation in association with soil profile development has been the major agent in preserving the spatial relationships between the different lithologic units. Slow tectonic uplift of the low relief coastal plain has also assisted in preserving the stratigraphy by distancing the barrier topography from the site of marine erosion.

References

Blackburn, G. (1966). Radiocarbon dates relating to soil development, coastline changes and volcanic ash deposition in southeast South Australia. *Aust. J. Science*, 29, 50-52.

Blackburn, G., Bond, R.D. and Clarke, A.R. (1965). Soil development associated with stranded beach ridges in south-east of South Australia. *C.S.I.R.O. Soil Publication*, 22.

Bloom, A.L., Broecker, W.S., Chappell, J., Matthews, R.K. and Mesolella, K.J. (1974). Quaternary sea level fluctuations on a tectonic coast: new Th^{230}/U^{234} dates from Huon Peninsula, New Guinea. *Quat. Res.*, 4, 185-205.

Boutakoff, N. (1963). The geology and geomophology of the Portland area. *Geol. Survey Victoria, Memoir*, 22.

Bretz, J.H. (1960). Bermuda: A partially drowned, late mature Pleistocene karst. *Geol. Soc. America, Bull.* 71, 1729-1754.

Broecker, W.S. and van Donk, J. (1970). Insolation curves, ice volumes and O^{18} record in deep sea cores. *Rev. Geophysics and Space Physics*, 8, 169-98.

Broecker, W.S., Thurber, D.L., Goddard, J., Ku. T.L., Matthews, R.K. and Mesolella, K.J. (1968). Milankovitch hypothesis supported by precise dating of coral reef and deep-sea sediments. *Science*, 159, 297-300.

Chappell, J. (1974). Geology of coral terraces on Huon Peninsula, New Guinea: A study of Quaternary tectonic movements and sea level changes. *Geol. Soc. America, Bull.* 85, 553-70.

Chappell, J. and Veeh, H.H. (1978). Late Quaternary tectonic movements and sea-level changes at Timor and Atauro Island. *Geol. Soc. America, Bull.* 89, 356-368.

Colwell, J.B. (1977). Heavy minerals in late Cainozoic sediments of southeastern South Australia. *Bureau of Mineral Resources, Record*, 1976/89.

Cook, P.J., Colwell, J.B., Firman, J.B., Lindsay, J.M., Schwebel, D.A. and von der Borch, C.C. (1977). Late Cainozoic sequence of Southeast of South Australia and Pleistocene sea level changes. *Bureau of Mineral Resources, Jour. Geol. Geophys.* 2, 81-88.

Fairbridge, R.S. and Teichert, C. (1953). Soil horizons and marine bands in coastal limestones of Western Australia. *J. Roy. Soc. N.S.W.*, 86, 68-86.

Firman, J.B. (1969). Quaternary Period. *In*: Parkin, L.W. (Ed.) *Handbook of South Australian Geology*. Geol. Survey of South Australia, Adelaide.

Firman, J.B. (1973). Regional stratigraphy of surficial deposits in the Murray Basin and the Gambier Embayment. *Geol. Survey of South Australia, Report*, 39.

Hossfeld, P.S. (1950). The Late Cainozoic history of the southeast of South Australia. *Trans. Roy. Soc. South Australia*, 73, 232-279.

Matthews, R.K. (1973). Relative elevation of late Pleistocene high sea level stands: Barbados uplift rates and their implications. *Quat. Res.* 3, 147-153.

Nelson, C.G. (1978). Temperate shelf carbonate sediments in the Cenozoic of New Zealand. *Sedimentology*, 25, 737-771.

Noye, B.J. (1967). A limnological study of the Coorong. *Horace Lamb Centre for Oceanographical Research, Research Paper, No.* 11.

Schwebel, D.A. (1978). *Quaternary stratigraphy of the south-east of South Australia*. Unpubl. Ph.D Thesis, Flinders University of South Australia.

Shackelton, N.J. and Opdyke, N.D. (1973). Oxygen isotope and palaeomagnetic stratigraphy of equatorial Pacific core V28-238: Oxygen isotope temperature and ice volumes on a 10^5 and 10^6 scale. *Quat. Res.* 3, 39-55.

Shackelton, N.J. and Opdyke, N.D. (1976). Oxygen isotope and palaeomagnetic stratigraphy of Pacific core V28-239 late Miocene to latest Pleistocene. *Geol. Soc. America, Memoir*, 145, 449-463.

Sprigg, R.C. (1952). Geology of the southeast province South Australia, with special reference to Quaternary coastline migrations and modern beach developments. *Geol. Survey of South Australia, Bulletin*, 29.

Thom, B.G., Bowman, G.M., Gillespie, R., Temple, R. and Barbetti, M. (1981). *Radiocarbon dating of Holocene beach-ridge sequences in South-east Australia*. Dept. of Geography, University of N.S.W., Duntroon, Monograph, No. 11, 36 pp.

Veeh, H.H. (1966). Th^{230}/U^{234} and U^{234}/U^{238} ages of Pleistocene high sea level stands. *J. Geophys. Res.*, 71, 3379-3386.

Von der Borch, C.C. (1965). The distribution and preliminary geochemistry of modern carbonate sediments of the Coorong area, South Australia. *Geochemica et Cosmochemica Acta*, 29, 781-799.

Von der Borch, C.C. (1977). Stratigraphy and formation of Holocene dolomite carbonate sediments of the Coorong area, South Australia. *Sedimentology*, 46, 952-966.

Ward, W.C. (1975). Petrology and diagenesis of carbonate eolianites of northeastern Yucatan Peninsula, Mexico. *Amer. Assoc. Pet. Geol's, Studies in Geology*, 2, 500-571.

Yaalon, D.H. and Laronne, J. (1977). Internal structure in eolianites and palaeowinds, Mediterranean coast, Israel. *J. Sed. Pet.*, 41, 1059-1064.

14

Quaternary Emergent Shorelines of Tasmania

A. R. BOWDEN

Australian Groundwater Consultants
Melbourne, Vic.

E. A. COLHOUN

Department of Geography
University of Tasmania
Hobart, Tas.

Introduction

Tasmania with its northern Bass Strait islands and southern submarine rises forms the continental continuation of south-eastern Australia (Fig. 1). This insular region has been influenced by Tertiary and Quaternary sea-level changes.

Former studies of Tasmanian Quaternary shorelines assumed that stable or quasi-stable tectonic conditions prevailed (Lewis, 1935; Edwards, 1941; Davies, 1959; Jennings, 1959, 1961; Chick, 1971; Kershaw and Sutherland, 1972). However, recent studies indicate the need to interpret emergent shore-lines in terms of middle and late Quaternary uplift (Bowden, 1978, 1981; van de Geer *et al.*, 1979). The principal evidence bearing on the latter approach is discussed here.

Evidence of Emergent Shorelines

The shores of Tasmania are subject to a temperate climate, and are dominated by westerly swell-waves and storm waves generated in the Southern Ocean. The tidal range is under two

COASTAL GEOMORPHOLOGY
IN AUSTRALIA
ISBN 0 12 687880 3

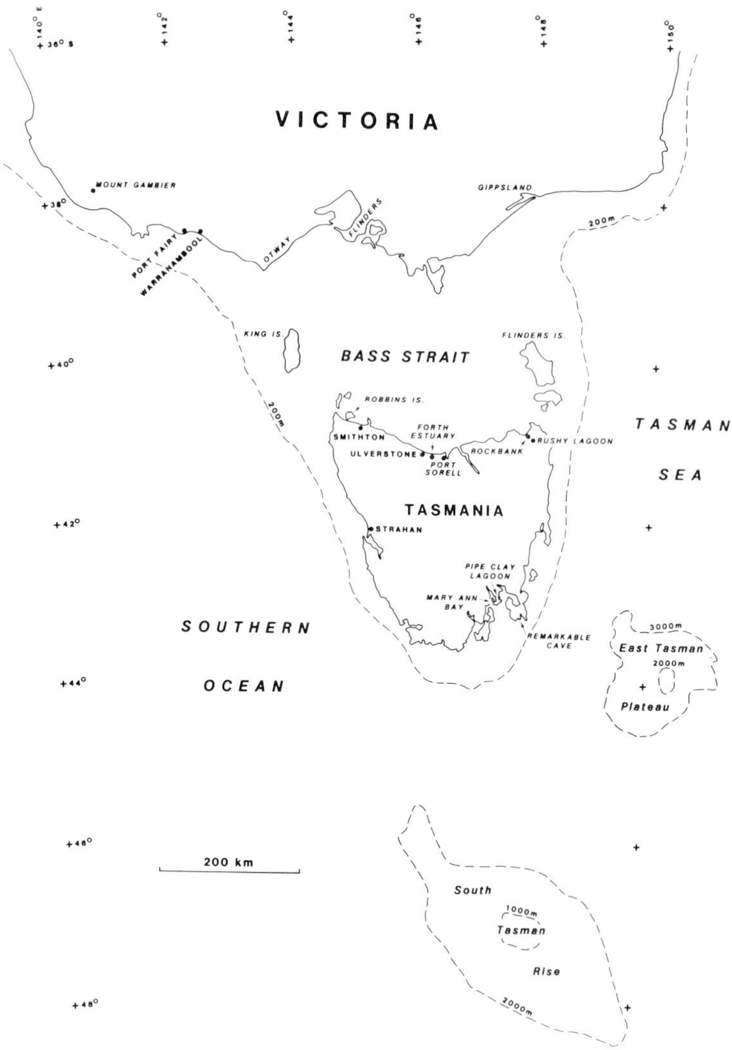

FIGURE 1. Tasmania and the southeast Australian coast.

metres except on the northwest coast where it approaches four metres. Consequently, wave action is concentrated within a narrow vertical range (Davies, 1964, 1972).

Erosional marine landforms in Tasmania include high cliffs, slope-over cliffs, natural arches, geos, rock platforms and sea caves at and above present sea level. Even though there is an abundance of inherited forms fashioned during the higher sea level conditions of interglacial times, it is difficult nevertheless to relate such features with any precision to the heights attained by former sea levels.

Depositional landforms and emergent bodies of marine sediment are common on the north, east and west coasts of Tasmania. The forms and deposits include sloping marine sand plains which are particularly extensive in the northeast. The east coast exhibits many old emerged bay floors with bayhead beaches, while estuarine muds are common in the drowned valleys of the southeast. On the west coast, narrow terraces of sand and gravel, some underlain by rock platforms, are frequent. Extensive beach-ridge plains (Fig. 2) occur in several localities, and are particularly well developed on the seaward sloping plain of the far northwest coast around Smithton and Robbins Island (van de Geer, 1981). Dune complexes related to interglacial emergent marine deposits occur in many places but are rarely exposed because they have been overridden by Holocene dunes. Sometimes the old coastal dunes are difficult to differentiate from terrestrial dunes, some of which have been derived from interglacial marine deposits during the Last Glacial Stage (Fig. 3) (Bowden, 1981, 1982).

Published data on emergent coastal landforms and sediments have been confined to King Island, the northwest coast, the northeast coast, Flinders Island and the southeast coast with only occasional observations elsewhere (Jennings, 1959, 1961; van de Geer, 1981; Chick, 1971; Bowden, 1978; Kershaw and Sutherland, 1972; Davies, 1959). There is no comprehensive map of emergent coastal features, though a general inventory of the coastal environment has been made (Macartney, 1979). These studies demonstrate that correlation of rocky with erosional and depositional shoreline features on an altimetric basis is not yet possible. Given the difficulties of determining the relationship between shoreline forms and any sea-level stand during an interglacial-glacial cycle, it is clear that the most important elements to determine are the maximum level and age of any transgression; these are best determined from depositional features. The height values are best related to present sea level for the region concerned because quasi-stationary mean sea level surfaces may differ significantly from the geoidal surface of datums (Everard, 1979).

FIGURE 2. Last Interglacial beach-ridge plain of Robbins Island in northwestern Tasmania. Landward limit of the sloping plain is c. 15m elevation.

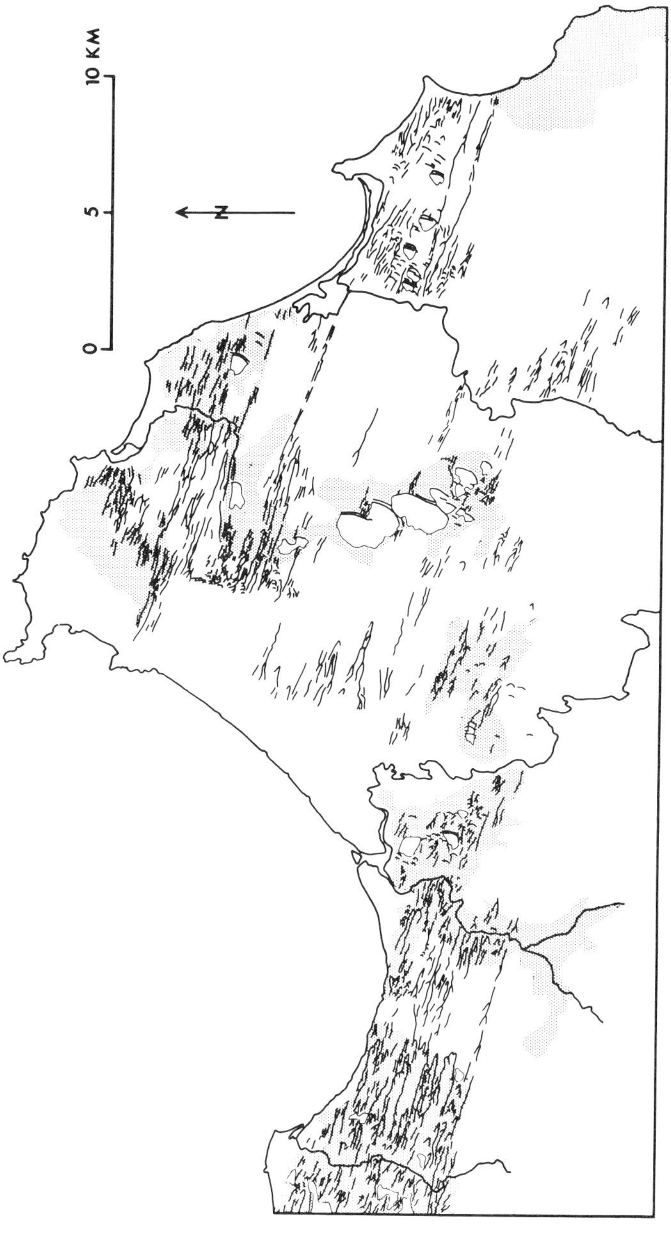

FIGURE 3. Linear dunes, lunettes, and interglacial marine sands of northeastern Tasmania.

The sequence of interglacial deposits that occurs in northeastern Tasmania is the most comprehensive of any studied to date, and it is used here as the basis for comparisons of former sea levels within Tasmania. The principal evidence which indicates the upper marine limits for the last three interglacial stages in Tasmania is presented, and the conclusions are used to support a hypothesis of regional tectonic uplift.

Interglacial Marine Deposits at Stumpys Bay

Tracts of the Stumpys Bay Sand Formation (Bowden, 1981) occur on the east coast of northeastern Tasmania and have been largely protected from deflation by the granite hills to windward. The original morphology of the sand plains is preserved almost intact and low beach ridges which are sub-parallel to the coast characterize the surface.

Environment of Deposition

At Stumpys Bay (Fig. 4) the Stumpys Bay Sand forms a narrow plain 2 to 3 km wide and 6 km long. The plain is wedged between the sea and the prominent rise of Mount William. The western boundary of the formation is marked by a fairly well defined break of slope at an altitude of 32.5 m above high water mark (HWM). Drilling demonstrates that the formation has a maximum thickness of 8 m and is underlain by granite. The cross section (Fig. 5) from the sea to the upper limit shows that the formation wedges out at the break of slope on its landward margin, and that it has been truncated by Holocene coastal deposits to seaward. An important feature of the plain is that some well preserved beach ridges lie sub-parallel to the coast (Fig. 6). Several small streams originate in the Mount William area and traverse the plain which is otherwise largely devoid of natural surface drainage.

The overall morphology of the Stumpys Bay plain and the sub-parallel beach ridges provide evidence for a marine origin. The ridges are well drained, elongate, sandy rises which commence with noticeable breaks of slope of the seaward, steeper flanks. The landward flanks form wide, very low angle surfaces. The swales between the ridges are very poorly drained and often contain small lagoons. The plain slopes seaward with a gradient of approximately 12.5 m/km.

Six ridges are shown on the cross profile of Stumpys Bay (Fig. 5). They occur over a horizontal distance of around

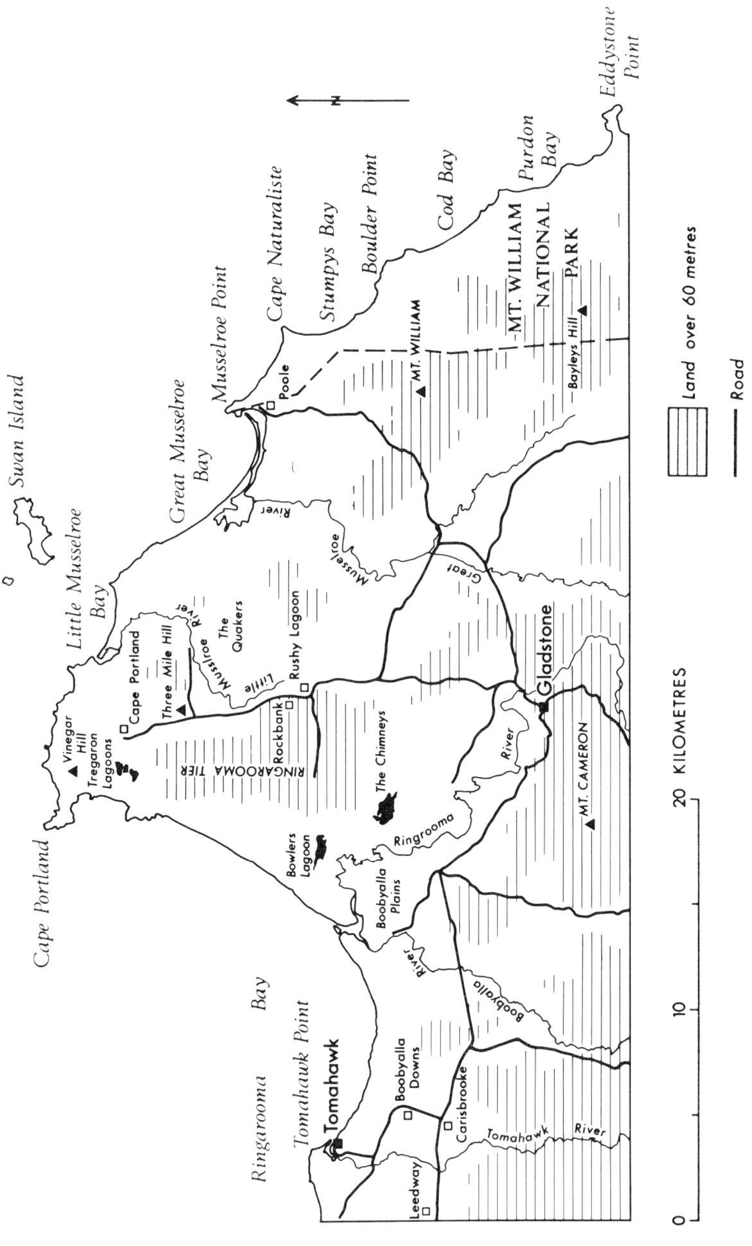

FIGURE 4. Northeast coastal plain of Tasmania.

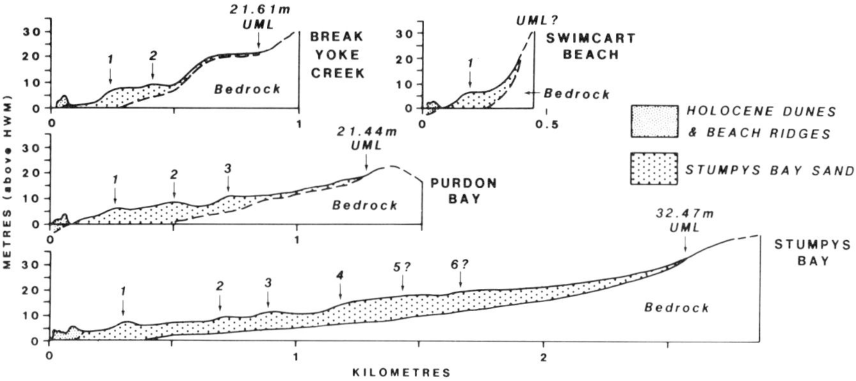

FIGURE 5. Cross sections of Stumpys Bay and adjacent bays to the south. (UML is the upper marine sand limit).

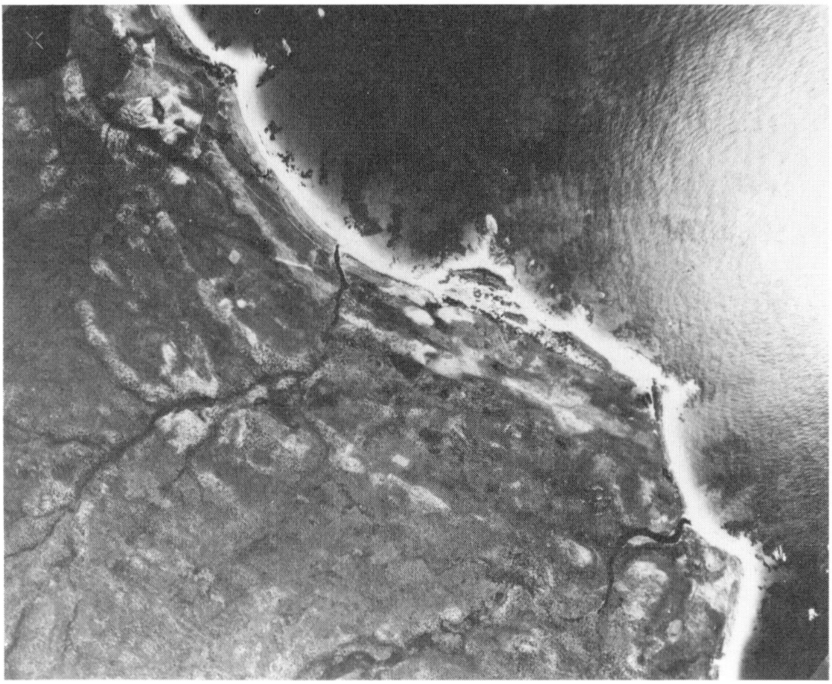

FIGURE 6. Stumpys Bay area showing beach-ridge crests parallel with the coast.

2 km. The top 40-80 cm of the ridges consist of coarse, very well rounded, quartz sand which is characteristic of a beach origin. The fossil content of sponge spicules also indicates a marine origin. The spicules are well preserved considering their delicate structure (Bowden, 1981). It seems unlikely that they have been reworked from another sediment. The above evidence clearly demonstrates that the sand plain at Stumpys Bay is a marine deposit.

Three auger holes were drilled in the plain to determine the sequence of deposition and to ascertain if the sands were deposited during a single interglacial. Interpretation of the grain size characteristics provides strong evidence of one marine transgression that involved landward migration of a barrier beach system and foredune which was backed by a lagoon. During the subsequent regression, probably represented by the top 1.5 m of beach sand, beach ridges were constructed at decreasing elevations on the sand plain. Since the beach sands are continuous from sea level to 32.5 m the sequence therefore indicates deposition during a transgression to approximately 32 m followed by regression to below present sea level. No evidence of peat layers or fossil soil horizons, which could indicate subaerial deposition and weathering, was encountered during drilling at Stumpys Bay. This indicates that only one interglacial cycle of marine transgression is represented.

Age

No material which can be radiometrically dated has been found at Stumpys Bay. However, several indirect lines of evidence suggest that the sequence is of Last Interglacial age.

The sediments are not a product of the Holocene marine transgression because of their vertical extent, and their truncation by marine sands and dunes of Holocene age. The age of the Stumpys Bay Sand is therefore older than Holocene.

The degree of sponge spicule surface solution provides a line of evidence which supports the conclusion that the Stumpys Bay Sand is substantially older than Holocene in age. Scanning Electron Microscope examinations of spicule morphology at 1000X were made of samples from the present beach at Tomahawk (Fig. 7 (i)), from sands of Holocene age which are probably 3000 to 6000 years old (Fig. 7(ii)), and from the Stumpys Bay Sand at both Tomahawk (Fig. 7(iii)) and Stumpys Bay (Fig. 7(iv)). The features shown depict typical surface forms due to solutional etching. Very little difference in surface morphology was detected within each sample or between samples from the same depositional unit but strong contrasts occurred between units.

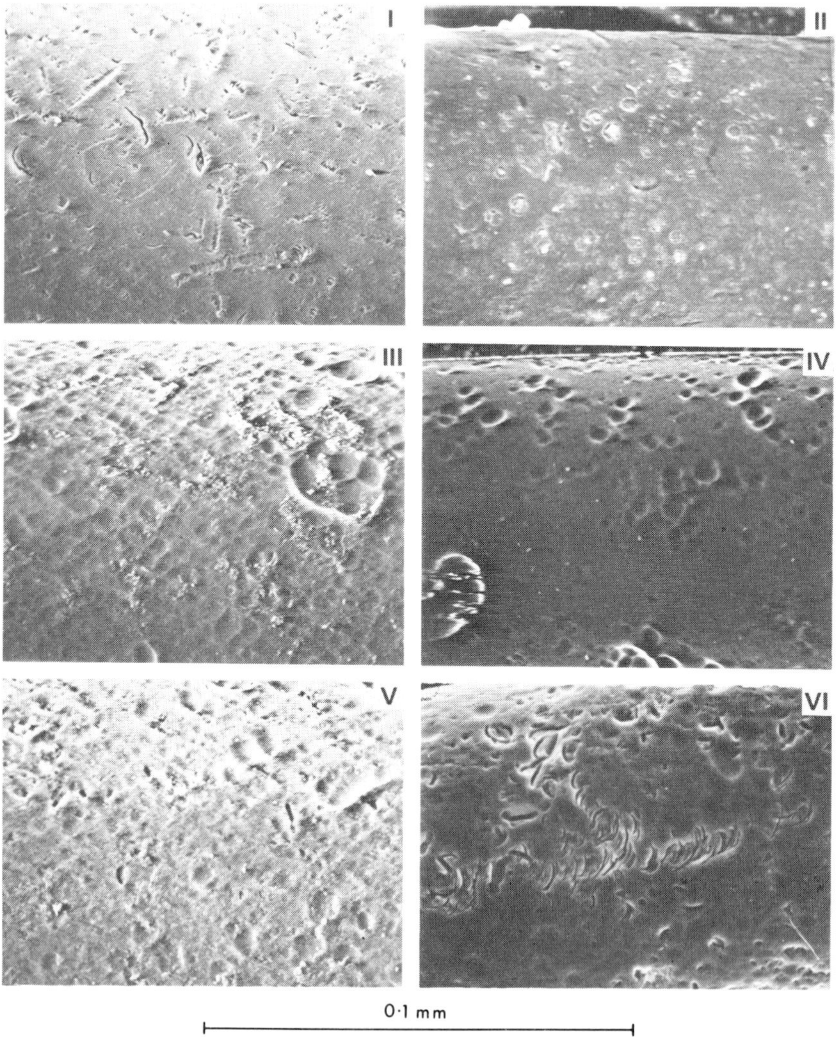

FIGURE 7. Micromorphology of megascleres from:
(i) Modern beach at Tomahawk
(ii) Holocene beach sands at Tomahawk
(iii) Last Interglacial marine sands at Tomahawk
(iv) Stumpys Bay sand at Stumpys Bay
(v) Rockbank sands at Rockbank
(vi) Ringarooma sands at Rockbank

Spicules from the present beach still retain most of their original surface, but are in the process of being modified by solution processes. Shallow crescentic solution crevasses (Krinsley and Doornkamp, 1973) are present and many are arranged in an *en echelon* pattern. They average 3.0 to 5.0 μm in length but extend to 10 μm, and are 0.5 to 1.0 μm in width. The *en echelon* solution crevasses occur as bands up to 30 μm in length, but most are approximately 10 μm, and are 1.0 μm to 3.0 μm in width.

Holocene spicules are more corroded (Fig. 7(ii)). A much lower proportion of the original spicule surface remains, and small, shallow solution pits 1 μm to 4 μm in diameter have been formed.

Spicules from the older Stumpys Bay Sand show an even greater degree of solution and therefore indicate a substantially greater age (Figs. 7(iii) and (iv)). In this case the original spicule surfaces have been entirely corroded away, and the solution pits have been englarged and deepened considerably, to give a 'raindrop impact' impression.

The increase in degree of surface solution with increase of age of the sponge spicules is very marked. This method has provided a good comparative index of relative age in northeast Tasmania because the spicules were all obtained from siliceous deposits below the water table, but differences would be expected if the spicules are extracted from deposits influenced by markedly different groundwater and oxidation conditions. However, the comparison indicates that the Stumpys Bay Sand is of pre Holocene and therefore of interglacial age.

Marshall and Thom (1976) have given clear evidence that the highest sea level during the Last Interglacial was 4-6 m on the tectonically stable central and northern New South Wales coast. Marine deposits at about this elevation also occur in Gippsland, Victoria (Jenkin, 1968). The Stumpys Bay Sand extends unbroken from below present sea level to over 32 m with no development of a pronounced shoreline at 4-6 m, a feature to be expected if the Stumpys Bay Sand is older than Last Interglacial in age.

Shackleton and Opdyke (1973) have shown that the deep sea oxygen isotope record indicates that interglacial sea levels on stable coasts over the last 800,000 years were within ±10 m of present sea level. The Stumpys Bay Sand is less than 800,000 years old, as indicated by the beach ridges which, although composed of unconsolidated sand are well preserved and are characterized by relatively weak soil profile development. As the beach sediments attain 32 m and are clearly of interglacial age it appears that the area must have experienced uplift since they were deposited. If any interglacial high sea level had occurred since deposition of the Stumpys Bay Sand, it would be reasonable to expect either erosional or depositional evidence for such an occurrence. The lack of such evidence leads to the conclusion that the Stumpys Bay Sand Formation was deposited during the Last Interglacial.

Other Evidence for Sea-Level in the Last Interglacial

Several small embayments of sand with beach-ridge topographies occur to the south of Stumpys Bay. They are all very similar in geographic setting and sediment composition to Stumpys Bay, and appear to be the products of marine transgression during the Last Interglacial. A prominent feature of the embayments is that the bedrock coast and hinterland backing the embayments becomes steeper with increasing distance southwards. A further feature is that the upper limit of depositional features becomes progressively lower towards the south.

Purdon Bay 9 km south of Stumpys Bay is backed by a small beach-ridge plain which is 2.5 km long, 1.4 km wide and has a seaward slope of approximately 16.5 m/km. The plain is composed of medium to coarse grained well rounded beach sands. Three broad-crested beach ridges lie sub-parallel to the coast, as shown by the profile of Figure 5. The sand wedges out at an indistinct break of slope at 21.4 m above HWM.

At Break Yoke Creek, approximately 29 km south of Stumpys Bay, the beach-ridge sand plain is about 1 km long and 0.8 km wide. It has a surface gradient of around 27 m/km towards the east. Surficial deposits consist of well rounded, coarse grained sand. The single beach ridge is wider than the beach ridges in Purdon and Stumpys bays, but it possesses a double crest (Fig. 5). This suggests that the beach ridge at Break Yoke Creek is a composite feature formed by superimposition of one ridge on another. The rounded beach sands wedge out sharply at a pronounced break of slope at 21.6 m above HWM.

Swimcart Beach is 40 km south of Stumpys Bay and is backed by a narrow sand plain 1 km long and 0.4 km wide. It has a surface slope of approximately 33.3 m/km (Fig. 5). Only one beach ridge is present. The upper marine limit of the sand plain has not been determined because aeolian sands are banked against the steep flank of the confining landward hill.

Table 1 is a comparison of the elevations of the upper marine limits and beach ridges, and the approximate surface gradient for each marine plain. The beach ridges are numbered from lowest to highest. The three main features of this table are:

1. The striking similarity in elevation of the three lowest beach ridges.

Table 1. Eastern Beach-Ridge Plains

		Stumpys Bay	Purdon Bay	Break Yoke Creek	Swimcart Beach	Mean	SD
Beach ridge elevation (m above HWM)	1	7.2	6.5	7.6	7.1	7.1	0.4
	2	9.5	9.2	9.4	Absent	9.4	0.1
	3	11.5	11.2	Absent	"	11.4	NA
	4	15.0	Absent	"	"		NA
	5	18.7	"	"	"		NA
	6	20.0	"	"	"		NA
Upper limit of beach sand (m above HWM)		32.5	21.4	21.7	Not obtainable		
Surface gradient (m/km)		12.5	16.5	27.0	33.3		
Distance south from Stumpys Bay (km)		0	9	29	40		

NA = Not applicable

2. The discordant upper marine limit elevations between Stumpys Bay and the remaining two areas.
3. The increasing surface slope in a southward direction.

Although the datum for each levelled transect was HWM and therefore heights could be ± 0.3 m in error, the strong correlation of the three lowest beach ridges over a distance of 40 km indicates that the shorelines which they represent have not been appreciably warped since deposition. However, if only the elevational differences of the upper limits of marine deposition were taken into account, there would appear to be some tectonic warping along the coast. This apparent paradox can be explained by variation of coastline sensitivity to modification by marine processes.

Two principal factors govern coastline sensitivity in northeastern Tasmania. The first is the amount of weathered granite mantle which is available for reworking by a transgressive sea. Steep surface slopes are less likely to develop a thick weathered layer than gentle slopes due to the more rapid removal of weathered products. Thus, wave trimming has a greater effect on weathered mantles than on bare rock or thinly mantled rock surfaces.

The second factor is the gradient of the surface which is being transgressed by the sea. Depositional features, particularly those which are the product of short term oscillations, such as beach ridges, are less likely to be evident on a high gradient coast than on a low gradient coast. Horizontal 'compression' of shore location on a steeply sloping coast during transgression and subsequent regression would not show the effects of short period variations in shore location as well as on a low gradient coast, where minor sea level fluctuations would result in larger horizontal displacements of the shore.

With respect to both weathering mantle thickness and surface gradient the embayments become increasingly less sensitive to coastal modification south of Stumpys Bay. Thus, of the six beach ridges recognized at Stumpys Bay, only three occur at Purdon Bay, and two at Break Yoke Creek where they occur in close juxtaposition. The principal conclusion is that the different elevations of the apparent limits of marine deposition are partly due to increased steepness of the surface. This has resulted in poor preservation of features on steep coasts during the transgression which appears to have attained a maximum of 32 m.

An additional factor which may partly explain the restricted occurrence of the highest levels of marine deposition is the difference in the nature of the transgression oscillations. The apparent upper marine limits at Purdon Bay and Break Yoke Creek are very similar, being 21.4 m and 21.6 m respectively.

At Stumpys Bay there is a subtle, but definite, break of slope at 21.3 m to 21.6 m above HWM. The 21 to 22 m level is therefore a very pronounced level and its presence suggests that during the Last Interglacial, sea level was at or below 21 to 22 m for a substantial portion of the transgression/regression. The transgression to 32 m was probably a relatively brief event which only left evidence on coasts that were very sensitive to marine modification. The sequence at Stumpys Bay is therefore the most complete and suggests that a brief transgression to 32 m was followed by a major period of beach-ridge construction at levels below 21 to 22 m during the Last Interglacial.

An abundance of emergent marine landforms and deposits extend to 20 to 22 m above HWM elsewhere in Tasmania. These have many similarities with the Quaternary marine features described from northeastern Tasmania.

A marked marine level at 18 to 21 m and another at 30 to 37 m is recorded on Flinders Island (Sutherland and Kershaw, 1971; Kershaw and Sutherland, 1972). A prominent level at 20 m is recorded on King Island (Jennings, 1959, 1961).

In northern Tasmania, upper marine limits occur at an elevation of 20 to 22 m in the Port Sorell area (Colhoun, 1976). In northwestern Tasmania, shelly sands underlie freshwater peat deposits that have been radiocarbon dated to more than 52,000 years BP (GrN-9342) in the Smithton area (van de Geer, 1981). Maximum elevations of approximately 20 m have been observed in the Ulverstone area (Chick, 1971). Near the mouth of the Forth River, marine sands at 17 m transgress glacial outwash terrace deposits of pre-Last Glacial age (Colhoun, 1976). The landward limit of beach ridges at Robbins Is. in 15m (Fig.2).

In southeastern Tasmania, marine interglacial deposits have been strongly deflated at several localities in the South Arm. This has resulted in the formation of lagoons, lunettes and dunes most of which are of Last Glacial age. At Pipe Clay Lagoon a thin peat accumulated on the damp floor of a deflation hollow in marine sands. The deflation hollow was later filled by aeolian sands. The peat has been radiocarbon dated to between 25,380 ± 640 years BP (SUA-151) and 19,810 ± 360 years BP (SUA-153) showing the marine sands to be older (Colhoun, 1977). At Mary Ann Bay similar marine sands with an abundant shallow-water shell fauna occur at 21 m above HWM (Colhoun *et al.*, 1982a).

In western Tasmania, marine interglacial deposits at 22 m at Strahan Airport occur inland of the Holocene Ocean Beach barrier and dune system (Davies, 1960; Banks *et al.*, 1977). They were formed by the reworking of sands and gravels from either glacial outwash (David, 1926), or from deposits of Tertiary age (Scott, 1960).

The widespread evidence shows that sea level attained at least 20 to 22 m throughout Tasmania. The evidence from the northeastern embayments suggests that such levels may be only apparent limits on insensitive coasts with a possible limit at approximately 32 m relating to a brief maximum transgression being recorded on the sensitive coast at Stumpys Bay prior to a more prolonged stillstand at about 20-22 m.

Older Marine Deposits

Two suites of marine deposits occur in the Rockbank area (Fig. 4) which occurs west of the head of the Little Musselroe River and west of Rushy Lagoon. They have been informally termed the Rockbank and Ringarooma deposits (Bowden, 1981).

Rockbank Deposits

The Rockbank deposits are exposed in two sections at the eastern foot of Ringarooma Tier. The deposits are up to 4 m thick and extend to approximately 49 m above HWM, where they wedge out against silty clays. The deposits consist of siliceous, moderately well rounded and moderately sorted unconsolidated sands. They are well bedded and low angle cross-bedding is common. Sponge spicules were found in most samples and are similar to those found in the Stumpys Bay Sand. Most spicules were megascleres, but some were microscleres. The abundance of sponge spicules is convincing evidence for a marine origin.

The age of the Rockbank deposits is difficult to ascertain but some conclusions can be drawn from their stratigraphic and topographic relations. The higher elevation of the upper marine limit (49 m) compared with the adjacent Stumpys Bay Sand (32.5 m) indicates that the Rockbank deposits are older. Although no sections occur at the contact, the surface of the Rockbank deposits has a slope of 20 to 25 m/km. It appears to pass below the surface formed on the Stumpys Bay Sand which has a gradient of 10 to 14 m/km.

The similar composition of the Stumpys Bay Sand and the Rockbank deposits permits comparison of the weathering of contained sponge spicules. Comparison of a scanning electron micrograph at 1000x of a typical sponge spicule from the Rockbank sands (Fig. 7(v)) with typical spicule surfaces from the Stumpys Bay Sand (Figs. 7(iii) and (iv)) shows that the 'raindrop impact' solution features are still apparent in the Rockbank deposits, but that the surface has been much more weathered and contains deep solution pits. This suggests a greater age for the Rockbank than for the Stumpys Bay sands.

The Rockbank and the underlying Ringarooma deposits are separated by a sandy peat layer. The sandy peat is probably a thin swamp deposit developed on, and largely derived from, the underlying silty clays of the Ringarooma deposit. Pollen analysis of a sample of the sandy peat showed that aquatics were highly represented with *Isoetes* being most abundant. Megaspores of *Selaginella uliginosa* also occur. A freshwater swamp is indicated. In addition, pollen of Gramineae and Compositae were very common but tree pollen was absent. The largely herbaceous pollen assemblage is similar to other assemblages formed during cold stages of the Quaternary in Tasmania. No Tertiary pollen types are present. The sandy peat seems to indicate that the Rockbank and Ringarooma marine deposits probably belong to different interglacial stages.

The height of the Rockbank deposits indicates that they are older than the Last Interglacial Stumpys Bay Sands. The pollen assemblage in the sandy peat shows that they are of Quaternary age. The sponge spicule weathering, although more advanced than Last Interglacial in age, still retains features characteristic of young deposits. This suggests that they may belong to the Second Last Interglacial which is equated with oxygen isotope Stage 7 of Shackleton and Opdyke (1973).

Ringarooma Deposits

The Ringarooma like the Rockbank deposits, have been partially protected from deflation by Ringarooma Tier to the west. They extend as a surface which slopes eastward at 40 to 45 m/km from a height of 71 m above HWM. This upper limit is marked by a conspicuous break of slope. The lower limit is a moderately well defined break of slope at the upper limit of the Rockbank deposits. The maximum thickness has not been determined.

The Ringarooma deposits consist of silty to clayey sand. The fine grained composition suggests a lagoonal facies. The deposits contain abundant sponge spicules (megascleres and microscleres) of marine origin. Their surface morphology shows that they have been more heavily altered by solution than the Rockbank deposits. The scanning electron micrograph at 1000X of Figure 7(vi) represents the surface of a typical spicule which has much stronger solution pitting than the spicule in Figure 7(v). However, caution is necessary due to possible differential solution rates in sediments of dissimilar composition.

The height of the Ringarooma deposits and the difference in degree of sponge spicule solution indicates that they are probably considerably older than the Rockbank deposits. The interstratified sandy peat deposit suggests that the two marine deposits belong to two interglacial stages that are separated by a glacial stage. If the Rockbank deposit represents oxygen isotope Stage 7 then it is likely that the Ringarooma deposits represent the Third Last Interglacial Stage or oxygen isotope Stage 9 of Shackleton and Opdyke (1973), or older.

Ringarooma, Rockbank and Stumpys Bay Sand deposits are marine deposits which have been successively superimposed at lower levels. The upper limits of each level are approximately 71 m, 49 m and 32 m. Each suite of deposits appears to belong to a separate interglacial stage of high sea level and the relative heights attained by the transgressions have been decreasing.

Study of the nature and occurrence of the Stumpys Bay Sand and the older marine deposits has produced a partial sea level curve for the Late Quaternary in northeastern Tasmania (Fig. 8). There are few records of high Quaternary sealevels elsewhere in Tasmania with which this curve may be compared. There are records of a boulder beach at 37-46 m and bedrock surfaces of probable marine origin at 69 m on King Island (Jennings, 1959, 1961). Rock benches have also been recorded at 61-75 m on Flinders Island which may be of marine origin (Kershaw and Sutherland, 1972). These features seem to be too high to correlate with the suggested 30-32 m Last Interglacial limit from northeastern Tasmania, and may well belong to the earlier interglacial stages suggested to equate with isotope Stages 7 and 9. However, being storm beaches and rock benches in widely separated locations they make determination of specific marine limits and dating difficult.

Shoreline Comparisons and Deduced Uplift Rates

In many studies of Quaternary emergent shorelines it has proved possible to relatively date and correlate some shorelines associated with the Last Interglacial, while correlation of older shorelines has proved more difficult. This is partly because of the difficulty of obtaining 'absolute' ages on marine deposits in temperate coastal areas, and partly because the maximum of the Last Interglacial transgression was the highest eustatic level attained in middle and late Quaternary times (Shackleton and Opdyke, 1973).

Accordingly, Last Interglacial shorelines associated with isotope Stage 5e are the highest usually recorded on stable coasts, and on unstable coasts they form the reference level from which uplift rates and ages of higher shorelines may be estimated. The age of the Last Interglacial maximum, between 120 and 130 ka BP, has now been radiometrically ascertained at so many tropical sites (e.g. Hawaii, Ku *et al.*, 1977; Bermuda, Harmon *et al.*, 1981) that it can be extended to emerged shorelines of temperate areas. This is especially the case where the stratigraphy and dating of the overlying deposits can be shown to belong to the Last Glacial Stage. Thus, assuming that the 30-32 m shoreline of northeastern Tasmania is correctly correlated with the Last Interglacial maximum at 125 ± 5 ka BP and isotope Stage 5e, then this shoreline and the lower 20-22 m shorelines are much higher than their counterparts in Victoria and New South Wales, and on stable coasts elsewhere in the world where levels attain 2-9 m (Veeh, 1966).

At Warrnambool and along the Otway coast of Victoria, Gill (1976) has recorded a prominent shoreline at 7.5 m. Gill and Amin (1975) have obtained an age of 125 ka BP for this shoreline by $^{230}Th/^{234}U$ dating of molluscs. However, the validity of the method has been questioned (Kaufman *et al.*, 1971). Marine sediments of the Inner Barrier on the stable New South Wales coast indicate a Last Interglacial sea level at 5 ± 1 m above present (Thom *et al.*, 1981b). $^{230}Th/^{234}U$ dates on corals from similar deposits at Newcastle Bight and Evans Head on the north coast suggest an early peak at 142-143 ka BP and a later peak at 114-127 ka BP. The later peak seems to correlate with isotope Stage 5e (Marshall and Thom, 1976).

If the correlation of the Last Interglacial eustatic levels from apparently stable Australian mainland areas and elsewhere with the northeastern Tasmanian 30-32 m shoreline is correct, then the uplift rates for Tasmania can be calculated between 0-125 ka BP assuming a constant rate. If +6 m is taken as the control datum for 125 ka BP then the uplift rate for Tasmania is about 0.21 m/ka. However, if the 20-22 m level dates to 125 ka then a lower uplift rate of 0.1 m/ka would apply.

The upper limit of the Rockbank deposits at approximately 49 m is interpreted to represent the Stage 7 sea-level maximum and the respective elevation and age of the Ringarooma deposits is approximately 71 m and Stage 9. Shackleton and Opdyke (1973) considered that the Stage 7 sea level attained -10 m and that the Stage 9 eustatic level was very close to present sea level. The ages of the assumed sea-level peaks correlated with Stage 7 and 9 from Shackleton and Opdyke are 220 ka and 325 ka.

Figure 8 graphs the present altitude of the *upper* marine limits minus the estimated eustatic sea level for the periods

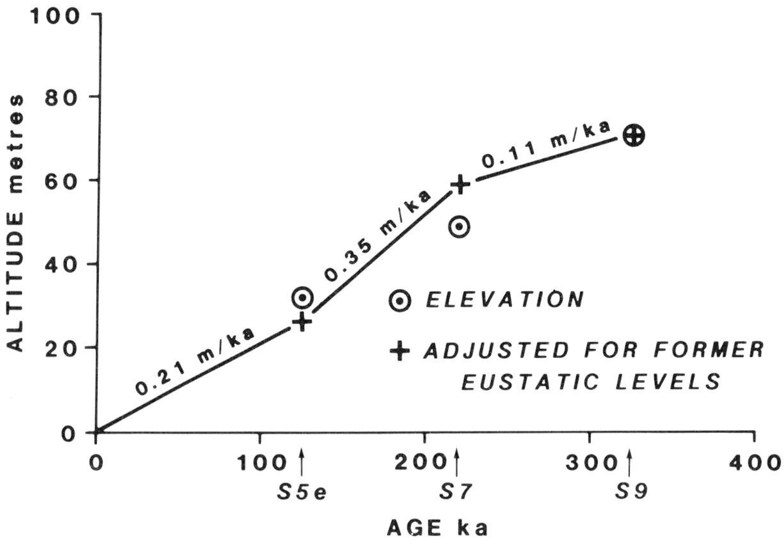

FIGURE 8. Elevation, age and uplift rates of marine limits having regard to former eustatic levels. Interglacial isotope stages are indicated.

concerned versus age. The resultant slopes reflect the uplift rates for the various intervals. The mean uplift rate over the last 325 ka is shown to have been 0.22 m/ka. The graph suggests that the rate of uplift may have risen from 0.11 m/ka during the interval between Stage 9 and Stage 7, to 0.35 m/ka from Stage 7 to Stage 5, and then dropped to 0.21 m/ka after Stage 5e. The general uplift rate of 0.22 m/ka over the longer period agrees closely with the better known rate over the last 125 ka. This supports the estimated ages of the Rockbank and Ringarooma deposits.

Implications for Holocene Shoreline Levels

Opinion is divided as to whether or not eustatic sea level has been higher than present during the Holocene in Australia (Fairbridge, 1961; Gill and Hopley, 1972; Thom *et al.*, 1972; Thom and Chappell, 1975; Cook and Mayo, 1978; Thom *et al.*, 1981a; Hopley, 1983). It has been claimed that several phases of higher sea level have occurred in Victoria (Gill, 1982) yet no traces of higher levels have been found on the coast of New South Wales (Thom *et al.*, 1969, 1972; Thom *et al.*, 1981a). The variations in recorded Holocene high sea levels seem to be regional which suggests regional crustal adjustments rather than general eustatic changes.

It has been suggested that sea level may have been 0.6-0.9 m higher than present during the mid-Holocene in southeastern Tasmania (Davies, 1959, 1961), but it has not been possible to prove unequivocally a high eustatic stand. However, even if eustatic sea level has neither risen nor fallen since the mid-Holocene, the uplift rates suggested here imply emergence of the land and relative lowering of sea level by up to 1.26 m for Tasmania. Although the similarity of these values might seem to indicate uplift as the cause of the relative emergence, it is difficult to be sure that Holocene marine sediments within 1 m of HWM are emergent, in the absence of knowledge of the effects of other factors such as reduced barometric pressures, onshore wind regimes, changing beach profiles, and the range of extraordinary high spring tides during storm conditions.

On King Island, Jennings (1959) has described cobble beaches up to about 4 m and marine platforms up to 3 m above HWM of spring tides. These features are not presently forming and are regarded as having emerged since the mid-Holocene. Kershaw and Sutherland (1972) describe numerous marine deposits of Holocene age on Flinders Island occurring between 3 and 4.5 m above mean low water spring tides. They explain their elevation as a result of emergence since the mid-Holocene. There are also several local occurrences of cobble ridges, estuarine deposits and dune-beach ridges at elevations of 1 to 3 m above HWM on all Tasmanian coasts. Unfortunately, there are very few radiocarbon dates that relate to such Holocene marine deposits. The few that exist are of late rather than mid-Holocene age. Without extensive further research the occurrence of a possible higher mid-Holocene eustatic stand can neither be accepted nor rejected.

Comparison of Uplift Rates with Other Regions

The uplift rates suggested by the emergent shorelines of Tasmania are moderate when compared with rates detected elsewhere. A logarithmic scale of uplift rates is useful for such comparison:-

Quasi-stable	0 to 0.005 m/k
Slow	0.005 to 0.05 m/ka
Moderate	0.05 to 0.5 m/ka
Rapid	0.5 to 5.0 m/ka
Very rapid	5.0 m/ka +

The moderate uplift rates suggested for Tasmania are higher than for any part of the southeastern Australian coast between western Victoria and northern New South Wales. Gill (1976) records a widely represented shoreline in western Victoria at 7.5 m as having an age of 125 ka BP. He regards a higher shoreline at 21 m at Port Fairy as possibly belonging to an earlier phase of the Last Interglacial. In New South Wales Thom *et al.*, (1981b) have suggested that the Inner Barrier beach systems represent a sea level of 5 ± 1 m above present. During the last 125 ka the southeastern Australian coastline is regarded as having been stable (Gill, 1976; Thom and Chappell, 1975) with the possible exception of Gippsland where a slow uplift rate of 0.06 m/ka has been suggested (Jenkin, 1968, 1976). Tasmanian rates are less than in southeastern South Australia where uplift adjacent to Mt. Gambier has apparently increased from the slow rate of 0.075 m/ka around 7000 ka BP to the rapid rate of 0.8 m/ka after about 300 ka BP (Sprigg, 1979).

Tasmanian uplift rates seem comparable with those of South Westland, New Zealand, where the Sandine 2 terrace attains 24-32 m (Nathan and Moar, 1975). However, they are generally less than the 0.2-0.4 m/ka rate associated with the Last Interglacial Awatuna Formation, further north in Westland. The Tasmanian rates are much less than the rapid Holocene uplift rate of 1.9 m/ka of the Rapahoe raised beach of North Westland (Suggate, 1965). The Tasmanian rates are also much less than the rapid rates of about 0.5 to over 4.0 m/ka recorded for North Island, New Zealand (Ota, 1977; Ghani, 1978). Such comparison serves only to demonstrate the regional variability of both short and long term uplift rates as emphasised recently from northwestern South Island by Williams (1982).

Viewed more widely, the 0.21 m/ka moderate uplift rate for Tasmania is only slightly exceeded by the rate of 0.24 m/ka calculated for the Nestor terrace on the tectonic coast of San Diego, California, and by the 0.23-0.43 m/ka calculated for the Barbados III terrace (Kern, 1977; Matthews, 1973). However, it is very much less than the rapid uplift of 0.94-2.56 m/ka for the Huon Peninsula in New Guinea and of 1-2 m/ka for Kikai in the Ryukyu Island arc of Japan, both of which are located near plate boundaries (Bloom *et al.*, 1974; Ota *et al.*, 1978). It should be noted, however, that these uplifted complexes are markedly terraced in contrast to the more uniform slope of each different aged surface in Tasmania (Fig. 5).

Factors contributing to Late Quaternary Uplift of Tasmania

Factors that have contributed to the elevation and sloping character of the middle and late Quaternary shorelines of Tasmania and the Bass Strait islands are difficult to identify.

They should explain why the interpreted 125 ka BP Tasmanian shorelines are higher than shorelines of the same age in Victoria and New South Wales, and why they often slope seawards (Fig.5).

The difference in elevation between Tasmanian and mainland shorelines is not likely to be due to glacial isostatic adjustments because last glaciation ice was confined to the central and southwestern mountains of Tasmania. The ice volume is unlikely to have caused crustal depression beyond the limits of glaciation which did not extend to the coast.

It also seems unlikely that the height differences are related in any measure to hydro-isostatic loading of the shelf during the Holocene transgression because, like New South Wales, the shelves adjacent to the eastern and western coasts of Tasmania are narrow and do not carry a great water load. Yet the 30-32 m shoreline at Stumpys Bay in northeastern Tasmania and other records of high sea level all contrast with the 5 ± 1 m sea-level limit in New South Wales. In addition, the shallow epi-continental sea of Bass Strait is common to the Victorian, the Bass Strait island and the northern Tasmanian coastlines, and yet there are marked height differences of the inferred 125 ka BP shoreline in areas located on different structural units.

Geophysical investigations indicate differences in crustal thickness and stability on either side of Bass Strait. Johnson (1972) indicates an average crustal thickness of 45 km for southern Victoria, 20-25 km for Bass Strait and 30-35 km for Tasmania. Webster and Richardson (see Shirley, 1980) estimate crustal thickness in northern Tasmania at 27-32 km. This suggests that the thinner crust in Bass Strait may be regarded as a 'weak linkage' of the Tasmanian peninsula with the mainland continental crust. The region surrounding Bass Strait consists of horsts and grabens that have experienced differential movements during the Tertiary. The structural units appear to have different degrees of instability. In the west, the Otway Basin appears to be subsiding relative to the adjacent Warrnambool-Cape Otway Platform which is quasi-stable (Fig. 9). The widespread occurrence of the higher, Last Interglacial shorelines in Tasmania suggests that Tasmania has been uplifted relative to the Bass, Torquay, Gippsland and Port Phillip Bay basins during the late Quaternary.

There is some geophysical evidence to support the hypothesis that Tasmania has experienced more late Quaternary uplift than the Victorian coast. The Otway, Bass and Gippsland basins have inward facing normal faults in their seismic basements and the basins contain thick late Mesozoic and Cainozoic sediments as shown in Figure 9 (Boeuf and Doust, 1975; Jones and Holdgate, 1980; Anfiloff, 1982). The region is undergoing crustal compression from a direction of approximately 120° and

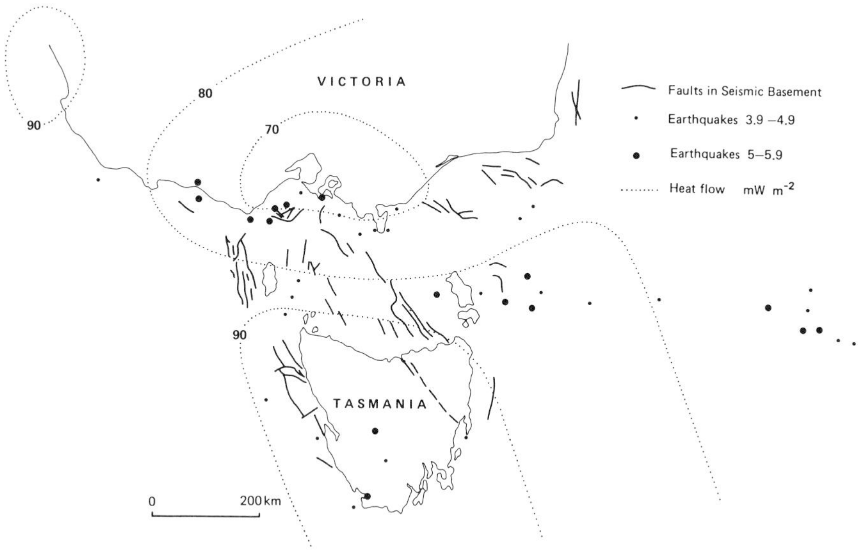

FIGURE 9. Selected structural and geophysical characteristics of southeastern Australia. (sources: Bureau of Mineral Resources Earth Science Atlas, 1980; Geology of Australia - Map scale 1:2,500,000, 1976; and Cull, 1982).

is seismically active probably along fault lines (Denham *et al.*, 1979). Records of earthquakes with magnitudes over 3.5 show that the depression between the King Rise and the Cape Otway Platform probably coincides with an active fault as does the east-west lineament at 40°S that extends from 148°E in the Bass Basin beneath Flinders Island to 157°E in the Tasman Sea. Between 1960 and 1978 over 800 earthquake epicentres were located in Tasmania and Bass Strait with some 30 occurring in the southern part of the Bass Basin (Robinson, 1974; Shirley, 1980). Seismic events are also concentrated near the edge of the continental shelf west and east of Tasmania and may be associated with either submarine slides or faulting. The epicentres of these earthquakes are not shown on Figure 9.

Northeastern Tasmania has not recently been reported as being seismically active, but between 1883 and 1885 the area experienced over 2500 earthquake events with up to 10 quakes per day and 2 quakes estimated at an intensity of 8 (Carey, 1960; Shirley, 1980). These earthquakes may have originated on the W-E lineament at 40°S that bisects Flinders Island.

A 20 km long submarine cliff with its base at -45 m has been recorded off the Otway Coast (Gill *et al.*, 1980) and a submarine cliff at -55 to -65 m has been recorded off the southern part of the west coast of King Island (Jennings, 1959). Although both partly reflect lowered Pleistocene sea levels, their locations suggest that they may be situated on fault scarps related to recent tectonism. Gill (1982) suggests that the cliff off the Otway Coast is located on the Torquay Fault which coincides with a line of submerged volcanoes.

Tasmania has high geothermal values of over 90 m Wm^{-2} These compare with 90-120 m Wm^{-2} for the late Cainozoic volcanic province of western Victoria, and with a decreasing gradient of heat flow from 70 m Wm^{-2} from the Victorian coast northwards to northern New South Wales and southern Queensland, which have under 50 m Wm^{-2} and are about world average (Cull and Denham, 1979; Cull, 1982). The heat flow values for Tasmania are not corrected for the effects of glaciation and are therefore minimal. This could indicate that Tasmania lies over a southward moving convection plume and that deep intrusion or underplating of the crust could be occurring as the Australian Plate moves northwards. This could cause uplift through thermal expansion of the crust and magmatic upwelling. In the southeastern highlands of mainland Australia, where underplating is considered to have caused continuous uplift during the last 90 m.y., the average rate is estimated at 0.01 m/ka (Wellman, 1982), which is less than one twentieth the rate required to uplift the 125 ka BP shoreline in northeastern Tasmania. Although the high heat flow values must be associated with greater than average surface radioactivity there is no evidence to demonstrate that either deep intrusion or underplating are occurring. There are no known Quaternary volcanics in Tasmania, although thermal springs in the Smithton area indicate a contemporary high heat flow in northwestern Tasmania (Colhoun *et al.*, 1982b).

None of the factors considered can be demonstrated to have either singly or jointly caused the emergence of the shorelines for which only a limited geomorphological record exists. Until further investigations are undertaken and better dating techniques become available, it will be difficult to ascertain fully, the relationships between shoreline heights, shoreline ages and Quaternary tectonism.

References

Anfiloff, V., (1982). Elevation and gravity profiles across Australia: some implications for tectonism. *BMR Journ. Austr. Geol. Geophys.* 7, 47-54.

Banks, M.R., Colhoun, E.A., and Chick, N.K., (1977). A reconnaissance of the geomorphology of Central Western Tasmania. *In* Banks, M.R. and Kirkpatrick, J.B. (eds.) *Landscape and Man*. Roy. Soc. Tasmania, Hobart., 29-54.

Bloom, A.L., Broecker, W.S., Chappell, J.M.A., Matthews, R.K., and Mesolella, K.J., (1974). Quaternary sea level fluctuations on a tectonic coast. New $^{230}Th/^{234}U$ dates from the Huon Peninsula, New Guinea. *Quat. Res.* 4(2), 185-205.

Boeuf, M.G. and Doust, H., (1975). Structure and development of the southern margin of Australia. *APEA Journ.* 15(1), 33-43.

Bowden, A.R. (1978). Geomorphic perspective on shallow groundwater potential, coastal northeastern Tasmania. *Aust. Water. Res. Council Tech. Paper*, 36, 84 pp.

Bowden, A.R., (1981). *Coastal Sands of Northeastern Tasmania: Geomorphology and Groundwater Hydrology.* Ph.D. thesis, Department of Geography, University of Tasmania.

Bowden, A.R., (1983). Relict terrestrial dunes: legacies of a former climate in coastal northeastern Tasmania. *Z. Geomorph. Supplementband* 45, 153-174.

Carey, S.W., (1960). Tasmania University Seismic Net. *Publication 84, Geology Department University of Tasmania.*

Chick, N.K., (1971). Fossil shorelines of the Ulverstone district, Tasmania. *Pap. Proc. Roy. Soc. Tasm.* 105, 29-40.

Colhoun, E.A., (1976). The glaciation of the lower Forth Valley northwestern Tasmania. *Austr. Geogr. Studies* 14, 83-102.

Colhoun, E.A., (1977). A sequence of late Quaternary deposits at Pipe Clay Lagoon, southeastern Tasmania. *Pap. Proc. Roy. Soc. Tasm.* 111, 1-12.

Colhoun, E.A., Turner, E. and van de Geer, G., (1982a). Pleistocene marine molluscan faunas from four sites in Tasmania. *Pap. Proc. Roy. Soc. Tasm.* 116, 91-96.

Colhoun, E.A., van de Geer, G. and Mook, W.G., (1982b). Stratigraphy, pollen analysis, and paleoclimatic interpretation of Pulbeena Swamp, northwestern Tasmania. *Quat. Res.* 18, 108-126.

Cook, P.J. and Mayo, W., (1978). Sedimentology and Holocene history of a tropical estuary (Broad Sound, Queensland). *Bur. Min. Res. Geol. Geophys. Bull.* 170.

Cull, J.P., (1982). An appraisal of Australian heat flow data. *BMR Journ. Austr. Geol. and Geophys.* 7, 11-21.

Cull, J.P. and Denham, D., (1979). Regional variations in Australian heat flow. *BMR Journ. Austr. Geol. and Geophys.* 4, 1-13.

David, T.W.E., (1926). Pleistocene glaciation near Strahan, Tasmania. *Austr. Assoc. Adv. Sci.* 17, 91-103.

Davies, J.L., (1959). Sea level change and shoreline development in southeastern Tasmania. *Proc. Roy. Soc. Tasm.* 93, 89-95.

Davies, J.L., (1960). Quaternary strandlines in Tasmania. *In:* Summary of the Report of the ANZAAS Sections C and P Committees, *Aust. J. Science* 23, 75.

Davies, J.L., (1961). Tasmanian beach ridge systems in relation to sea level change. *Pap. Proc. Roy. Soc. Tasm.* 95, 35-40.

Davies, J.L., (1964). A morphogenic approach to world shorelines. *Zeitschrift fur Geomorphologie* 8, 127-142.

Davies, J.L., (1972). *Geographical Variation in Coastal Development.* Oliver and Boyd Ltd., Edinburgh.

Denham, D., Alexander, L.G. and Worotnicki, G., (1979). Stresses in the Australian crust: evidence from earthquakes and *in situ* stress measurements. *BMR Journ. Austr. Geol. and Geophys.* 4, 289-295.

Edwards, A.B., (1941). The north west coast of Tasmania. *Proc. Roy. Soc. Vic.* 53, 233-267.

Everard, C.E., (1979). Marine geomorphology - mean sea level. *Progress in Physical Geography* 3(4), 587-596.

Fairbridge, R.W., (1961). Eustatic Changes in sea level. *Physics and Chemistry of the Earth,* 4, Pergamon, New York, 4, 99-185.

Ghani, M.A., (1978). Late Cenozoic crustal movements in the southern North Island, New Zealand. *N.Z. Journ. Geol. Geophys.* 21(1), 117-125.

Gill, E.D., (1976). Warrnambool-Port Fairy District. *In* Douglas, J.G. and Ferguson, J.A. (eds.) *Geology of Victoria Geol. Soc. Aust. Spec. Publ.* 5, 299-304.

Gill, E.D., (1982). Australian sea levels in the last 15,000 years - Victoria. *In* Hopley, D. (ed.) *Australian Sea Levels in the Last 15,000 years. Occasional Paper 3, Monograph Series,* Dept. of Geography, James Cook University, 59-63.

Gill, E.D. and Amin, B.S., (1975). Interpretation of the 7.5 m and 4 m Last Interglacial shore platforms in southeast Australia. *Search* 6(9), 394-396.

Gill, E.D. and Hopley, D., (1972). Holocene sea levels in eastern Australia - a discussion. *Marine Geology* 12, 223-233.

Gill, E.D., Segnit, E.R. and Hunt, I., (1980). Pleistocene submerged cliff off the Otway coast of Victoria. *Proc. Roy. Soc. Vict.* 91(1), 43-51.

Harmon, R.S., Land, L.S., Mitterer, R.M., Garrett, P., Schwarcz, H.P. and Larson, G.J., (1981). Bermuda sea level during the Last Interglacial. *Nature* 289, 481-483.

Hopley, D., (1983) (ed.). Australian sea levels in the last 15,000 years: a review. *Occasional Paper 3, Monograph Series*, Dept. of Geography, James Cook University, 104 pp.

Jenkin, J.J., (1968). The geomorphology and Upper Cainozoic Geology of southeast Gippsland, Victoria. *Geol. Survey Memoir Victoria*, 27, 1-147.

Jenkin, J.J., (1976). Western Port and southern Gippsland. *In* Douglas, J.G. and Ferguson, J.A. (eds.) *Geology of Victoria. Geol. Soc. Aust. Spec. Publ.* 5, 315-325.

Jennings, J.N., (1959). The coastal geomorphology of King Island, Bass Strait, in relation to changes in the relative level of land and sea. *Rec. Queen Victoria Museum, Launceston*, N.S. 11, 39 pp.

Jennings, J.N., (1961). Sea level changes in King Island, Bass Strait. *Z. Geomorph. Supplementband* 3, 80-84.

Johnson, B.D., (1972). *Coastal structure studies in Tasmania*. Ph.D. thesis, Department of Geology, University of Tasmania

Jones, H.A. and Holdgate, G.R., (1980). Shallow structure and Late Cainozoic geological history of western Bass Strait and the west Tasmanian Shelf. *BMR Journ. Austr. Geol. and Geophys.* 5, 87-93.

Kaufman, A., Broecker, W.S., Ku, T.-L and Thurber, D.L., (1971). The status of U-series methods of mollusc dating. *Geochim. et Cosmochim Acta* 35, 1155-1183.

Kern, J.P., (1977). Origin and history of upper Pleistocene marine terraces, San Diego, California. *Bull. Geol. Soc. Amer.* 88(11), 1553-1566.

Kershaw, R.C. and Sutherland, F.L., (1972). Quaternary Geomorphology of Flinders Island. *Rec. Queen Vict. Museum* 43, 1-28.

Krinsley, D.H. and Doornkamp, J.C. (1973). *Atlas of Quartz Sand Surface Textures*. Cambridge University Press, London, 91pp.

Konishi, K., Schlanger, S.O. and Omura, A., (1970). Neotectonic rates in the central Ryukyu islands derived from ^{230}Th coral ages. *Marine Geology* 9, 225-240

Ku, T.-L., Kimmel, M.A., Easton, W.H. and O'Neil, T.J., (1974). Eustatic sea level 120,000 years ago on Oahu, Hawaii. *Science* 183, 959-962.

Lewis, A.N., (1935). Correlation of the Tasmanian Pleistocene raised beaches and river terraces in unglaciated areas. *Pap. Proc. Roy. Soc. Tasm.* (1934), 75-86.

Macartney, P.M., (1979). *Coastal Tasmania: Tasmanian Coastal Environment Study*. Tas. Cons. Trust Inc. Hobart, 2 vols.

Marshall, J.F. and Thom, B.G., (1976). The sea level in the Last Interglacial. *Nature* 263, 120-121.

Matthews, R.K., (1973). Relative elevation of late Pleistocene high sea level stands: Barbados uplift rates and their implications. *Quat. Res.* 3, 147-153.

Nathan, S. and Moar, N.T., (1975). Late Quaternary terraces between Ship Creek and the Whakapohai River, South Westland, New Zealand. *Journ. Roy. Soc. N.Z.* 5(3), 313-327.

Ota, Y., (1977). Quaternary studies in New Zealand with special reference to the study of sea-level records. *The Quat. Res. Tokyo* 15(4), 141-155.

Ota, Y., Machida, H., Hori, N., Konishi, K. and Omura, A., (1978). Holocene raised coral reefs on Kikai-jima (Ryukyu Islands) - an approach to Holocene sea level study. *Géogr. Rev. Japan* 51 (2), 109-130.

Robinson, V.A., (1974). Geologic history of the Bass Basin. *A.P.E.A. Journ.* 14, 45-49.

Scott, B., (1960). Comments on the Cainozoic history of western Tasmania. *Rec. Q. Vict. Mus. Launceston* 12, 1-10.

Shackleton, N.J. and Matthews, R.K., (1977). Oxygen isotope stratigraphy of dated interglacial coral terraces in Barbados. *Nature* 268, 618-620.

Shackleton, N.J. and Opdyke, N.D., (1973). Oxygen isotope and palaeomagnetic stratigraphy of Equatorial Pacific Core V28-238: oxygen isotope temperature and ice volumes on a 10^5 year and 10^6 year scale. *Quat. Res.* 3, 39-55.

Shirley, J.E., (1980). Tasmanian seismicity - natural and reservoir induced. *Bull. Seismol. Soc. Amer.* 70(6), 2203-2220.

Sprigg, R.C., (1979). Stranded and submerged sea-beach systems of southeast South Australia. *Sed. Geol.* 22(1-2), 53-96.

Suggate, R.P., (1965). Late Pleistocene Geology of the northern part of the South Island, New Zealand. *N.Z. Geol. Surv. Bull.* N.S. 77, 1-91.

Sutherland, F.L. and Kershaw, R.C., (1971). The Cainozoic geology of Flinders Island, Bass Strait. *Pap. Proc. R. Soc. Tasm.* 105, 151-175.

Thom, B.G., Hails, J.R. and Martin, A.R.H. (1969). Radiocarbon evidence against higher post-glacial sea levels in eastern Australia. *Marine Geology,* 7, 161-168.

Thom, B.G., Hails, J.R., Martin, A.R.H. and Phipps, C.V.G., (1972). Post-glacial sea levels in eastern Australia - a reply. *Marine Geology* 12, 233-242.

Thom, B.G. and Chappell, J.M.A., (1975). Holocene sea levels relative to Australia. *Search* 6(3), 90-93.

Thom, B.G., Bowman, G.M., Gillespie, R., Temple, R. and Barbetti, M., (1981a). Radiocarbon dating of Holocene beach-ridge sequences in south-east Australia. *Dept. Geogr. Univ. N.S.W., Duntroon, Monograph* 11, 1-70.

Thom, B.G., Bowman, G.M and Roy, P.S., (1981b). Late Quaternary evolution of coastal sand barriers, Port Stephens-Myall Lakes area, central New South Wales, Australia. *Quat. Res.* 15, 345-364.

van de Geer, G., (1981), *Late Quaternary Marine and Freshwater Swamp Deposits of Northwestern Tasmania.* Ph.D. thesis, Department of Geography, University of Tasmania.

van de Geer, G., Colhoun, E.A. and Bowden, A.R., (1979). Evidence and problems of interglacial marine deposits in Tasmania. *Geol. en Mijnbouw* 58(1), 29-32.

Veeh, H.H., (1966). Th^{230}/U^{234} and U^{234}/U^{238} ages of Pleistocene high sea level stand. *Journ. Geophys. Res.* 71, 3379-3386.

Wellman, P., (1982). The southeastern Highlands: A geological and environmental history from 100 million years ago to the present. *Vict. Nat.* 99, 110-120.

Williams, P.W., (1982). Speleothem dates, Quaternary terraces and uplift rates in New Zealand. *Nature* 298, 257-260.

INDEX

4 5 6 7 8 9 0 1 2 3
A B C D E F G H I J